AF517270

A MONOGRAPH FROM
THE WAITE AGRICULTURAL RESEARCH INSTITUTE
University of Adelaide
South Australia

SOIL AND PLANT ANALYSIS

Other Related Books

Title	Author
• A Manual on Conservation of Soil and Water	*USDA*
• Agriculture Chemical Users Manual	*Geoff Cowles*
• Agrochemicals in Plant Disease Management	*N.G. Ravichandran*
• Chemistry of the Soil *2nd Ed*	*F.E. Bear*
• Crop Production in Salt Affected Soils	*I.C. Gupta*
• Fundamentals of Organic Farming and Gardening: An Instructors Guide	*A. McGregor*
• Fundamentals of Soil Science	*A. Rathinasamy*
• Genesis and Management of Sodic (Alkali) Soils	*S.K. Gupta*
• Management of Saline & Waste Water in Agriculture	*S.K. Gupta*
• Principles in the Quantitative Analysis of Water Fertilizers Plants and Soils	*U.S.S. Ramulu*
• Research Methods in Plant Sciences: Allelopathy (Vol. 1)	*S.S. Narwal*
• Resource Conserving Techniques in Crop Production	*A.R. Sharma*
• Saline Wastelands Environment and Plant Growth	*B.K. Garg*
• Salinity Tolerance in Plants: Methods, Mechanisms and Management	*B.K. Garg*
• Salt Affected Soils: Reclamation and Management	*S.K. Gupta*
• Sanrakshit Kheti Ke Antargat Satat Fasal Prabhandhan	*K.V.R. Rao*
• Soil and Plant Analysis	*C.S. Piper*
• Soil and Plant Analysis Laboratory Manual *2nd Ed*	*John Ryan*
• Soil and Water Conservation in Semi-Arid Areas	*FAO*
• Soil Chemical Analysis	*M.L. Jackson*
• Soil Conservation *3rd Ed*	*N. Hudson*
• Soil Conservation Technical Handbook	*D.H. Hicks*
• Soil Dynamics in Tillage and Traction	*W.R. Gill*
• Soil Salinity Assessment	*FAO*
• Soil Survey Laboratory Methods Manual	*USDA*
• Soil Survey Manual *New Revised Ed*	*USDA*
• Soil-Water Engineering Field and Laboratory Manual	*T.J. Trout*
• Standard Methods for Analysis of Soil Plant and Water	*I.C. Gupta*

SOIL AND PLANT ANALYSIS

A Laboratory Manual of Methods for the Examination of Soils and the Determination of the Inorganic Constituents of Plants

C. S. PIPER D.Sc.

SENIOR CHEMIST
WAITE AGRICULTURAL RESEARCH INSTITUTE
UNIVERSITY OF ADELAIDE

Published by:
Scientific Publishers
5A, New Pali Road, P.O. Box 91
Jodhpur 342 001, India
E-mail: info@scientificpub.com
Website: www.scientificpub.com

Print : 2019

All Rights Reserved. No part of this publication may be reproduced or distributed in any form or by any means without the prior written permission of the publishers.

Disclaimer: Whereas every effort has been made to avoid errors and omissions, this publication is being sold on the understanding that neither the editors (or authors of chapters in edited volumes) nor the publishers nor the printers would be liable in any manner to any person either for an error or for an omission in this publication, or for any action to be taken on the basis of this work. Any inadvertent discrepancy noted may be brought to the attention of the publishers, for rectifying it in future editions, if published.

ISBN: 978-81-72336-20-2 (Hardbound)
ISBN: 978-93-86237-53-8 (E-book)

Second Indian Reprint for Asia, 2017
First Indian Reprint for Asia, 2011

Original Edition 1942

All Rights Reserved under
International Copyright Conventions

Printed in India

PREFACE

In 1928, in order to secure greater co-ordination between the various laboratories in Australia engaged in advisory or systematic work on soils, the methods then in use at the Waite Institute were published by Professor J. A. Prescott and the present author *(Methods for the Examination of Soils, C.S.I.R. Pamphlet No. 8)*. The methods which were given in that publication were those selected as the result of critical studies, carried out on Australian soils in the chemical laboratories of the Waite Agricultural Research Institute and tentatively adopted for use. They were published to serve as a basis for discussion, so that some measure of standardization of the methods in use in the various Australian laboratories could be reached.

Since that publication the present author has had many valuable discussions with chemists working throughout Australia and New Zealand, as well as in other parts of the world. Much additional experience in the chemical and physical examination of soils and the chemical analysis of plant material has also been gained by his colleagues in the Department of Agricultural Chemistry of the University of Adelaide and the Division of Soils of the Council for Scientific and Industrial Research. This additional experience has led to many amendments in the original methods, to make them applicable, generally, to the wider range of soils encountered, while new methods have been added, for determinations not previously included. As a result of the favourable reception accorded to the earlier publication, and in response to numerous requests, it has been felt desirable to compile a new edition of the methods in use, incorporating the additional material available. With the exception of two or three of the less common determinations, all of the methods included in the present publication are in actual use in the laboratories of the Waite Institute and can be recommended to give accurate and reliable values for a wide range of samples.

The experience gained in handling numerous soil samples, which have been collected from all parts of Australia by the C.S.I.R. Division of Soils, has been particularly valuable, since it has necessitated the adoption of methods which will give dependable values when applied to these widely varying soil types. Methods originally developed for a single soil type have often been found to be inapplicable, without suitable modification, to the wide range of soils encountered in Australia.

The science of pedology is rapidly growing. Methods for the examination of soils, at present in use, will change as the store of our knowledge increases. New methods will continue to be developed to enable the measurement, in the laboratory, of those soil properties, which will give a more accurate description and definition of each soil type, while older methods will be improved as developments take place in other branches of pure science.

The author wishes to express his appreciation to all his colleagues in the laboratories of the Waite Agricultural Research Institute for the many valuable discussions which he has had with them and for the assistance which they have so generously given. The compilation of the present edition of the methods in use in these laboratories would not have been possible without this whole-hearted co-operation. It has been impracticable to give personal reference in the text to many of the small modifications introduced into methods, since these have frequently been adopted as the result of discussions among several different workers on a problem, but as far as possible acknowledgement has been made. The author desires particularly to express his indebtedness to Professor J. A. Prescott, Director of the Waite Institute, for the active interest which he has maintained in this work and the encouragement and advice which he has so freely given throughout the whole time that it has been the author's privilege to work with him.

C.S.P.

Adelaide,
December, 1942.

CONTENTS

PART II

METHODS FOR THE DETERMINATION OF THE INORGANIC CONSTITUENTS OF PLANTS

LIST OF ILLUSTRATIONS

INTRODUCTION

The choice of the analytical method for any given determination depends upon several factors, particularly the purpose for which the analysis is required. The chemist, called upon to undertake the examination of soils or plant materials, without previous experience in this branch of analytical chemistry, is confronted with a vast array of methods in the literature, frequently with little indication of their applicability or relative value. The methods may be physical, chemical, or biological. In the final choice, accuracy and reproducibility of the values obtained should be of first importance and the results should be directly comparable with results obtained in other laboratories. The methods selected should be such that, with proper equipment and organization, large numbers of analyses can be readily carried out, if necessary, as for instance in soil surveys.

In many laboratories empirical methods have often been developed at some time or other, in connexion with particular investigations, in which rapidity has been a prime essential in the methods adopted. While these methods have frequently yielded valuable data in the particular problems for which they were first proposed, they have too often been adopted by other workers for entirely different soil types or used under entirely different conditions. It is not, therefore, surprising that under such conditions they often gave erroneous and conflicting values. Empirical methods should only be used when the conditions can be completely standardized and it is known that there are no disturbing factors affecting their accuracy.

In the following pages a selection of methods is given for the chemical and physical examination of soils and the chemical determination of the inorganic constituents of plants.

It is desirable that the results of all soil analyses should be expressed on an oven-dry basis, instead of the air-dry basis, before publication. With the exception of the moisture constants, which have always been expressed on an oven-dry basis, to facilitate direct comparison with each other, it has been a frequent practice in the past to express results as a percentage of the air-dry soil. Since oven-drying changes many of the properties of a soil, it is seldom desirable or permissible to use oven-dried samples for the individual determinations and the results obtained in the laboratory will continue to be determined on an air-dry basis. Their recalculation to the moisture-free basis is, however, a simple operation.

The exchangeable ions should always be expressed in terms of milligram-equivalents per 100 g. of oven-dry soil, usually abbreviated to m.e.%. The method of expression in milligram-equivalents is now universally accepted. The milligram-equivalent, or milli-equivalent as it is sometimes called, is the chemical equivalent or combining weight of the element or radical, expressed in milligrams, i.e. it corresponds to the gram-equivalent weight divided by 1,000. This method of expression facilitates the comparison between the amounts of the different ions present, for if such values as 0·040% Ca, 0·056% CaO, 0·024% Mg, and 0·046% Na are expressed in milligram-equivalents per cent., it is at once seen that they all represent the same chemically equivalent amount of the different substances, since each corresponds to 2 m.e.%. For the sake of uniformity, and to facilitate comparison with the exchangeable ions, values for water soluble salts should also be reported in milligram-equivalents per 100 g. of soil.

The results of plant analyses are always expressed on the basis of the oven-dry material. It is desirable to express the results in terms of the individual elements, rather than the corresponding oxides.

In many fertility problems the information gained from a study of the soil can be considerably enhanced by the analysis of the plants growing on that soil. Although several chemical methods have been proposed at different times for the determination of the availability of the plant nutrients in the soil, none of these methods gives results sufficiently reliable to meet with general acceptance. After all there are many factors which must be considered in connexion with the general question of availability. The absolute concentration of the nutrient in the soil solution is not the only important factor involved. The rate at which further supplies of the nutrient are released by the soil, when this static equilibrium is disturbed, is also important. The latter is undoubtedly affected by the total amount of the plant nutrient, the chemical and physical form in which it exists in the soil, the soil reaction, and by many other factors, which are not so well known. It is improbable that any simple form of chemical analysis can simulate the conditions existing in the soil throughout the period of active plant growth and so give a reliable measure of availability under all conditions. For this reason many of the methods used to assess the availability of the plant nutrients are biological, rather than chemical, and make use of the plant itself to measure availability. Such methods include the well-known Neubauer, Mitscherlich, and *Aspergillus* methods. It has been found impracticable to include these biological methods in the present monograph.

The analysis of a range of plant species growing on a given soil type can prove a useful guide to the relative availability of the different mineral elements in that soil. Plant analysis has been found particularly valuable in regard to the availability of the trace elements such as manganese, copper, zinc and boron.

The determination of the mineral constituents of plants is of importance in connexion with many fertilizer investigations. It has become of even greater importance in recent years, on account of the many animal nutrition problems associated with deficiency diseases.

Up to the present, interest has been mainly centred in the amounts of the different elements present in plants, these being an indication of their relative availability in the soil. Little is as yet known of the functions of many of the elements, particularly the trace elements. This will require more knowledge on their distribution in different parts of the plant. For this purpose the present methods may not always be sufficiently sensitive, particularly when the size of the sample available makes the absolute amount of the element to be determined extremely small. The present methods are frequently capable of determining extremely minute amounts but, for some purposes, it will be necessary to develop even more sensitive micro-methods and use even greater refinements in technique.

Throughout the methods given in the following pages, when reference is made to concentrated reagents the following strengths are understood:

Concentrated sulphuric acid	S.G. 1·84
Concentrated nitric acid	S.G. 1·42
Concentrated hydrochloric acid . .	S.G. 1·18
Concentrated ammonia	S.G. 0·91

When reference is made to diluted reagents the numbers used indicate the proportions of reagent and water used for diluting it. For example, dilute sulphuric acid (1 + 4) means one part of concentrated sulphuric acid mixed with 4 parts, by volume, of water.

PART I

CHAPTER I

THE COLLECTION AND PREPARATION OF SOIL SAMPLES

Field Methods.

The purpose for which the soil sample is required primarily determines the method of sampling. Where the sample is to be representative of a given area of land it is necessary to take a number of samples scattered uniformly over the field or block to be examined. Such composite samples are required in sampling experimental plots, but are of little or no value for soil survey purposes. Here the basis of sampling is the soil profile and no attempt should be made to secure a mixed sample representative of a given area. The sample is intended to represent soil conditions at some particular point on the map and this position must be carefully selected so as to be truly representative of the soil formation which has been defined from previous bores over the area. To determine the nature and magnitude of the natural variations from type a number of independent samples should be collected.

Under Australian conditions the ordinary post-hole auger of commerce has proved a most useful sampling tool. It can be obtained in sizes down to three inches, but the four-inch size is the most useful. Its combined cutting and digging edge makes this tool the most serviceable for general use in hard ground. The stem should be marked at intervals of three inches so that the depth of sampling can be ascertained at a glance. The main objection to the use of this auger is the fact that the diameter at the top is slightly larger than at the bottom, thus causing slight contamination of the lower samples. In soil surveys this auger is extensively used to determine the nature of the soil and the character of the deeper layers of the profile. It is used to select the site for taking the type sample.

The cylindrical borers and screw augers, shown in most catalogues, are only serviceable on light or moist soils. In the case of screw augers the pitch of the screw should be sufficiently narrow to bring up the sample. When composite samples are required from light moist soils a semi-cylindrical sampling tool, with a cross section similar to a cheese tester, is very useful.

Where only small samples are required, as for moisture determinations or bacterial numbers, particularly in a growing crop, the Fränkel borer may be recommended if the ground is reasonably moist. This borer is pushed or hammered down in a closed condition and filled by a clockwise rotating movement, which opens the receptacle and scrapes the soil into the opening; a half turn in the opposite direction closes the opening again, and the borer can be withdrawn.

Method for Taking Composite Samples.

By means of a 4-inch post-hole auger or, in moist soils, a cylindrical borer, sample a number of sites representative of the plot or area to be investigated. Sample the surface mulch separately from the rest of the soil. This is important when nitrates or other soluble salts are to be determined. Take the remaining samples to represent 6- or 9-inch layers as desired. Combine the soil from the appropriate depth in each hole, break up any lumps into smaller pieces averaging about half an inch in diameter, and mix thoroughly on a piece of canvas or sacking. From this material take a "grab sample," that is, spread the material into a layer, then take small portions of soil, at random, so that the sample taken (about 2 lb.) is representative of the original material. Place in a calico bag or, if moisture is to be determined, in a tin with a tightly fitting lid. Label clearly with the location and depth of sample and transport to the laboratory.

Method for the Collection of Type Samples in Soil Surveys.

Considerable care must be devoted to the selection and collection of these samples since they form the basis for all subsequent laboratory description of the soils. These samples should be taken from the face of a freshly dug pit, sufficiently deep to enable the different horizons in the upper part of the profile to be easily examined and clearly differentiated. This method of sampling also avoids contamination of the sample with soil from the overlying horizons. Except in the case of the surface layer, no attempt should be made to obtain a sample representing the average of the whole of a horizon, since the limits of each horizon are diffuse, gradually merging both in physical and chemical properties into the horizons above and below it. The soil sample taken should represent that part which is typical of the horizon, and this is usually obtained by marking the full extent of the horizon and sampling a layer several inches in depth near the centre. Deeper samples of the C horizon, below the bottom of the pit, may be taken by means of a post-hole auger.

The following is the method used by the Soils Division of the Council for Scientific and Industrial Research:

Carefully choose the site of the pit, taking into consideration ground cover, micro-relief, degree of erosion, surface drainage, proximity to trees (if in open country), and all other factors likely to affect the soil in comparison with the normal type it is intended to represent. It is desirable to put down a trial hole with a post-hole auger to ensure that the site is satisfactory. By means of suitable bearings, fix the position of the pit on the map.

As far as possible dig the pit so that the sun shines on one end at the time of sampling. The depth of the pit will vary with soil type and sampling requirements. If possible make the depth sufficient to expose the C horizon of the soil. In deep soils use an arbitrary depth of 3-5 feet. The main consideration in these soils is that the B_2 horizon should be penetrated. Sample layers deeper than the bottom of the pit by means of a 4-inch post-hole auger.

Clean one face of the pit carefully with a spade and note the succession and depth of each horizon. Pick the surface with a knife or edge of the spade to show up structure and define colour and compactness. Before sampling clear away all litter and plant material on the land surface, but do not remove grass roots or organic material embedded in the A horizon.

Take the first sample to represent the whole of the surface horizon, to the depth of the first distinct change in colour or texture. Sample each horizon below this by cutting steps in the pit face and removing a portion of the layer, typical of the horizon, without attempting to include its full depth. For example, take a band, about 4-5 inches thick, out of the central part of a horizon 8 inches thick. By discarding an inch or two at the junctions of the horizon the transitional phases are effectively excluded from the sample.

Place each sample, together with a slip of paper identifying the soil, in a numbered calico bag for transport to the laboratory. Two to three pounds of soil constitutes a suitable amount.

In deep profiles take further samples for general description of the underlying layers and for pH and salt determinations, by boring into the bottom of the pit with a post-hole auger until the C horizon, or other definite substratum, is reached. Record a detailed description of the profile, and samples taken, in the field notebook or on a form such as that shown on the next page. Designate the different soil horizons A_0, A_1, A_2, B_1, B_2, B_3, and C, as usual, and use the following terms in the description of some of the other features:

Boundary: Clear, gradual, or diffused. Straight, wavy, or irregular.

Lime and Gypsum: Slight, low, medium, or high.

Stone and Rubble: Slight, low, medium, or high.

Structure: Granular, cloddy, nutty, blocky, columnar, or platey. (The type of cracking should be noted.)

Consistence: Dry: Loose, friable, brittle, compact, hard,or cemented. *Moist:* Soft, plastic, sticky, stiff, or cemented.

Preparation of the Soil Sample in the Laboratory.

When the sample reaches the laboratory break up any large lumps and spread it out in a large aluminium tray, or on a stout sheet of brown paper, to become air-dry. If desired, keep one or two of the larger lumps to illustrate structure and colour of the original soil. When air-dry, grind the main sample in a mortar under such conditions that the aggregate particles are crushed but no actual grinding or break-down of the ultimate particles of the soil occurs. Effective grinding can be carried out by pounding and rubbing the soil with a wooden pestle in a large iron mortar or, in the case of heavy clay soils, by grinding the soil in an end-runner grinding mill, fitted with a pestle of hard wood and a rubber lining to the mortar.

SHEET USED BY SOIL SURVEYORS FOR THE FIELD DESCRIPTION OF SOIL PROFILES

Soil No.............................. Date.............................. State..............................
County.............................. Parish.............................. Survey..............................
Portion ....
Genetic Type.. Soil Type..............................
Geographical Landscape ....
.... Elevation..............................
Geology ....
Micro Relief ....
Drainage..
Native Vegetation..
Culture ....
Remarks ....
....
....

Bag No.	Horizon	Depth ins.	Boundary	Colour	Texture	Lime Gypsum	Stone Rubble	Structure	Consistence	
									Dry	Moist

C.S.I.R. Division of Soils Collected by..............................

Sift the soil through a sieve with round holes 2 mm. in diameter and return the coarse material to the mortar for further crushing. Repeat the screening and crushing of this residue until all aggregate particles are fine enough to pass the sieve and only stones and organic residues remain on it. Before discarding, weigh this coarse material to determine the proportion of "stones and gravel" in the original sample.

Weigh the "fine earth" which has passed through the 2 mm. sieve, mix thoroughly, and store it in a suitable sample bottle. For identification, place a label in the bottle as well as on the outside.

For special determinations, for which it is necessary to weigh out small quantities of soil, e.g., organic carbon and calcium carbonate, it is desirable to use material that has been ground more finely than the main sample, so as to reduce sampling errors. For these determinations grind a portion of the sample until as much as possible will pass through a screen having round holes 0·5 mm. in diameter. Mix any hard material remaining on the screen back into the finely ground sample.

On account of the importance which is now attached to traces of copper, zinc, and other heavy metals in soils, copper and brass utensils must be avoided during the grinding and handling of the soil sample. This applies particularly to the use of brass sieves. Excellent sieves can now be obtained in stainless steel in certain sizes. Aluminium or ordinary iron (not galvanized) are useful alternatives, although the former is somewhat soft to withstand rough handling and the latter is liable to rust unless kept dry.

Nitrate, nitrite, and ammonia determinations must be carried out on soils straight from the field. Air-drying of these samples is not permissible. The procedure for the handling of these soils is given in Chapter XI.

CHAPTER II

HYDROGEN ION CONCENTRATION, CONDUCTIVITY, AND WATER SOLUBLE SALTS

These determinations are considered together since soil reaction, chlorides and conductivity are most conveniently carried out on the one soil suspension. Mechanical shaking of the suspension is necessary to obtain equilibrium between the soil and water for the conductivity determination, and it is also desirable for the determination of soil reaction. For many soils its adoption gives more reproducible pH values than those given by the one minute shaking so frequently used. Total soluble salts are determined by a conductivity method, chlorides by electrometric titration of an aliquot of the suspension, and hydrogen ion concentration potentiometrically on the balance of the suspension. Filtration is unnecessary for any of these determinations. This combination of all three determinations is most convenient when large numbers of soils require investigation. However, for the convenience of those who have to make occasional determinations only, alternative methods are given.

Up till recently soil reaction has always been determined in dilute aqueous or normal potassium chloride suspensions. The introduction of the glass electrode has made possible the determination of hydrogen ion concentration in soils of much lower moisture content than was formerly feasible. McGeorge (11) has developed the spear type glass electrode and determines pH in soils at field moisture content (p. 26).

Preparation of the Soil Suspension for the Routine Determination of Conductivity, Chlorides and Soil Reaction.

By means of an automatic pipette, transfer 100 ml. of aerated distilled water into a 200 ml. wide-mouthed hard glass bottle and add 20 g. of soil. Stopper, shake mechanically for one hour, and then place near the conductivity apparatus to attain constant temperature. Prepare each soil suspension in duplicate and carry out the determinations of conductivity (p. 32), chloride (p. 44), and pH (p. 15 or p. 19) in the order named. Unless conductivity and chlorides are determined before determining the hydrogen ion concentration, some potassium chloride is certain to find its way into the suspension and lead to high salt values.

HYDROGEN ION CONCENTRATION

The hydrogen ion concentration of soils can be determined either by measuring the potentials developed in a suitable electrical half cell or by the colours given with suitable indicator solutions. The problem of determining soil reaction electrometrically falls into two distinctly separate parts, namely:

the preparation of a soil: water suspension in which the pH is to be determined, and

the correct measurement of pH in a properly prepared soil: water suspension.

The latter includes those difficulties commonly referred to as electrode errors. The proper determination of soil reaction is so greatly affected by analytical details that the technique used must be standardized, if the results obtained in different laboratories are to be strictly comparable.

ERRORS IN THE PREPARATION OF THE SUSPENSION.

For most European soils contact for one minute between the soil and water before making a determination has been sufficient to give the equilibrium value. Many Australian soils fail to reach equilibrium with the same rapidity and, for these soils, mechanical shaking of the soil and water is essential, if reproducible values are to be obtained. Equilibrium is nearly always reached by shaking for one hour, and this period has been adopted for all soils. Many of the electrode drifts, observed when the suspension is made without mechanical shaking, can be traced to lack of attainment of equilibrium between the soil and suspension.

Particular care must be taken with light sandy soils, and other weakly buffered soils, to avoid alteration of the pH during the preparation of the suspension. All distilled water used should be brought into equilibrium with the atmosphere by bubbling air through it for 24 hours. The air should be drawn from outside the laboratory building. Ordinary distilled water contains variable amounts of carbon dioxide and this would affect the hydrogen ion concentration, particularly in the case of light soils. Poor quality glassware also introduces errors through the solution of significant amounts of alkali. All glass apparatus, including the bottles used for shaking, should be made from borosilicate or other non-alkali glass.

Alkaline suspensions become less alkaline on standing, due to the absorption of carbon dioxide from the atmosphere. To avoid this error, suspensions should be covered, if left standing for any length of time.

The hydrogen ion concentration varies considerably with changes in the soil: water ratio and it is most important that the ratio adopted be standardized. One part of soil to five parts of water was used by most workers with the hydrogen electrode, and this ratio is still widely used in America with other electrode systems. Biilmann and Jensen (4) used a 1 : 1 soil: water ratio with the quinhydrone electrode and the Soil Reaction Committee of the International Society of Soil Science (8) recommended the adoption of a ratio of one part of soil to 2·5 parts of water. A 1:5 soil: water ratio has been in general use for soluble salt determinations. Since mechanical shaking has been found to be essential in the preparation of the suspension for pH measurements it is preferable to combine the two determinations (soil reaction and total soluble salts), standardizing on the ratio of one part of soil to five parts of water for both. This ratio has been adopted for all determinations by the C.S.I.R. Division of Soils.

The pH value determined with a 1:5 soil: water ratio does not differ greatly from that at a 1:1 ratio. In general it is about 0·1 to 0·3 pH units higher in the more dilute suspension. In acid soils this increase in pH value is largely a dilution effect. In neutral and alkaline soils the increase in dissociation, resulting from the dilution of the soluble salts by the increased volume of water, is probably the most important factor in giving a higher value, for Puri and Asghar (12) have shown that even small amounts of neutral salts appreciably alter the pH of the suspension.

Some analysts prefer to determine soil reaction in suspensions made with a normal solution of potassium chloride instead of water. Values obtained in such potassium chloride suspensions are generally about 1·5 units lower than in aqueous suspensions. The pH values determined in N potassium chloride suspensions appear to be less influenced by changes in biological and meteorological conditions and thus measure a more permanent characteristic of the soil. The seasonal fluctuations are largely associated with variations in the salt content of the soil. The concentration of potassium chloride used masks any effects due to the small and variable amounts of salts naturally present. However, the conditions in potassium chloride suspension are purely artificial, and there is little to commend the practice.

ERRORS IN THE MEASUREMENT OF SOIL REACTION.

Soil reaction should be determined at a temperature as close to 18-20° C. as possible. If determined at other temperatures, this should be clearly stated. According to Best (2) the errors in determinations at other temperatures are small except when the quinhydrone electrode is used at alkalinities beyond its working range. Fluctuations in temperature during a series of determinations are to be avoided, since they necessitate frequent restandardization of the reference electrode system.

In all electrometric methods in which hydrogen ion concentration is measured in an aqueous medium, serious errors can be introduced by the junction or contact potentials between the suspension and the reference half cell. The design of the electrode system is important in minimizing this error. In some arrangements a potassium chloride-agar tube is used to make the connexion; in other systems a saturated solution of potassium chloride is used. When an agar bridge is employed, as in the set-up described on p. 16, the tube should dip just below the surface of the suspension and be well removed from the platinum electrode. When not in use the agar tubes should be stored with their ends dipping into a saturated solution of potassium chloride, to maintain the concentration of this salt in the agar jelly. Before use and between each determination, the tip of the tube should be rinsed with water and lightly wiped with a piece of filter paper, to remove excess of potassium chloride. If the agar jelly tends to break away from the end of the tube a small length of the glass tube should be cut off so as to maintain a plane and unbroken surface to the gel.

With the glass electrode, better and more reproducible results have been obtained by making the junction with a saturated solution of potassium chloride contained in a tube terminating in a ground glass cap. Provided that a fresh drop of potassium chloride solution is released and the ground glass cap is rinsed with water and dried with a scrap of filter paper between each determination, a good junction is obtained. The Morton type of glass electrode vessel (often supplied with the Cambridge Valve Potentiometer), although very convenient for many pH determinations, is not applicable to the determination of pH of soils, on account of the serious error due to the junction potential, which develops at the contact between the soil suspension and the saturated potassium chloride solution.

At different times the hydrogen electrode, quinhydrone electrode, antimony electrode, and glass electrode have been used for the determination of soil reaction. Soil suspensions are highly complex in composition, and it is not to be supposed that each of these electrode systems will give equally valid results for all soils. Each electrode has limitations and some of the known sources of error are discussed, together with a brief description of the four electrode systems, in the following paragraphs.

The Hydrogen Electrode is the ultimate standard for the determination of pH in solutions of known composition, if there are no disturbing substances present. It gives an absolute value of the hydrogen ion concentration in the absence of substances which act as electrode poisons, or substances which can react catalytically with hydrogen in the presence of platinum black. Values obtained by it for all soils are not, therefore, to be taken as fundamentally correct, unless it is known that disturbing substances are absent. The hydrogen electrode vessel must be so designed that the equilibrium is not upset by loss of carbon dioxide from the soil. Suitable electrode vessels have been described by Crowther (6) and Heintze and Crowther (7). The advent of the glass electrode has reduced the need for the hydrogen electrode and for fuller information on its use the reader is referred to the method described in the report of the Soil Reaction Committee of the I.S.S.S. (8).

The Quinhydrone Electrode was first introduced for soil work by Biilmann (3). It has been very extensively used, particularly for routine determinations, on account of its simplicity and wide application. Reliable results are obtained for most soils when certain well-recognized exceptions are excluded. In the first place the quinhydrone electrode is not applicable to alkaline soils with a pH greater than 8·5. Such soils give erroneously low values since one of the products of dissociation of quinhydrone is hydroquinone, which is acid in nature. Erroneous values are also given by soils containing the higher oxides of manganese. In the presence of quinhydrone, manganese dioxide is reduced and the divalent manganese so formed decreases the amount of metal ion unsaturation of the soil. With these soils rapid drift in the potential of the electrode occurs, especially during the first 30-60 seconds after adding the quinhydrone. Values as much as 1·5 pH units too high have been noted for some manganiferous soils. In its second report, the Soil Reaction Committee of the International Society of Soil Science recommends that, before using the quinhydrone electrode, its applicability to the particular soil should be tested by comparing the value obtained 10 seconds after adding the quinhydrone with the value measured after 60 seconds. If rapid drift is detected the soil should be noted as subject to quinhydrone drift. When such drift occurs, the 10 second value is more nearly correct but, for these soils, it is preferable to determine pH values by means of the glass electrode.

The quinhydrone electrode is also susceptible to certain electrode poisons. In particular, care must be taken to keep the electrode free from all traces of mercury; a binding screw should be used instead of mercury for making electrical connexion to the electrode. Small capillary cracks in the glass, produced by flaming the electrode, and which allow mercury when present to reach the electrode, have been a frequent source of error.

Another source of error with the quinhydrone electrode is an adaptation lag which is shown when suspensions of differing reaction are measured in succession. The potential developed deviates from the true potential in the direction of the last made determination and slowly drifts towards the true equilibrium. To avoid this adaptation lag the electrode should be carefully wiped with a small piece of filter paper between determinations. Duplicate determinations should be made in succession and not be separated from each other by other determinations, either of soil or buffer solutions.

The Antimony Electrode was introduced for the determination of the reaction of those alkaline or manganiferous soils which give rise to large quinhydrone errors. The values determined by it require empirical standardization, and are subject to temperature and salt errors. However, the development of convenient potentiometric equipment for use with the glass electrode has simplified the correct determination of reaction in these soils. There is now no justification for the use of the antimony electrode.

The Glass Electrode has become possible for routine determinations of soil reaction since the development of simple thermionic-valve potentiometers capable of convenient and accurate measurement of the small potentials produced at the glass-liquid interface. The production of electrodes of relatively low resistance has also contributed to the practical success of the measurements. In its second report, the Soil Reaction Committee of the I.S.S.S. (9) showed that the glass electrode gave satisfactory values for the pH of all soils tested. It is probably the most accurate method for the determination of pH in soils generally.

There should be a clear realization of the magnitude of the small currents produced in the glass electrode system. Failure to appreciate this leads to the operation of a modern valve potentiometer as a mechanical instrument instead of a piece of physical apparatus and this practice may allow certain inaccuracies to escape detection. The current drawn from the glass electrode system must be kept very small (less than 10^{-10} amp.) otherwise the electrode will polarize, giving rise to low potentials. The resistance of the glass electrode varies from 1-100 megohms, depending on the type and size of the electrode. With such a resistance in circuit the insulation must be kept in a high state of efficiency, otherwise current may leak across the insulating material, when the full potential will not be impressed on the terminal of the valve potentiometer. In general, all leads should be kept as short as possible and be supported only at the electrode and the potentiometer. The insulation of the negative electrode in the arrangement shown on p. 22 is most important; only special insulating material, such as amber or amberoid, is sufficiently good for this purpose. The insulation should be kept clean and dry. Moisture and dust on its surface can account for sufficient leakage to lead to low values.

An asymmetry potential is developed at the glass-liquid interface but this does not introduce any error, since it is balanced out during the standardization of the system against buffer solutions of known value.

To obtain good results by the glass electrode system the glass electrode used should be checked, each day that it is used, against two buffer solutions differing by about 2 pH units and, if the difference in potentials developed is more than 1·5 mv. (0·03 pH units) from the theoretical difference, the electrode should not be used. For some unknown reason, possibly due to the development of an extremely minute pinhole in the membrane, an occasional electrode will fail to develop the theoretical potential difference after long periods of use.

The outside stem of the glass electrode should be kept perfectly clean and dry, as it otherwise presents an easy leakage path for the current, thus leading to low potentials. The stem should be frequently wiped with a piece of soft filter paper. Between determinations the electrode should always be handled by the same part of the stem, so as to leave the rest of the stem unsoiled by the fingers.

Colorimetric Methods are useful where occasional or

approximate results only are desired. The selection of the indicator is of the highest importance and indicators, such as methyl red, which are absorbed by some soils, should not be used in soil suspensions. Colorimetric methods are capable of giving results to an accuracy of about ± 0·3 pH units. As with all indicator methods, however, a salt error occurs if much soluble salt is present.

Kuhn's method, in which the soil suspension is flocculated by means of barium sulphate, is described on p. 28 and a method which is suitable for obtaining approximate values in the field, by the use of a mixed indicator and a standardized colour plate, is given on p. 29.

Soil Reaction: Quinhydrone Electrode Method.

Reagents:

Quinhydrone. Dissolve 100 g. of ferric ammonium alum in 300 ml. of distilled water. Heat to 65° C., pour with stirring into a solution of 25 g. of hydroquinone in a 100 ml. of distilled water, previously heated to the same temperature. Cool the mixture, filter off the fine needles of quinhydrone on a Buchner funnel, wash with cold water, then dry between sheets of filter paper at room temperature.

Veibel Buffer Solution pH 2·03. Dissolve 6·710 g. of potassium chloride in water, add 100 ml. of 0·1N hydrochloric acid and dilute to 1 litre. Store in a stoppered pyrex reagent bottle.

3·5N Potassium Chloride. Dissolve 261 g. of potassium chloride in water and dilute to 1 litre.

Buffer Solutions for Standardizing. Any of the buffer solutions given on p. 20 may be used.

Arrangement of the Apparatus:

Platinum Electrodes. Two electrodes are required. Weld a length of platinum wire to a piece of platinum foil, about 8 x 14 mm., and seal it into a glass tube. Attach a binding screw at the top for making the electrical connexions. Do not use mercury for this purpose. Before use, and at intervals, clean the electrodes with a hot mixture of chromic and sulphuric acids. Wash well to remove the last traces of acid and heat to redness in an alcohol flame for a few seconds, to ensure constancy and correctness of potential.

Agar-Potassium Chloride Tubes. Soak 8 g. of agar in 200 ml. of water and heat in an autoclave for half an hour at a pressure of 0·75 atmospheres. Then add 52 g. of potassium chloride and sufficient neutral red (about 0·05 g.) to give a distinctive colour to the gel, so that tubes filled with it will be easily distinguished from the corresponding agar-potassium nitrate tubes used in the chloride determination. Keep the mixture hot in a water bath and fill several suitably shaped glass tubes (about 5 mm. bore) by siphoning the warm jelly into them. The tubes used must be clean and quite dry, otherwise a mobile film of moisture may arise between the walls of the tube, and the agar will then act as a siphon. Close the outer ends of the tubes with a short piece of rubber tubing and a clip and leave the other end of the tubes dipping into the gel until cold. When cold remove the rubber tubing and, if necessary, cut off a short length of the glass tube from each end to leave plane agar surfaces. When not in use keep the tubes in a wide-mouthed stoppered bottle containing a saturated solution of potassium chloride, so as to maintain the chloride concentration and prevent the agar drying out.

Potentiometer. A simple potentiometer, giving readings to 0·001 v., is quite convenient for the measurement of the potentials produced by the quinhydrone electrode. Many modern instruments are calibrated to give readings directly in pH units, compensating automatically for the temperature and the potential of the reference half cell.

Figure 1 shows a convenient set-up for the quinhydrone electrode and potentiometer. Partly fill a suitable electrode vessel, A, with Veibel buffer solution, add about 0·1 g. of quinhydrone, insert a stopper carrying a bright platinum electrode, then shake for a few seconds. Release the screw clip on the rubber tube sufficiently to allow the side arm to fill, then close the clip again. Insert the side arm through the stopper of the connecting vessel, B, so that it dips below the surface of the 3·5N potassium chloride solution contained in it. In making this connexion press gently on the rubber tube on the electrode vessel, A, and then release it a little, thus forcing two or three drops of solution out of the side arm and sucking back a corresponding amount of potassium chloride, to make a good liquid junction within the side arm. Introduce an agar-potassium chloride bridge through the other hole in the stopper of B and adjust the position of this tube so that its free end will dip just below the surface of the

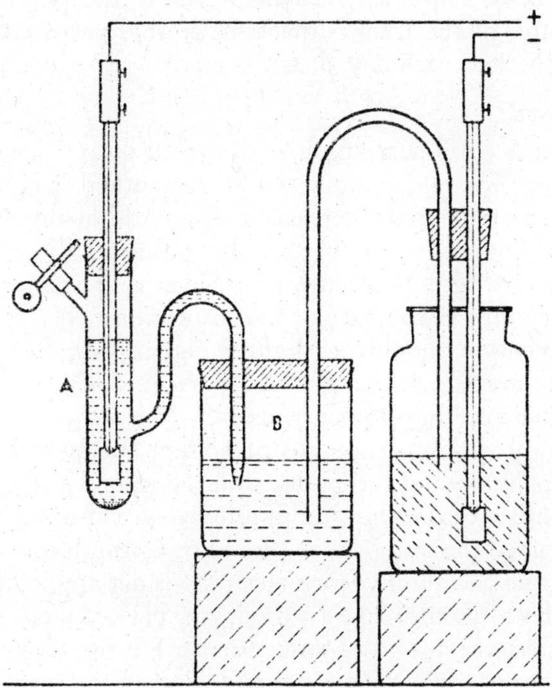

Fig. 1. The quinhydrone electrode assembly.

soil suspension to be measured. Attach the second platinum electrode to the agar tube by the rubber stopper as shown, and adjust its position so that the electrode will reach nearly to the bottom of the soil suspension and be as far removed as possible from the free end of the agar tube. Connect the electrode system to the potentiometer and standardize the latter against a standard Weston cell.

Check the standardization of the electrode system by placing 25-30 ml. of a buffer solution, approximating in value to that of the soils to be determined, in a beaker, add about 0·1 g.

of quinhydrone, stir and place it so that the agar bridge and electrode dip into it as shown. Measure the potential developed. If it deviates by more than 2 mv. from the theoretical value (obtained from the formula below) ascertain the cause of the fault before making any determinations.

The Veibel half cell maintains a steady potential for 2-3 days. It is, however, recommended by the Soil Reaction Committee of the I.S.S.S. that it be freshly prepared and the set-up checked each day that it is used.

Method:

Prepare a soil suspension as described on p. 7 or take the soil suspension remaining in the wide-mouthed shaking bottle after the conductivity determination. Add about 0·1 g. of quinhydrone and shake vigorously. As rapidly as possible place the bottle in position with the agar connecting tube just dipping below the surface of the suspension. The surface of the suspension should be slightly higher than the level of solution in vessel B, to prevent potassium chloride slowly siphoning over into the suspension. Measure the potential, preferably within 10 seconds of adding the quinhydrone to the suspension, and record it as the 10 second value. After 60 seconds measure the equilibrium value. If rapid drift has not occurred the latter value should be used. Rapid drift indicates that the quinhydrone electrode is not applicable to the particular soil under investigation, as explained on p. 12.

Remove the bottle of suspension, rinse the electrode and tip of the agar tube with water and wipe off the surplus moisture from both by means of small pieces of filter paper. Then make the determination on the duplicate suspension. Duplicate determinations should be carried out in succession, to minimize errors due to adaptation lag of the electrode (p. 13).

At the end of each day's determinations check the electrode system against a suitable buffer solution.

Calculation of the Results:

When measured with the set-up described above, the pH of the suspension is given by the expression

$$\text{pH} = 2 \cdot 03 + \frac{\text{observed E.M.F. (in volts)}}{\text{R.T.}}$$

T is the absolute temperature and R has the value 0·0001983. For soils the observed values of E.M.F. are always negative in sign. R.T. has the following values at different temperatures.

t°	R.T.	t°	R.T.
12	0·0565	22	0·0585
14	0·0569	24	0·0589
16	0·0573	26	0·0593
18	0·0577	28	0·0597
20	0·0581	30	0·0601

A table can be constructed so that the pH can be read off directly from the observed value of E.M.F. at any temperature, but as most potentiometers are now made so that readings are given directly in pH units such a table is not included in the present work.

Soil Reaction: Quinhydrone Electrode with Calomel Half Cell.

The quinhydrone electrode can be used with a standard calomel half cell as the reference half cell in place of the Veibel quinhydrone half cell. A suitable calomel half cell is described on p. 21. The calomel half cell is frequently preferred when the electrode system is kept permanently set up. However, such an arrangement offers little advantage over the one described, since the Veibel half cell is very easily set up each day that it is required. Moreover, when a saturated calomel half cell is used the polarity of the combination changes at about pH 7·8, so that it is necessary to include a reversing switch in the circuit, if soils more alkaline than this are encountered. Since the change of polarity occurs at pH 2·03 with the Veibel electrode the whole range of soil reaction can be covered without reversal of polarity. Values for the potential of the calomel half cell at different temperatures can be obtained from published tables.

Soil Reaction: Glass Electrode Method.

Ordinary potentiometers are not sufficiently sensitive to measure the small currents produced by the glass electrode. For the measurement of these minute currents the thermionic-valve potentiometer has completely superseded the quadrant electrometer because of its convenience in operation. Various reference half cells have been used with the glass electrode. These include quinhydrone, silver: silver chloride and calomel half cells. Two saturated calomel half cells balanced against each other, as originally used by Kerridge, are preferred for soil work. If Veibel's buffer solution of pH 2·03 is placed inside the glass electrode the whole range of pH above this value is covered without reversal of polarity of the system. This simplifies the insulation requirements; only the calomel electrode, negative in respect to the glass electrode, then needs specially high insulation. The set-up is very convenient for manipulation and errors due to junction potentials are eliminated.

Standard Buffer Solutions:

A useful range of buffer solutions is covered by the following:

Veibel Solution pH 2·03. Dissolve 6·710 g. of potassium chloride in water, add 100 ml. of 0·1N hydrochloric acid and dilute to 1 litre.

0·05M Acid Potassium Phthalate pH 3·97. Dissolve 10·207 g. of recrystallized potassium hydrogen phthalate and dilute to 1 litre.

Sorensen's Phosphate Buffer Solutions. Dissolve the appropriate amounts of potassium dihydrogen phosphate (KH_2PO_4) and sodium phosphate ($Na_2HPO_4.2H_2O$), as indicated in the table below, in water and dilute the solution to 1 litre.

pH of solution	KH_2PO_4	$Na_2HPO_4.2H_2O$
6·24	7·262 g.	2·375 g.
7·17	2·723 g.	8·313 g.
7·73	0·908 g.	10·688 g.

Sodium Borate Solution pH 9·16. Dissolve 4·768 g. of sodium borate ($Na_2B_4O_7.10H_2O$) and 3·728 g. of potassium chloride in water. Dilute to 1 litre.

The above buffer solutions are all easily prepared and do not require the use of standard sodium hydroxide solution. For those who prefer them, Clark and Lub's series of buffer solutions may be used. Another useful series of buffer solutions is that devised by McIlvaine and given by Clark *(The Determination of Hydrogen Ions. 3rd. Ed. p. 214)*. From two stock solutions (0·2M disodium phosphate and 0·1M citric acid) buffer solutions covering the range pH 2·2–8·0 can readily be prepared.

Arrangement of the Apparatus:

Calomel Half Cells. Calomel half cells are described as normal, tenth normal, or saturated depending on the concentration of potassium chloride solution used in the cell. For routine work the saturated calomel half cells are preferred. Since the two calomel half cells are opposed in the set-up used for the glass electrode, their absolute potential is unimportant so long as it is steady. Any difference between the two cells is automatically balanced out during the standardization. Once assembled, the calomel half cells will last indefinitely.

To prepare a saturated calomel electrode, clean the electrode vessel A (Fig. 2) and rinse it with water. Add mercury so as to form a layer about 7-10 mm. deep in the bottom of the cell, sufficiently deep to make contact with the platinum wire inserted later. Place a layer of a mixture of calomel and mercury, to a depth of about 10 mm., over the pool of mercury. Electrolytic calomel as purchased is quite suitable. Then fill the cell to a point above the side arm with a saturated solution of calomel and potassium chloride. Fill the side arm at the same time. (The calomel-potassium chloride solution is prepared by shaking a saturated solution of potassium chloride with a little electrolytic calomel and allowing it to stand over the excess calomel.) Insert the glass stem carrying the platinum wire and see that it makes contact with the pool of mercury in the bottom. The calomel electrode is then ready for use. Fill the reservoir bulb at the top with a saturated solution of potassium chloride for flushing out the side tube and renewing the liquid junction at the ground glass cap between determinations. When not in use leave the lower stem of the electrode vessel dipping beneath the surface of water in a small beaker, to prevent sticking of the ground glass cap.

Glass Electrodes. The Kerridge pattern recessed type of glass electrode, as shown in Fig. 2, is strongly recommended. The stem, 17-18 cm. long, is made from soft soda glass tubing, 13 mm. in diameter. The bulb, 22-28 mm. in diameter, is blown from Corning 015 special electrode glass.

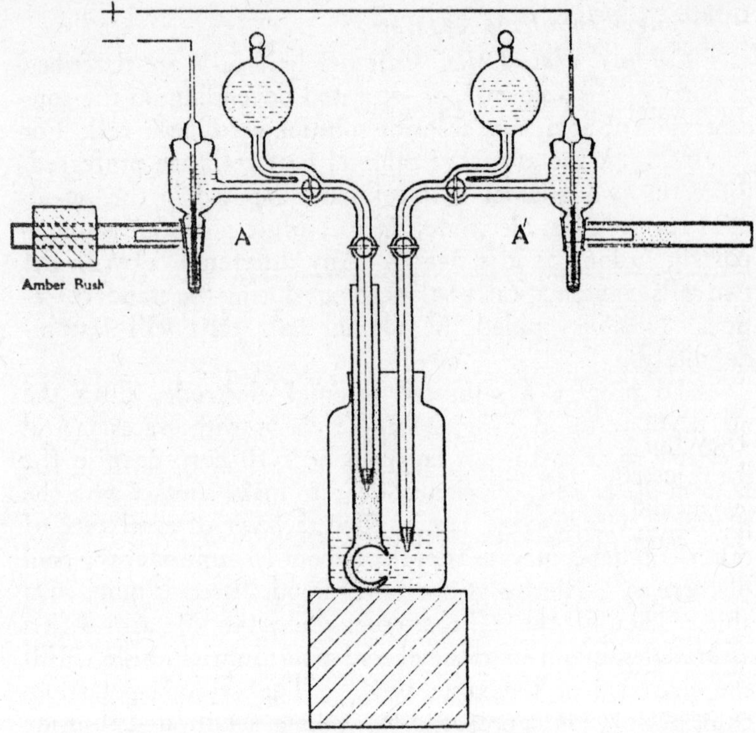

Fig. 2. The glass electrode assembly.

The recessed bulb, 14–20 mm. in diameter constitutes the thin membrane and is well protected from mechanical damage. Well-made electrodes have an electrical resistance of 5-10 megohms at 15° C. The resistance falls very rapidly with rise in temperature.

In the second report of the Soil Reaction Committee of the I.S.S.S., no significant differences were noted between the Kerridge pattern and the MacInnes and Dole type electrodes.

The Kerridge pattern electrode has, however, a much lower resistance and this reduces the possibility of electrical leakage across the insulators.

To clean new electrodes soak them in tenth normal hydrochloric acid for two or three days and rinse with water. If new electrodes are required at short notice, cleaning with dilute hydrochloric acid, filling with water and placing in a beaker of boiling water for 15 minutes will frequently be found sufficient. When not in use keep all electrodes filled with water (or Veibel solution) and leave standing in a beaker of water. Test all new electrodes for linearity in response of potential to change in hydrogen ion concentration by the use of two or three buffer solutions, covering the range of soil reactions to be determined. Do not use an electrode that deviates from theoretical responses by more than 1 mv. up to pH 8·5 or 3-4 mv. up to pH 10.

Since electrodes sometimes fail after long periods of use they should be checked, by determining the pH of two separate buffer solutions on each occasion that they are used. If they fail to develop the full potential difference they should be rejected. The outside stem of the electrode must be kept clean and dry, otherwise the leakage currents across it will be a cause of low potentials, particularly in solutions of high pH.

Valve Potentiometer. Any of the commercial instruments, incorporating an electrometer-type valve, is suitable. Most instruments are now designed to give readings either in millivolts or directly in pH units. In the latter case the effect of temperature is eliminated automatically during the standardization of the instrument. A large capacity accumulator, kept in good condition, is essential to ensure stability of the potentiometer circuit.

Insulation and Shielding. Particular care must be paid to the insulation of the negative calomel electrode. The electrical lead from it to the potentiometer should be as short as possible and must not touch any other object. The terminal on the potentiometer is suitably insulated, but the calomel half cell must be supported in its clamp by means of an amber bush. Ebonite or bakelite is not sufficiently good insulation at this point. The insulation must be kept clean and free from moisture. Normal insulation is sufficient for all other parts of the circuit.

With electrodes of resistance less than 10 megohms shielding is generally unnecessary, but, if desired, an earthed metal sheet can be attached to the bench, beneath the potentiometer and electrode vessels. High resistance electrodes increase the requirements for insulation and shielding.

Fig. 2 shows the most convenient arrangement of the glass electrode assembly. Standardize the potentiometer against a Weston Cell as usual. Then set up and standardize the electrode system as follows:

Fill the glass electrode with fresh Veibel solution and clean and dry the outside stem with a piece of filter paper. Rinse the bulb of the electrode with a buffer solution of about pH 6, to be used for the standardization, and stand the electrode in a small beaker containing some more of this buffer solution.

Turn all stopcocks on the calomel electrode vessels to the open positions in respect to the potassium chloride reservoirs. Momentarily loosen the ground glass cap on each, so as to release two or three drops of potassium chloride solution and renew the liquid junction on the ground surface. Dry the lower stems with a small piece of filter paper and close the lower stopcocks to prevent accidental leakage of potassium chloride solution.

Raise the beaker and glass electrode so that the stem of the negative calomel electrode dips below the surface of the Veibel solution in the glass electrode and the ground glass cap of the positive calomel electrode is just beneath the surface of the buffer solution in the position shown in Fig. 2. Place a block of wood or paraffin wax under the beaker to support it in this position. Set the pH dials of the potentiometer to correspond with the known value of the buffer solution and standardize the instrument in accordance with the directions supplied with it. Check the standardization firstly against a duplicate portion of the buffer solution and then against a second buffer solution of about pH 8.

Method:

Prepare a soil suspension as described on p. 7 or take the soil suspension remaining in the wide-mouthed shaking bottle after the conductivity determination.

When the potentiometer has been standardized, remove the glass electrode from the buffer solution and thoroughly rinse the outside with distilled water from a wash bottle. Then dip the bulb in a beaker of aerated water and flick off any remaining drops. Wipe the upper part of the stem carefully, as before. Place in the wide-mouthed bottle containing the soil suspension and stir gently.

Rinse the stem and ground glass cap of the positive calomel reference electrode with water and dry lightly with a piece of filter paper. Turn the lower stopcock to the on position, loosen the ground glass cap and allow a few drops of potassium chloride solution to escape, in order to renew the liquid junction. Replace the cap and close the stopcock. Wipe the outside of the cap with a piece of filter paper as before. Bring the wide-mouthed bottle containing the soil suspension and glass electrode into position, supporting it on the paraffin block as before, and make the pH measurement as nearly as possible 60 seconds ($\pm$ 15 seconds) later.

After recording this value, remove and rinse the glass electrode as before, wiping the stem if necessary, and place in the duplicate suspension. Rinse the ground glass cap of the positive calomel half cell and renew the junction as described above, before proceeding to the next determination. There is no need to renew the potassium chloride junction at the ground glass cap of the negative half cell between each determination, but it is essential in the case of the positive half cell.

At the end of a series of determinations check the standardization of the electrode system against one of the standard buffer solutions.

During determinations, the stopcocks of the calomel electrodes are normally kept in the following positions (after the initial flushing with potassium chloride at the beginning of each day).

Negative Calomel Electrode.

Upper stopcock—connexion between electrode chamber and lower stem.

Lower stopcock—off position.

Positive Calomel Electrode.

Upper stopcock—connexion between potassium chloride reservoir and lower stem. Stopper of reservoir should also be left loose.

Lower stopcock—off position during measurements but turned on to renew potassium chloride junction between each determination.

On account of the small currents involved, the electrical circuit is established even though the stopcocks are in the off position.

Calculation of the Results:

Most valve potentiometers are calibrated to give pH values directly, when standardized against a buffer solution. However, if the potentiometer is calibrated in millivolts only, the pH of the suspension is given by the following expression:

$$\text{pH} = \text{pH of standard buffer solution} - \frac{E_{(\text{suspension})} - E_{(\text{buffer})}}{\text{R.T.}}$$

$E_{(\text{suspension})}$ and $E_{(\text{buffer})}$ are the observed potentials of the soil suspension and the buffer used for standardizing respectively. These potentials are usually negative in sign. Values for R.T. are given on p. 19.

Soil Reaction: Determination at Field Moisture Capacity.

McGeorge (11) has applied a very rugged type of glass electrode to the determination of pH values in soils at low moisture contents. Determinations can be made so long as the soil is able to supply sufficient moisture to make a continuous moisture film contact between the glass electrode and the reference calomel electrode. The determination is usually carried out at a moisture content approximating to the moisture equivalent. In general the pH value of neutral and alkaline soils decreases as the soil : water ratio varies from 1 : 10

to that at the moisture equivalent. Acid soils, on the other hand, show a decrease in pH value between the soil:water ratios of 1:10 and 1:1, but with the further narrowing of the ratio to that at the moisture equivalent the pH value tends to increase again.

The spear type of glass electrode was first used for this determination but many plain bulb type electrodes, supplied commercially, are now sufficiently rugged for use with stiff soil pastes. Chapman, Axley, and Curtis (5) recommend the Beckmann bulb type glass electrode, in conjunction with the Beckmann sleeve type calomel electrode, as a reference half cell. They do not recommend making the determination at moisture contents below that of moisture equivalent. At this moisture level stable readings are obtained, provided that the electrodes are "conditioned" or worked around, by pushing them in and out of the moist soil several times, before making the reading. Particular care must be taken to press the moist soil firmly around the electrode, for an imperfect coverage of the glass electrode with the soil moisture film leads to erratic and unreliable values. To obtain more stable and reproducible values they prefer to make the determination at the moisture content corresponding to the sticky point, or in the case of heavy clays, at a moisture content slightly wetter than this. The proper moisture content, corresponding to this point, can be more readily determined than that corresponding to the moisture equivalent, while the pH values at the former moisture content differ only slightly from those at the moisture equivalent. Their method is as follows:

Transfer 100 g. of air-dry soil to a beaker and moisten it with successive small portions of water, mixing and working the soil with a porcelain spatula until the soil reaches the desired moisture content, corresponding to that of the sticky point. If the soil is very heavy add a little more water to facilitate good coverage of the glass electrode.

To make the determination rinse the glass and calomel electrode assembly with aerated distilled water and remove the excess water by wiping it with strips of filter paper. Then insert the assembly in the soil paste and withdraw it three or four times, inserting it at different points each time. This is necessary to condition the electrode and ensure reproducible values. After the fourth insertion press the soil firmly around the electrodes, by means of a glass rod, and leave them in undisturbed contact with the soil while the pH value is read by means of a suitable valve potentiometer.

Soil Reaction: Kuhn's Colorimetric Method (10).

When a soil suspension is shaken vigorously with very pure barium sulphate, the latter flocculates the soil colloids and leaves a clear and colourless solution. If an indicator, which is not absorbed by the soil, is present its colour will denote the soil reaction. The amount of barium sulphate necessary to clear a suspension depends on the amount of colloids present so that, for loams and heavy clay types, it is necessary to reduce the quantity of soil used. However, for reliable results as much soil as possible, compatible with obtaining a clear solution, should be taken. Crowther (8) recommends the adoption of Clark and Lub's sulphonphthalein indicators so as to avoid errors introduced by methyl red and similar basic indicators, through ionic exchange with the soil. Crowther also recommends the use of colour discs, such as are supplied with the Lovibond or Hellige comparators. These serve as permanent colour standards and are more convenient than the preparation of numerous buffer solutions. If colour discs are used they should be re-marked so that the pH values are adjacent to the colour plates to which they refer.

Reagents:

Indicator Solutions. The indicator solutions described on p. 126 are suitable for this method.

Method:

For sandy soils place a layer of barium sulphate (*pro Röntgen*), 1 cm. thick, in the bottom of a dry test tube. A long narrow test tube is most suitable. Then add a layer of soil about 3 cm. deep and fill with carbon dioxide-free water to a depth of 9-10 cm. Add a suitable indicator solution, depending on the pH range expected, and fill the tube to a total depth of 15 cm. with water. Close with a paraffined cork and shake vigorously until about half a minute after the contents of the tube are thoroughly mixed. Vigorous shaking is necessary to mix the indicator intimately with the soil particles. Place the tube aside to settle. If the correct proportions have been taken the upper part of the suspension becomes clear within a few minutes, and the colour of the indicator can be seen. Compare this colour against suitable colour standards, either glass colour discs or a range of freshly prepared buffer solutions to which the indicator has been added.

For loams and clay soils use less soil and more barium sulphate, but do not exceed the limits of 4-5 cm. for the depth of the two layers.

If flocculation occurs too rapidly the supernatant liquid may not be properly cleared. This is due to the use of too large a proportion of barium sulphate and it is necessary to repeat the determination using more soil and less barium sulphate.

If the suspension clears too slowly it implies that too little water has been added. In this case pour off some of the suspension, add more water and indicator solution and shake again.

Soil Reaction: Waite Institute Hydrionmeter.

The development of permanent colour standards on porcelain makes possible the determination of soil reaction by a simple colorimetric method, with sufficient accuracy for many purposes. The method is very useful in that preliminary determinations of pH can be carried out in the field and values obtained to within about half a pH unit.

The Waite Institute Hydrionmeter consists of a shallow porcelain plate with two circular bands showing a series of colours corresponding to the colour changes of the two indicator solutions used. The colours are surrounded by the white background of the plate. The outer band shows the range of colours given by Indicator No. 1 at reactions of pH 4, 5, 5·5, 6, 7, and 8, while the inner circle shows the colours produced by Indicator No. 2 at pH 4, 4·5 and 5. Indicator No. 1 consists of a mixture of methyl red and brom thymol blue adjusted to approximately pH 6. Indicator No. 2 is a solution of brom cresol green adjusted to approximately pH

4·5 and is used for testing soils more acid than pH 5. It is not absorbed by acid soils, as is the case with methyl red.

Indicator Solutions:

Indicator No. 1. Dissolve 0·4 g. of methyl red in 85 ml. of 96 per cent. alcohol and 15 ml. of 0·1N sodium hydroxide. Filter the solution. Dissolve 0·4 g. of brom thymol blue in 94 ml. of alcohol and 6·4 ml. of 0·1N sodium hydroxide. Mix these two solutions and dilute fivefold with water. Store in a stoppered pyrex reagent bottle.

Indicator No. 2. Dissolve 0·5 g. of brom cresol green in 100 ml. of 96 per cent. alcohol, add 7·1 ml. of 0·1N sodium hydroxide and dilute to 125 ml. with water. For use dilute with water to one-tenth of this strength.

Method:

Take about 1-2 g. of soil, from a representative sample, and place it on the porcelain plate. Add Indicator No. 1 solution, drop by drop, until there is just enough surplus liquid to enable the colour of the liquid at the edge of the wet soil to be seen. When the colour shows no sign of further change (about one minute) compare it with the standard colours. Stirring is recommended if it does not cause undue turbidity.

For strongly acid soils the determination should be repeated with Indicator No. 2 as this indicator is not adsorbed by the soil and gives more reliable results.

WATER SOLUBLE SALTS

The determination of water soluble salts is of special importance in a semi-arid country, particularly where irrigation is practised. Of the various constituents, chloride and nitrate ions are unaffected by the ratio of soil to water, but this ratio is important with respect to carbonate and bicarbonate ions, and has a further bearing on the relative proportions of the cations, owing to cation exchange phenomena. It is therefore necessary to adopt some conventional relationship between the weight of soil and the volume of extraction water. A ratio of 1 part of soil to 5 parts of water is widely adopted for the extraction of soluble salts.

Two methods are available for the determination of total

soluble salts. In the gravimetric method, described on p. 36, the suspension is filtered through a Chamberland filter candle and an aliquot is evaporated to dryness and weighed. For routine determinations on a large scale the conductivity method (p. 32) is extensively used. The conductivity of dilute suspensions depends chiefly on the ions present and use is made of this property in obtaining an approximate value for the amount of total soluble salts in a soil suspension. The conductivity is measured and referred to that of a standard solution of potassium chloride, measured under the same conditions. By this means errors due to measurements at different temperatures are largely eliminated, provided that both soil suspension and potassium chloride are measured at t' : same temperature. Since conductivity depends on temperature, the size of the electrodes, their distance apart, and the shape of the conductivity cell, reference to the conductivity of a standard potassium chloride solution, determined at the same time, enables the conductivity of the soil suspension to be expressed as a specific conductivity with sufficient accuracy for the purposes required.

Conductivity measurements must be made within a few hours of the preparation of the suspension, for its conductivity changes considerably on standing overnight, due to micro-organic activity. The increase of conductivity on standing has been proposed as a measure of soil fertility. Hard glass bottles should be used to avoid solution of alkali from the glass during the preparation of the suspension. All suspensions should be made with good quality distilled water which has been aerated, to bring it into equilibrium with the carbon dioxide of the atmosphere.

Values obtained by the conductivity method are frequently calculated and expressed as total soluble salts. Even if the conductivity values are standardized against values determined by the gravimetric method, approximations only are possible because the conductivity of any solution depends on the relative composition of the salts present. This will vary somewhat from soil to soil, even in any group of soils from one locality. The specific conductivity of a 1 : 5 soil suspension is, however, a fundamental value, capable of exact measurement. It is, therefore, preferable to report this conductivity value itself, rather than a value for total soluble salts derived from it on the assumption of an absolute correlation between these values. Although it has been customary to think in terms of some material value, such as percentage of total soluble salts, the practical understanding of specific conductivities would soon follow if these values were commonly reported. After all, the relative values on the pH scale are now widely understood.

If the relationship given on p. 36 is accepted, namely that the percentage of total soluble salts corresponds to 375 x the specific conductivity of a 1:5 soil suspension at 20° C., then the following scale shows the approximate relationship between the two sets of values.

TOTAL SOLUBLE SALTS

0	0·1	0·2	0·3	0·4	0·5	0·6	0·7	0·8	0·9	1·0%
0	0·27	0·53	0·80	1·07	1·33	1·60	1·87	2·13	2·40	2·67 mhos $\times 10^{-3}$

SPECIFIC CONDUCTIVITY OF A 1:5 SOIL SUSPENSION AT 20° C.

When much gypsum is present it will not be brought into solution completely in a 1:5 soil: water suspension. However, the amount of total soluble salts, whether determined conductometrically or gravimetrically, loses significance in such soils. Complete solution of the gypsum would only be required for the determination of water soluble sulphates and in such a case the soil should be extracted by shaking with dilute hydrochloric acid, before precipitating sulphates as barium sulphate, in the filtrate.

The Determination of Specific Conductivity.

Conductivity is measured by means of a Wheatstone bridge, preferably of the dial reading pattern. To avoid polarization at the electrodes alternating current must be used. This is generally obtained from an induction coil. A telephone receiver is then used to determine the point of minimum sound. More modern equipment makes use of a thermionic-valve oscillator and amplifier. If a bank of condensers is embodied in the circuit, background noises in the telephones can be eliminated and the point of minimum sound made much sharper. A further refinement replaces the telephone receiver with a cathode ray tube (magic eye) and the point of balance can then be detected visually. Such an arrangement is very useful and reduces fatigue when large numbers of determinations have to be carried out.

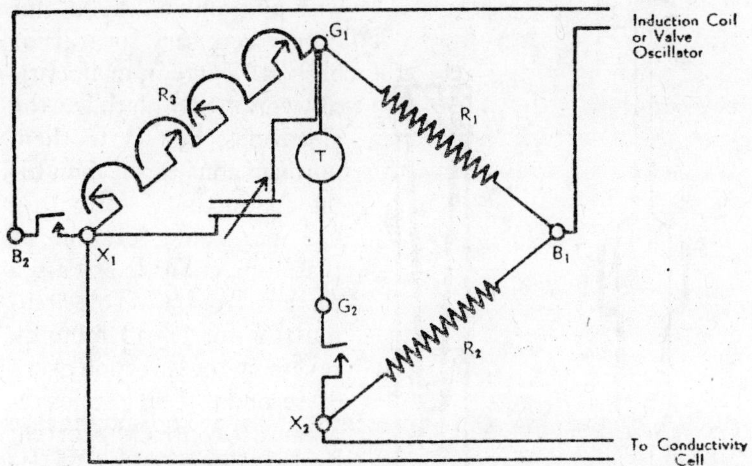

Fig. 3. The arrangement of the apparatus for the determination of the conductivity of soil suspensions.

Arrangement of the Apparatus:

Fig. 3 shows the necessary connexions to the Wheatstone bridge. R_1 and R_2 represent the ratio arms. Connect the leads of the induction coil to the terminals B_1 B_2 and the telephone receiver across G_1 G_2. Connect the conductivity cell across the terminals X_1 X_2, using short lengths of stout copper wire for this connexion.

If a thermionic-valve oscillator is used instead of the induction coil connect it across the terminals B_1 and X_1, so as to eliminate the tapping key in this circuit. Likewise connect the valve amplifier across G_1 and X_2. The telephone receiver or the electron ray tube is incorporated in the valve amplifier unit. A variable condenser is desirable to balance out the capacity of the conductivity cell. Connect it in parallel with the variable resistance R_3, that is, across the terminals X_1 G_1.

The Conductivity Cell may be either the pipette type (Fig. 4 A) or dipping type (Fig 4 B). The electrodes are platinum, about 20 mm. x 10 mm., firmly sealed to the glass at each corner, to maintain a constant position in relation to each other. If desired the electrodes may be plated with platinum black, but this is only necessary for soils of high salt content, or in critical work. To platinize the electrodes, first clean them in warm aqua regia, then fill the conductivity cell with a 0·3 per cent. solution of platinum chloride. Pass a current from a 4-volt accumulator for 10–15 minutes, reversing the direction every 30 seconds. Use a series resistance to control the current so that there is only a gentle evolution of gas. Then rinse the electrodes with water and remove the last traces of chloride by immersing them in 1 per cent. sulphuric acid and passing a current between them for 10 minutes, reversing its direction as before. For routine work such a plating will last for a considerable time, if the electrode vessels are kept full of water when not in use.

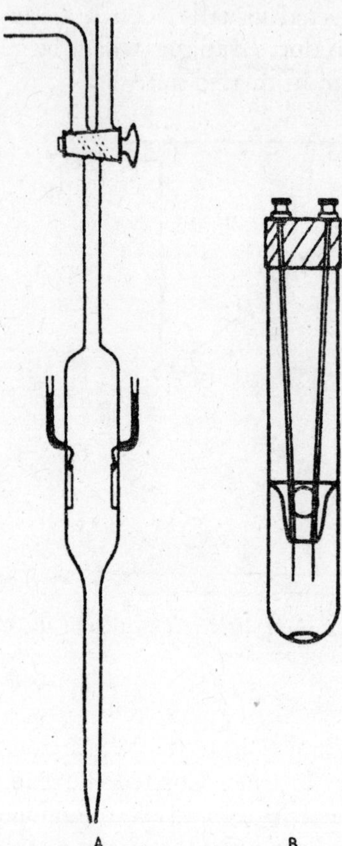

Fig. 4. Pipette type and dip type conductivity cells.

Reagents:

0·005N Potassium Chloride. Dissolve 0·7456 g. of potassium chloride in aerated distilled water and dilute to 2 litres.

Method:

Set the ratio arms of the Wheatstone bridge to a suitable ratio. For most soil suspensions the best results are obtained at the 1,000:1,000 ratio. At the point of balance the dial reading of the variable arm, R_3, will then give the value of the unknown resistance directly in ohms.

Rinse and fill the pipette electrode with the soil suspension, prepared as described on p. 7, or if a dip-type electrode is used, immerse it in the suspension. Adjust the variable resistance of the Wheatstone bridge until the point of minimum sound is reached or the cathode ray tube shows a sharply defined sector. As the null point is approached adjust the variable condenser to assist in obtaining a sharp balance at the null point.

To obtain the resistance of the soil suspension (in ohms) multiply the dial setting of the Wheatstone bridge by the ratio arm settings $\frac{R_2}{R_1}$.

Between determinations rinse the electrode with water. Allow to drain for a few seconds before starting the next determination. Before and after each series of determinations measure the resistance of the standard potassium chloride solution, carefully rinsing the electrodes with two lots of the solution.

Calculation of the Results:

The specific conductivity (in reciprocal ohms or mhos.) of a 1:5 soil suspension at 20° C. is given by the expression

$$\frac{R_{(KCl)}}{R_{(suspension)}} \times \frac{1}{\rho}$$

where $R_{(KCl)}$ = Observed resistance of 0·005N potassium chloride

$R_{(suspension)}$ = Observed resistance of the soil suspension

and ρ = Specific resistance of 0·005N potassium chloride at 20° C.

0·005N potassium chloride has a specific resistance of 1540 ohms at 20° C.

The above calculation gives the specific conductivity at 20° C., with sufficient accuracy, provided that the conductivity of the soil suspension is measured at a temperature within a few degrees of 20° C. The temperature of the standard potassium chloride solution used for comparison must of course be the same as that of the soil suspension at the time of measurement.

The results obtained by this method should be reported as "Specific Conductivity of a 1:5 Soil Suspension at 20° C." However, when some approximation of the total soluble salts is required this may be obtained by multiplying the values for specific conductivity by 375, a value derived from actual correlations of the specific conductivities with the amounts of total soluble salts determined gravimetrically in a large number of Australian soils. The correlation is only fair as it is largely influenced by the composition of the salts, particularly in soils of low salt content. Soils containing unusual combinations of salts will require a different factor. For this reason it is much more desirable to report the absolute value of specific conductivity rather than a value for total soluble salts derived from it by an imperfect correlation. Closer approximations can be obtained for any group of soils if the curve connecting specific conductivity and total soluble salts is determined for a representative number of samples. Joseph and Martin obtained the total soluble salts in Sudan soils by multiplying the specific conductivity at 30° C. by 250, a value derived from the known conductivity of a solution of equal proportions of sodium chloride and sulphate. This corresponds to a factor of 300 if the specific conductivity at 20° C. is used.

The Determination of Water Soluble Salts.

PREPARATION OF THE SOIL EXTRACT: To 1,000 ml. of aerated distilled water in a large bottle or shaking cylinder, add 200 g. of air-dry soil and shake in a mechanical shaking machine for one hour. Allow the heavier particles to settle and decant the supernatant liquid into a tall 500 ml. cylinder. Insert a Chamberland filter candle, connected to a 1 litre filter flask by means of rubber and glass tubing, and filter by applying suction to the filter flask. The filtration apparatus is illustrated in Fig. 5. Reject the first 50-100 ml. of filtrate which passes through the candle filter. This serves to rinse the filter. The rate of filtration varies considerably with different soils. With very alkaline soils complete filtration may take some hours, since a thick deposit of finely dispersed clay collects on the outside of the candle filter. This may be removed by disconnecting from the vacuum, removing the filter

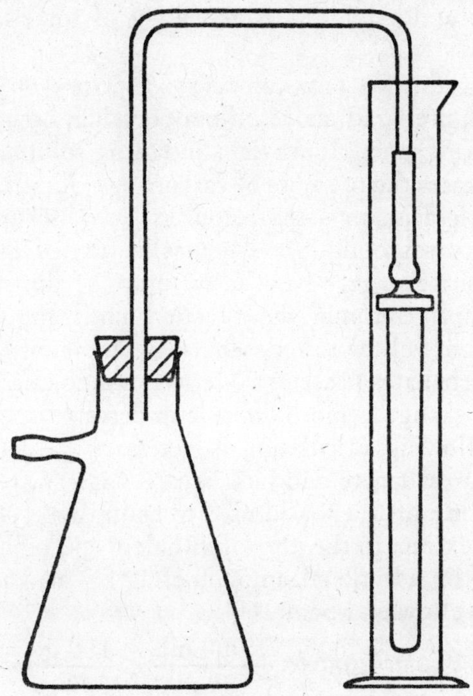

Fig. 5. Filtration of a soil suspension by means of a filter candle.

from the suspension and blowing gently into it through the rubber tube connexion. The slight back pressure will cause the coating of clay to glide off and leave a clean filtering surface. This operation may be repeated at intervals to hasten filtration. Before each filtration the filtering surface of the candle can be renewed by lightly rubbing the surface with sandpaper.

Use aliquot portions of the filtrate for the determination of total soluble salts, carbonates and bicarbonates, sulphates, and chlorides.

TOTAL SOLUBLE SALTS: Pipette 100 ml. of the clear filtrate into a tared platinum basin and evaporate over a water bath; when the volume is reduced to about 5 ml. add 2 ml. of 20 volume hydrogen peroxide, free from salts, to oxidize soluble organic matter, and evaporate to dryness. Dry in an oven at 110° C. for one hour, cool in a desiccator and weigh. The percentage of total soluble salts = weight of total soluble salts x 5.

CARBONATES AND BICARBONATES: Pipette 100 ml. of the soil extract, prepared above, into a porcelain basin and add 2 or 3 drops of phenolphthalein indicator solution. A red colour indicates the presence of carbonates. Titrate with 0·1N hydrochloric acid until the colour is just discharged. This end point corresponds to the neutralization of carbonates to the bicarbonate stage. Now add 2 drops of dimethyl yellow indicator and continue the titration until the colour just changes from yellow to red denoting complete neutralization of the bicarbonates present. Methyl orange can be used, but the colour change is more difficult to perceive.

The following calculation is necessary to determine the amounts of carbonate and bicarbonate. Let V_1 represent the volume in ml. of 0·1N acid used to neutralize 100 ml. of the above soil extract to the phenolphthalein end point and let V_2 represent the additional amount of 0·1N acid to reach the dimethyl yellow end point.

$$\%\ CO_3' = V_1 \times 0{\cdot}0060 \times \frac{1{,}000\ \text{ml.}}{100\ \text{ml.}} \times \frac{100\ \text{g.}}{200\ \text{g.}} = V_1 \times 0{\cdot}030$$

$$\%\ HCO_3' = (V_2 - V_1) \times 0{\cdot}0061 \times \frac{1{,}000\ \text{ml.}}{100\ \text{ml.}} \times \frac{100\ \text{g.}}{200\ \text{g.}}$$
$$= (V_2 - V_1) \times 0{\cdot}0305.$$

It is preferable to express the amounts of carbonate and bicarbonate present as milligram-equivalents per 100 g. of soil (m.e. %). These values are easily calculated since 1 ml. of a tenth normal solution corresponds to 0·1 milligram-equivalent.

$$\text{m.e.\%}\ CO_3' = \frac{V_1}{10} \times \frac{1{,}000}{100} \times \frac{100}{200} = \frac{V_1}{2}$$

$$\text{m.e.\%}\ HCO_3' = \frac{V_2 - V_1}{10} \times \frac{1{,}000}{100} \times \frac{100}{200} = \frac{V_2 - V_1}{2}.$$

If the standard acid is not exactly tenth normal use the appropriate factor.

CHLORIDES: The method for chlorides is given below.

To express the amount of chlorides present as milligram-equivalents per 100 g. of soil make the following calculation:

$$\text{m.e. \% Cl}' = \text{volume of standard silver nitrate} \times \text{normality factor} \times \frac{1{,}000}{\text{volume titrated}} \times \frac{100}{200}$$

SULPHATES: Measure the volume of the soil extract remaining after the preceding determinations have been carried out and transfer it to a beaker. Make just slightly acid to dimethyl yellow with hydrochloric acid, boil and precipitate the sulphates as barium sulphate by the addition of a slight excess of 5 per cent. barium chloride solution, run in slowly from a pipette. Boil for five minutes and leave to stand overnight. Filter through a 9 cm. Whatman No. 44 filter paper and wash completely with hot water. Ignite in a weighed crucible, cool in a desiccator and weigh as $BaSO_4$. Then

$$\% \ SO_4' = \text{Weight of precipitate} \times \frac{96}{233} \times \frac{1{,}000}{\text{Volume taken}} \times \frac{100}{200} = \frac{\text{Weight of precipitate}}{\text{Volume taken}} \times 206$$

$$\text{m.e. \% } SO_4' = \frac{\text{Weight of precipitate}}{0{\cdot}117} \times \frac{1{,}000}{\text{Volume taken}} \times \frac{100}{200} = \frac{\text{Weight of precipitate}}{\text{Volume taken}} \times 4{,}270$$

If the presence of silica is suspected purify the partly ignited precipitate by treatment with a little hydrofluoric acid and a drop of sulphuric acid. Carefully fume off the acids, complete the ignition, and weigh as $BaSO_4$.

The methods described under exchangeable cations can be used for the determination of calcium, magnesium, sodium and potassium in the water extract of the soil.

CHLORIDES

Chlorides in the soil can be easily and accurately determined by titration with standard silver nitrate either electrometrically (p. 40) or in the presence of potassium chromate as an indicator (p. 45). In the latter method it is necessary to obtain a clear extract of the soil, either by filtration or by flocculation with an excess of potassium chromate. The amount of chloride brought into solution does not depend on any equilibrium between the soil and water. Values obtained are therefore not influenced by the amount of water used or the time of standing.

Chlorides: Electrometric Titration Method.

In this simple and rapid method, due to R. J. Best (1), a silver wire, coated with silver chloride, dips into the unknown suspension and is connected to a quinhydrone half cell by means of an agar-potassium nitrate bridge. At the end point in the titration of a chloride solution with silver nitrate the chloride ion concentration is $1 \cdot 0 \times 10^{-5}$ at 25° C. and the theoretical potential of a silver-silver chloride electrode in this solution would be − 0·521 volts. If therefore this electrode is connected with a quinhydrone reference half cell giving a steady potential of 0·521 volts the combination will have zero potential at the end point of the titration. A quinhydrone half cell made from a buffer solution of pH 3·03 meets these requirements at a temperature of 25° C. At 16° the calculated pH of the reference half cell is 3·3. In practice the buffer solution can have any value between 3·0 and 3·3 without introducing an error greater than one-half drop of $\frac{N}{35 \cdot 5}$ silver nitrate in the usual volume titrated, for all temperatures between 16° and 25° C. Of course chlorides must be absent from the system to prevent diffusion into the silver-silver chloride half cell. The agar connexion is therefore made with potassium nitrate instead of potassium chloride, and sulphuric acid is used in place of hydrochloric acid in making the phthalate buffer of pH 3·0-3·3.

The end point is denoted by the reversal of the direction of the current, for the silver-silver chloride electrode is negative in respect to the quinhydrone electrode so long as free chloride ions remain in solution. As the titration proceeds and silver nitrate is added the potential difference decreases until, at the theoretical end point, it is zero. The further addition of silver nitrate solution causes the silver-silver chloride electrode to become the positive pole so that a reversal of current occurs at this point. With a suitable galvanometer in circuit this reversal of direction can be easily detected to within one drop of $\frac{N}{35 \cdot 5}$ silver nitrate.

Snyder (13) has pointed out that some soils show a drift at the end point unless the soil reaction is first adjusted to pH 2·0 with dilute sulphuric acid. This is, however, unnecessary in routine determinations since the error is only significant in alkaline soils of low salt content. Best *(priv. comm.)* considers that correct values will be obtained if such soils are first made slightly acid to methyl red, with dilute sulphuric acid. He attributes the slow drift to silver ions entering the exchange complex of the soil if it is not acidified.

Reagents:

Buffer Solution pH 3·2. Dissolve 10·21 g. of acid potassium phthalate in water, add 148 ml. of 0·1N sulphuric acid and dilute to 1 litre. Store in a stoppered pyrex bottle.

Quinhydrone. Prepare a supply as described on p. 15.

$\frac{N}{35{\cdot}5}$ *Silver Nitrate.* Dissolve 9·584 g. of silver nitrate in water and dilute to 2 litres.

0·01N Potassium Chloride. Dissolve 0·7455 g. of potassium chloride in water and dilute to 1 litre.

Arrangement of the Apparatus:

Fig. 6 shows a convenient set-up, although the motor stirrer is not essential. Hand stirring gives equally satisfactory results.

Silver-Silver Chloride Electrode. Thoroughly clean a piece of 16 gauge silver wire, about 12-15 cm. long, by successive treatments with alcohol, ammonia, and dilute nitric acid, washing with water between each operation. Deposit a layer of silver chloride on the lower portion of the wire by immersing it to a depth of 4-5 cm. in 0·1N hydrochloric acid, connecting it through a series resistance to the positive pole of a 2 volt accumulator, and passing a current of 3-4 milliamps between it and a platinum wire immersed in the same solution. When a good deposit of silver chloride has formed, after electrolysis for 15-30 minutes, disconnect the silver wire and rinse it with water. The silver chloride so deposited is brown in colour and adheres firmly to the wire.

When not in use keep the silver-silver chloride electrode

immersed in a test tube of water. With care an electrode will last for several years.

Agar-Potassium Nitrate Bridges. Prepare these exactly as described on p. 16 but using 51 g. of potassium nitrate instead of potassium chloride. Colour the agar gel with sufficient brom phenol blue to give a blue colour, thus readily distinguishing these tubes from the corresponding potassium

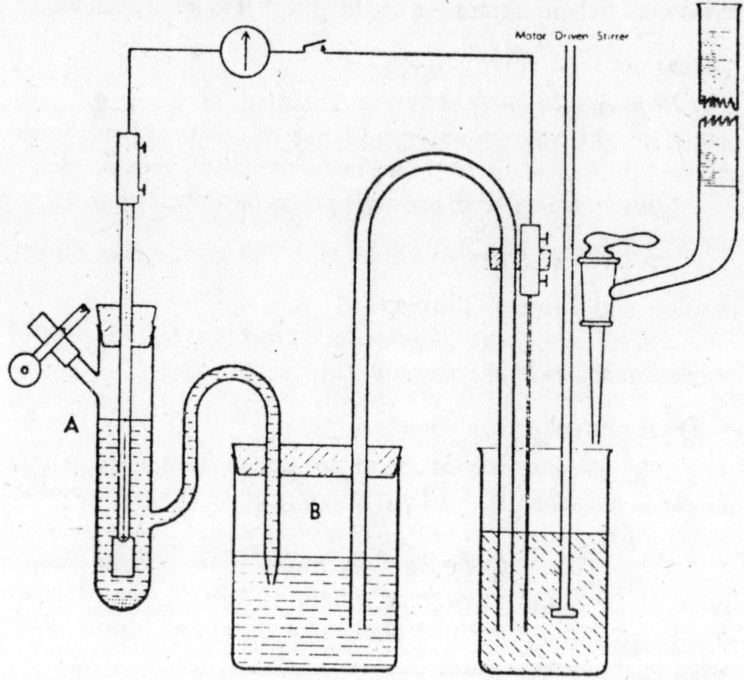

Fig. 6. R. J. Best's assembly for the electrometric determination of chlorides.

chloride tubes used for the determination of soil reaction by the quinhydrone electrode.

Quinhydrone Reference Half Cell. Partly fill a suitable electrode vessel A (Fig. 6) with the phthalate-sulphuric acid buffer solution of pH 3·2, add about 0·1 g. of quinhydrone, insert a stopper carrying a bright platinum electrode (p. 15), and shake for a few seconds. Release the screw clip on the rubber tube, just enough to allow the side arm to fill, and close the clip again. Insert the side arm through the stopper of the connecting vessel B so that it dips below the surface of the saturated potassium nitrate solution. Momentarily squeeze the rubber tube on the electrode vessel so forcing a few drops of solution out of the tip of the side arm and sucking back a corresponding amount of saturated potassium nitrate solution. This ensures a good liquid junction between the reference half cell and the potassium nitrate solution. Insert an agar-potassium nitrate bridge through the second hole in the stopper of B, and the half cell is ready for use.

Provided that potassium nitrate does not diffuse back into the quinhydrone half cell it will maintain a steady potential for 2-3 days. It is, however, preferable to fill it afresh each day that it is required.

Galvanometer. A moving coil pointer type galvanometer with an internal resistance of about 1,000 ohms and a sensitivity of about 0·3 milliamps per mm. scale division is very suitable.

Connect the galvanometer, through a tapping key, to the two electrodes, as in Fig. 6, and check the standardization of the set up, by titrating 20 ml. of 0·01N potassium chloride in 50 ml. of water with standard silver nitrate, exactly as described below for a soil suspension. If correctly set up the theoretical titration value will be obtained.

Method:

Transfer 5 g. or 10 g. of soil to a 100 ml. tall shaped beaker, add approximately 50 ml. of water, stir and leave to stand for 30 minutes. Place the beaker so that the silver-silver chloride electrode and the agar bridge tube dip into the suspension, depress the tapping key and note the initial deflection of the galvanometer to get a rough idea of the amount of standard silver nitrate to add. From a burette run in $\frac{N}{35 \cdot 5}$ silver nitrate in suitably small amounts at a time. Stir vigorously for 2-3 seconds after each addition, then depress the tapping key and note the deflection. Continue the titration until the pointer of the galvanometer just reverses direction. Warning of the end point is given by the decreasing deflections shown by the galvanometer pointer. The titration is reversible and if the end point is overshot it can be regained by back titration with standard potassium chloride. This, however, is seldom necessary.

Calculation of the Results:

One ml. of $\frac{N}{35\cdot5}$ silver nitrate is equivalent to 1 mg. of chlorine. The percentage of chlorine therefore corresponds to

$$\frac{\text{Volume of silver nitrate}}{1{,}000} \times \frac{100}{\text{Weight of soil taken}}$$

or $$\frac{\text{Volume of silver nitrate}}{10 \times \text{Weight of soil taken}}.$$

To convert chlorine values to sodium chloride multiply by 1·65.

Chlorides: Routine Electrometric Determination on a Prepared Soil Suspension.

The reagents are the same and the apparatus is set up exactly as described for the preceding method. The titration is carried out on an aliquot of the suspension prepared for the conductivity determination.

Method:

Transfer 25 ml. of the soil suspension, prepared as on p. 7, to a 50 ml. tall shaped beaker. If a pipette type conductivity cell is used in the conductivity determination it should be calibrated to deliver 25 ml. so that this volume can be transferred automatically after each conductivity measurement. Introduce the silver-silver chloride electrode into the beaker and carry out the titration exactly as previously described.

Calculation of the Results:

The percentage of chlorine in the soil corresponds to:

$$\frac{\text{Volume of silver nitrate}}{1{,}000} \times \frac{100}{25} \times \frac{100}{20}$$

$$= \frac{\text{Volume of silver nitrate}}{50}$$

The above calculation neglects the volume occupied by the soil in making the suspension. This does not significantly affect the values for low to moderate amounts of salt but for soils with a high salt content the determination should be repeated on a separate 5 g. portion of soil as in the method above.

Chlorides: Chromate Titration Method.

Pipette 50-100 ml. of the clear soil extract, prepared for the determination of water soluble salts (p. 36), into a porcelain basin and add 3-4 drops of a 1 per cent. solution of potassium chromate as indicator. Titrate the solution with $\frac{N}{35 \cdot 5}$ silver nitrate until the chloride is completely precipitated as silver chloride and the first faint tinge of red silver chromate persists. This colour change can be readily seen if a second basin, containing water and 3 or 4 drops of potassium chromate, is used as a standard for comparison. The colour change is more easily seen when the titration is carried out in a white basin instead of a beaker.

The percentage of chlorine in the original soil then corresponds to:

$$\frac{\text{Volume of silver nitrate}}{1{,}000} \times \frac{1{,}000}{\text{Volume of extract}} \times \frac{100}{200}$$

$$= \frac{\text{Volume of silver nitrate}}{2 \times \text{Volume of extract used}}.$$

REFERENCES

(1) Best, R. J.: *J. Agric. Sci.*, 1929, *19*, 533–540.

(2) Best, R. J.: *J. Agric. Sci.*, 1931, *21*, 337–365.

(3) Biilmann, E.: *J. Agric. Sci.*, 1924, *14*, 232–239.

(4) Biilmann, E. and Jensen, S. T.: *Trans. 2nd Comm. I.S.S.S.*, 1927, *B*, 236–274.

(5) Chapman, H. D., Axley, J. H. and Curtis, D. S.: *Proc. Soil Sci. Soc. Amer.*, 1940, *5*, 191–200.

(6) Crowther, E. M.: *J. Agric. Sci.*, 1925, *15*, 201–231.

(7) Heintze, S. G. and Crowther, E. M.: *Trans. 2nd Comm. I.S.S.S. (Budapest)*, 1929, *A*, 102–110.

(8) International Society of Soil Science. Soil Reaction Committee: *Soil Research*, 1930, *2*, 77–139 and 141–144.

(9) International Society of Soil Science. Soil Reaction Committee: *Trans. 3rd Int. Congress Soil Science*, 1935, *1*, 127–132.

(10) Kuhn, S.: *Z. Pflanz. Dung.*, 1930, *18A*, 309–314.

(11) McGeorge, W. T.: *J. Amer. Soc. Agron.*, 1937, *29*, 841–844.

(12) Puri, A. N. and Asghar, A. G.: *Soil Sci.*, 1938, *46*, 249–257.

(13) Snyder, E. F.: *Soil Sci.*, 1933, *35*, 43–48.

CHAPTER III

MECHANICAL ANALYSIS

The determination of particle size in soils is one of the most important of the physical determinations. Since field estimates of texture are generally based on the behaviour of the soil under the auger and during hand working, they should be checked and correlated with mechanical analyses at frequent intervals, particularly when commencing surveys in new areas.

Mechanical analysis consists essentially of two distinct operations, namely dispersion of the soil and grading the dispersed particles into size groups. Until a few years ago there was no uniformity either in the methods of dispersion or the size groups separated. The methods of dispersion in use in different laboratories varied widely from the simple boiling of the soil with water to those methods which aimed at securing complete dispersion by conversion of the clay to the sodium saturated condition after pre-treatment of the soil with hydrogen peroxide to oxidize organic matter. There were also many different groups of particle sizes separated. The existence of two such variables in the method made the comparison of mechanical analyses carried out in different laboratories difficult and uncertain. However, provided that ultimate dispersion had been attained, mechanical analyses carried out according to one system of particle size could be transposed to another system, with a reasonable degree of certainty, by the use of summation curves based on settling velocities.

In 1926, the First Commission (Soil Physics) of the International Society of Soil Science formulated tentative proposals for an international method of mechanical analysis and these proposals were adopted, with only minor modifications, at the Washington Congress in 1927. Details of two methods of dispersion were given, but it was recommended that "Method A," which aimed at securing the ultimate dispersion of the soil particles, should be adopted for fundamental work. The separation of four groups of particles, coarse sand, fine sand, silt and clay, with limiting diameters according to the Atterberg scale, was also recommended. Clay particles with a maximum diameter of 0·002 mm. were defined as those particles with a settling velocity of 10 cm. in 8 hours (0·000347 cm. per sec.) in water at 20° C., and it was agreed to use this definition of clay as the fundamental basis for the calculation, by Stokes's law, of the settling velocities corresponding to the other particle sizes in the Atterberg scale. It was later agreed, at Versailles in 1934 (2), that particles between 0·2 and 2 mm. in diameter should be separated by the use of accurate sieves with square apertures, while sedimentation methods were prescribed for all particles smaller than 0·2 mm. Elutriation methods were only permissible, as an alternative to sedimentation, for particles between 0·2 and 0·02 mm. in diameter. It was also decided at this latter conference that sodium hydroxide should be used for the dispersion instead of ammonia, which had been previously recommended. Alternative methods of dispersion were permissible, provided that it was known by experience that they yielded values substantially in agreement with the standard procedure.

The standardization effected by the International Society of Soil Science has been of great value in the efforts to secure uniformity in the methods for the mechanical analysis of soils. The recommendations were immediately accepted in Australia and Great Britain. The large number of analyses already carried out by the American system, made acceptance of the new standards slower in that country. However, the United States Bureau of Chemistry and Soils has now officially adopted the international system of particle sizes.

Stokes's law which connects the velocity of sedimentation with particle size is as follows:

$$v = \frac{2}{9} \cdot \frac{gr^2}{\eta} \cdot (D - d)$$

where v = velocity of sedimentation.
g = acceleration due to gravity.
r = radius of the spherical particles.
η = coefficient of viscosity of the fluid.
D = density of the spherical particle.
d = density of the fluid.

If all the constants are combined $v = Kr^2$. Since both η and d depend upon temperature, K will also depend on the temperature. In the international system the value of K has been indirectly defined, from the relationship connecting the particle size of clay with its settling velocity, as 34,700 at 20° C., v and r being expressed in cm. per sec. and cm. respectively.

In the application of Stokes's law to soil particles certain tacit assumptions are made. For instance it is assumed that all the particles behave as true spheres and that they all have the same specific gravity. Soil particles vary considerably in specific gravity, but, in the definition of clay by the International Society of Soil Science an average value has been accepted, and this is included in the above value for K.

All soil particles are not true spheres, so that Stokes's law is not strictly followed, but it is convenient to use it in the calculation of the assumed particle diameters, corresponding to each settling velocity. In the construction of summation curves practical considerations of scale make it impossible to plot particle diameters directly and the logarithm of particle diameter is sometimes used. It is preferable to use the logarithm of the settling velocity since this is the experimental value actually determined for the smaller particles, which constitute the important fractions in the definition of soil texture. It is, however, still necessary to make the assumption that the settling velocities of the larger sand particles follow Stokes's law.

Table 1 shows the particle size groups, and the limiting settling velocities in water at 20° C., for each group separated in the international, former British and former American systems of mechanical analysis. For the particle groups in which the separation is effected by sieving, and not based on an experimentally determined settling velocity, the latter value has been calculated from Stokes's law, using the two fundamental values defined in the international system, namely (1) particles of diameter of 0·002 mm. have a settling velocity of 0·000347 cm. per sec. at 20° C. and (2) the 70 mesh I.M.M. sieve prescribed, with an aperture 0·18 mm. square, just retains particles of 0·2 mm. diameter, the lower limit of coarse sand. It should be noted that the commonly accepted limiting

TABLE 1.

The Particle Sizes and Settling Velocities used in the International, Former British and Former American Systems of Mechanical Analysis.

System and Description of Particles.	Maximum Diameter of Particle. mm.	Method of Defining Upper Limit.	Settling Velocity at 20° C. cm. per sec.	Log. Settling Velocity.
International System				
Coarse Sand	2·0	2 mm. round hole sieve	347	2·54
Fine Sand	0·2	70 mesh I.M.M. sieve	3·47	0·54
Silt	0·02	Sedimentation: 10 cm./4·8 mins.	0·0347	$\bar{2}$·54
Clay	0·002	Sedimentation: 10 cm./8 hrs.	0·000347	$\bar{4}$·54
Former British System				
Fine Gravel	2·0	2 mm. round hole sieve	347	2·54
Coarse Sand	1·0	1 mm. round hole sieve	87	1·94
Fine Sand	0·2*	90 mesh I.M.M. sieve	2·1	0·32
Silt	0·04	Decantation: 10 cm./100 secs.	0·114†	$\bar{1}$·06
Fine Silt	0·01	Decantation: 7·5 cm./12·5 mins.	0·0114†	$\bar{2}$·06
Clay	0·002*	Decantation: 8·6 cm./24 hrs.	0·000114†	$\bar{4}$·06
Former American System				
Fine Gravel	2·0	2 mm. round hole sieve	347	2·54
Coarse Sand	1·0	1 mm. round hole sieve	87	1·94
Medium Sand	0·5	0·5 mm. round hole sieve	21·7	1·34
Fine Sand	0·25	U.S.Std. 60 mesh sieve	6·70	0·83
Very Fine Sand	0·10	U.S.Std. 140 mesh sieve	1·145	0·06
Silt	0·05	Tyler 300 mesh sieve	0·227	$\bar{1}$·36
Clay	0·005	Sedimentation: 10 cm./77 mins.	0·00217	$\bar{3}$·34
(Colloid)	0·002	Sedimentation: 10 cm./8 hrs.	0·000347	$\bar{4}$·54

* From the method used for defining the particle size these values should be 0·155 and 0·0014 mm. respectively.

† These values correspond to settling velocities of 0·1, 0·01 and 0·0001 cm. per second, respectively, at 15° C.

values for the diameters of fine sand and clay in the former British system, quoted in this table, differ from those theoretically calculated from the experimental values used in their separation. For the separation of fine sand in this system a 90 mesh I.M.M. sieve was commonly used, and the settling

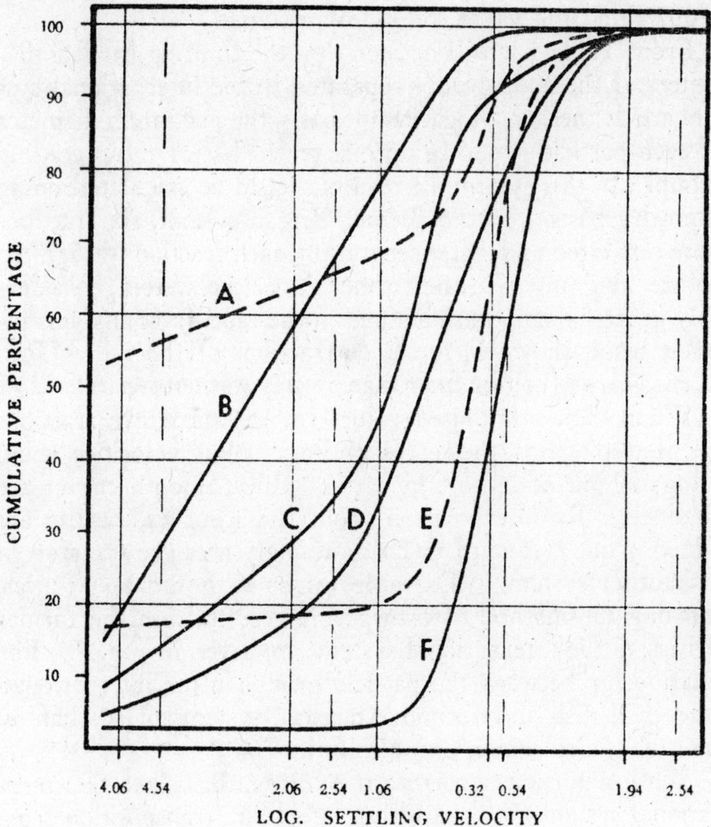

Fig. 7. Some typical mechanical composition or summation curves of soils

A. Heavy clay (Georgetown, S. Aus.).
B Silty loam (Booboorowie, S. Aus.).
C. Fine sandy loam (Waite Institute).
D. Sand to sandy loam (Kuitpo, S. Aus.).
E. Fine sandy loam (Pinnaroo, S. Aus.).
F. Sand (Copeville, S. Aus.).

The solid vertical lines correspond to the limiting settling velocities of the fractions in the former British System, while the broken lines represent the limiting settling velocities of the fractions in the International system.

velocity in the above table has therefore been calculated on the basis of the relative apertures of this screen to those of the 70 mesh screen used in the international method. Similarly the maximum diameter of the clay particles separated in this system, calculated from the time of sedimentation and the international value for K, corresponds to 0·0014 mm. instead of 0·002 mm., the value commonly assumed.

From Table 1 it will be seen that the limiting particle diameters of the four groups separated in the international system are defined on a logarithmic basis, the maximum diameter of each particle group being one-tenth that of the preceding group. In this system all results should be calculated on an oven-dry basis. In the former British system, six fractions were separated and, after separation, each fraction was ignited before weighing. In the former American system, considerably greater detail was obtained in the sand fractions, less interest being shown in the finer fractions of the soil. Until recent years a fraction finer than 5μ clay was not separated.

When the accumulated values for the successive fractions are plotted against the corresponding settling velocities, using a logarithmic scale for the latter values, smooth curves are obtained. By interpolation from these curves, results obtained in one system of mechanical analysis can be transferred to another system. Examples of some summation curves, showing the method of transposing results from the former British to the international system, are given in Fig. 7. The relationship between the particle groups in the international, former British and former American systems of mechanical analysis is also shown graphically in Fig. 8.

While interpolation from the former British to the international system can be readily carried out, transposition from the former American system is uncertain unless a fraction finer than 5μ clay has been separated. When interpolating from the former British system to the international system, allowance must be made for the difference between the ignited and oven-dry basis of the fractions. If hydrogen peroxide has been used in the pre-treatment of the soil, organic matter is largely oxidized or rendered soluble and the loss on ignition of the silt and clay fractions is mainly water of constitution.

By the use of average values for this loss a suitable correction can be applied. For Australian soils it has been found that the mean value of oven-dry international clay is 14 per cent. greater than the ignited value while the corresponding difference for silt is 7 per cent. Coarse and fine sand fractions seldom lose more than 0·5 per cent. on ignition. The values for all four fractions, interpolated from the former British system, can, therefore, be converted from the ignited to the oven-dry basis by the use of the above factors.

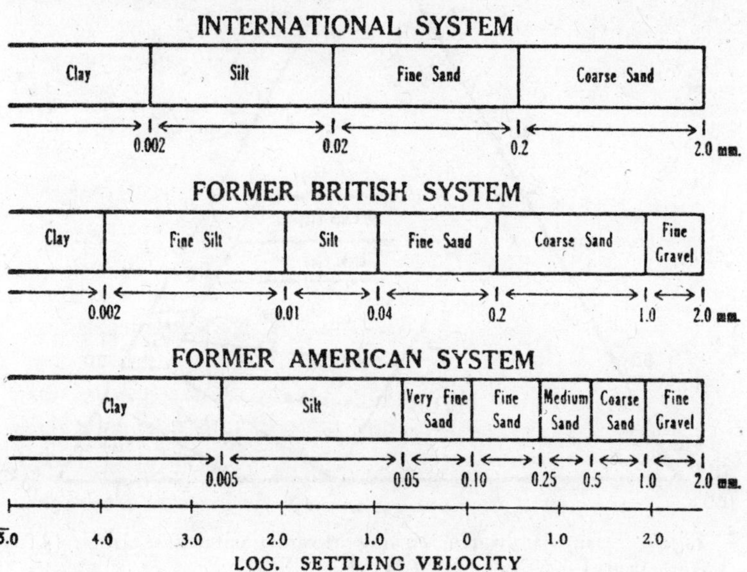

Fig. 8. A comparison of the fractions in the International, former British and former American methods of mechanical analysis.

In addition to summation curves, the results of mechanical analyses may be represented graphically by plotting the percentages of coarse and fine sand, silt and clay respectively in a triangular diagram, or all four fractions in a tetrahedron. For these diagrams the values for the mineral fractions of the soil are recalculated, before plotting, so as to give a summation of 100 per cent. Because of the influence of varying amounts of organic matter and calcium carbonate, the varying nature of the clay minerals (kaolinite or montmorillonite) and the exchangeable ions, and possibly the varying degree of microstructure in the soil, the correlation between mechanical analysis and field texture is imperfect. At the present time

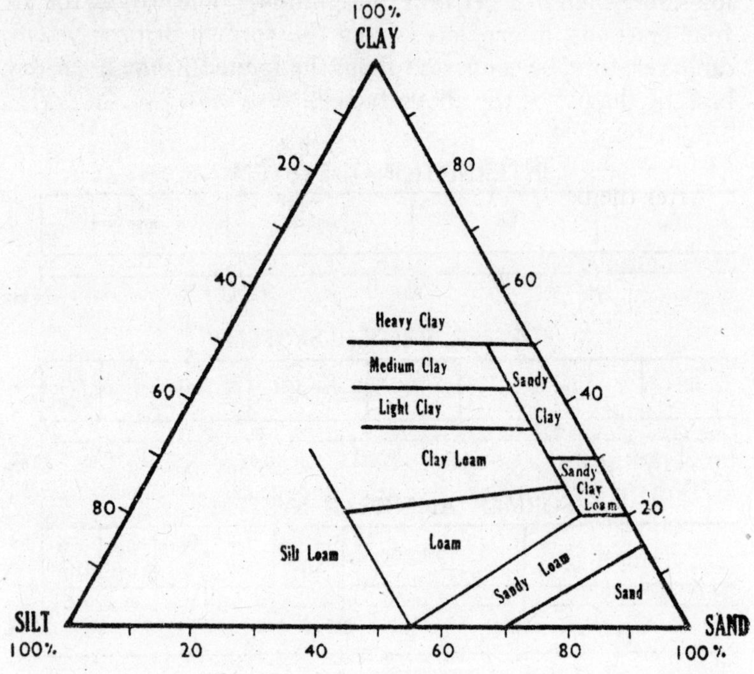

Fig. 9. Triangular diagram showing the various soil texture classes. (After J. A. Prescott, J. K. Taylor and T. J. Marshall (4)).

the best description of texture, that can be derived from the mechanical analysis, is that based on the triangular diagram of Prescott, Taylor and Marshall (4), shown in Fig. 9. This diagram applies only to mineral soils low in organic matter.

In the international "Method A" for mechanical analysis, the soil is treated with hydrogen peroxide, as originally proposed by Robinson (5) for humus soils. During this pretreatment a large proportion of the organic matter is oxidized and the subsequent dispersion of the clay is consistently better, being less subject to personal error. Oxidation of soil organic matter by hydrogen peroxide depends upon the reaction of the soil and is far from complete in calcareous and other alkaline soils. More complete oxidation can be effected by making the soil slightly acid before treatment with hydrogen peroxide but, for Australian soils, Hosking (3) has found that satisfactory dispersion is obtained after the use of the standard method of pre-treatment. Manganiferous soils lead to difficulties in the peroxide treatment since managanese dioxide decomposes hydrogen peroxide catalytically. For such soils Robinson (7) uses sodium bisulphite to reduce the manganese dioxide, before the addition of hydrogen peroxide.

After the peroxide treatment, the soil is treated with dilute hydrochloric acid, to dissolve carbonates and to remove all exchangeable metal ions. The excess of hydrochloric acid is then removed by filtration and washing, and the soil dispersed in the presence of a small amount of sodium hydroxide. The finer fractions (silt and clay) may be separated by a beaker method of decantation or determined by the pipette method of analysis, in each case using the sedimentation velocities defined by the International Society of Soil Science and shown in Table 1. If the determination is carried out at a temperature other than 20° C. the time of sedimentation must be corrected since, according to Stokes's law, velocity of sedimentation is inversely proportional to the viscosity of the fluid. Either the depth or the time of sedimentation may be altered to correct for the effect of temperature. In practice it is generally more convenient to keep the depth of sedimentation constant and to alter the time. Suitable times, appropriate to the usual range of temperature are given in Tables 2 and 3. Coarse sand is retained on a 70 mesh I.M.M. sieve while fine sand is separated from the finer fractions by decantation, as mentioned above.

Whenever sieves are used for the separation of soil fractions they should be handled carefully to avoid undue wear. Owing to the fineness of the wire, the mesh is easily strained and the size of the apertures altered. The wire gauze should be periodically examined, under a microscope, and replaced whenever serious irregularities develop. Sandy residues remaining on the screen should be rubbed as lightly as possible during sieving. Mechanical sieving is preferable to hand sieving, since variations due to personal error are minimized.

Beaker methods of mechanical analysis require no special equipment. All separations are made by repeated decantations after appropriate periods of sedimentation. Ultimate dispersion is secured since the decantations are continued until no further material remains in suspension at the end of the period of sedimentation. Soluble salts are removed completely in the first few decantations so that flocculation from this source is eliminated. Repeated puddling with a rubber pestle between decantations secures good mechanical dispersion. Instead of the standard pre-treatment with hydrogen peroxide and fifth normal hydrochloric acid, a variety of other preliminary treatments, such as N sodium chloride and N ammonium chloride, have been used. Since only one clay decantation can be made each day and from 12 to 24 decantations are usually necessary for the complete separation of clay, the beaker method is slow and tedious. On the other hand, beakers and bottles are the only equipment necessary and, if bench space is available, large numbers of analyses can be carried out at one time. Details of a beaker method, which gives results in accordance with the international system, are given on p. 75.

To avoid the tedious separation of clay and silt by decantation the pipette method of mechanical analysis was devised by Robinson (6). In this method a stable suspension of the fully dispersed soil is obtained and, by withdrawing samples at definite depths and times, by means of suitable pipettes, the change in concentration of the suspension with time, at a given depth, is obtained. If the times and depths of sampling correspond to the limiting velocities of silt and clay, the amounts of these two fractions are determined directly. Coarse sand is obtained by the use of a 70 mesh I.M.M. sieve, while the fine sand is obtained by removal of the finer fractions by a decantation method.

The accuracy of any pipette method depends primarily upon the effectiveness of dispersion and the stability of the suspension. Preliminary treatment of the soil with hydrogen peroxide and hydrochloric acid ensures that natural aggregations are broken down by the destruction or dissolution of the cementing agents, and good dispersion is readily secured by mechanical means, such as shaking, after thorough puddling of the soil paste with a rubber pestle, or agitation with a rapidly revolving propeller blade. However, the stability of the suspension is equally important. All flocculating agents must be removed before preparing the suspension. Whereas, in the beaker method, they are automatically removed during the course of the decantations, and complete dispersion is eventually obtained, in the pipette method no such separation occurs and it is essential to obtain complete dispersion, and a stable suspension, at the outset. Neutral salts, if present in sufficient quantity, would tend to flocculate the finer particles. For this reason it is preferable to leach the soil with two or three lots of hydrochloric acid, to ensure complete removal of all soluble calcium salts, and then with water, until only traces of free acid remain. The soil is more permeable to leaching with dilute acid but, as the excess of acid is removed, the permeability rapidly decreases.

When calcium carbonate is present in soils, sufficient additional acid must be used to ensure its complete solution in the initial treatment. It usually dissolves readily. If excessive frothing occurs the addition of one or two drops of capryl alcohol will break the froth. Soil carbonate sometimes occur as hard compact nodules or as dolomite and may be incompletely dissolved in the acid in one hour. In such cases the soil should be left in contact with the acid for a longer period, frequently overnight, to ensure complete solution. Gypsum soils also present certain difficulties in the pipette method, for, unless the gypsum is completely removed in the preliminary treatment, it will flocculate the subsequent suspension. Its complete removal is most readily assured by shaking the soil with dilute hydrochloric acid for several hours, before filtration and washing.

Suspensions made in the presence of a small amount of sodium hydroxide are more stable than those made with ammonia, and Robinson (7) found that for many soils dispersion was more complete. With most Australian soils the use of ammonia is satisfactory, provided that the soil is thoroughly leached after treatment with dilute acid. Its use is convenient, since no correction is necessary for the amount in solution, after evaporation of the pipette sample. However, a few heavy types of soil have been encountered, which failed to disperse completely in the presence of ammonia. For these, Walkley (10) found sodium hydroxide brought about satisfactory dispersion. It is therefore recommended that sodium hydroxide should be used for the dispersion of all soils. Because of the greater stability of sodium clay suspensions, washing of the soil after the acid treatment does not need to be so complete as in the case of ammonia dispersions. Washing until the filtrate is nearly neutral is usually sufficient.

When large numbers of mechanical analyses are required, the method given on p. 59 is recommended. This technique has been in use in these laboratories since 1925 and it has been found to give results strictly in agreement with the international method.

The pre-treatment with hydrochloric acid dissolves all calcium carbonate before determining the particle size groups. If it is desired to include the carbonate in the mechanical analysis of calcareous soils the soil can be pre-treated with sodium hypobromite as in Troell's Method (9). In this method the organic matter is oxidized by the hypobromite and the soil is left in the sodium saturated condition. After filtration and washing, it is dispersed in the presence of a small quantity of sodium hydroxide and the mechanical analysis carried out by the pipette method as usual. In many mallee soils it is found that a large part of the calcium carbonate falls within the limits of the clay fraction.

As a rapid method of mechanical analysis, Bouyoucos (1) introduced the hydrometer method. The method has been extensively used by engineers (8), particularly in connexion with the specifications of foundations for highways. For this purpose drastic treatment of the soil aggregates is avoided in preparing the suspensions and sodium silicate is commonly preferred as a dispersing agent. More complete dispersion can be obtained when sodium oxalate is used. The method is highly empirical but, for many soils low in organic matter and free from soluble salts, useful values for silt and clay can be obtained by the technique described on p. 77. The method is not recommended for soils containing calcium carbonate. Besides differing in the method of dispersion used, the hydrometer method differs from the pipette method in that the density of the suspension is determined by means of a hydrometer, instead of the concentration by weighing the amount of soil dispersed in a known volume. Whereas the depth of sampling is easily defined in the pipette method, the depth corresponding to the hydrometer reading is ill-defined, since a large volume of suspension is displaced by the hydrometer, and the effective depth varies with the length of stem immersed. The calculations necessary to determine the effective depth are very involved. They have been fully worked out by Thoreen (8). He considers that satisfactory values for particle diameters can be calculated from Stokes's law if the effective depth of immersion of the Bouyoucos hydrometer is taken as 0·42 times the total depth of immersion. Because of the other limitations of the method correction for the effective depth of immersion is seldom warranted in rapid routine analyses.

Mechanical Analysis: Pipette Method.

MOISTURE.

Transfer 10 g. of air-dry soil to a weighed silica dish, fitted with an aluminium cover. Place the dish (uncovered) in an oven at 105° C. and dry for 12-16 hours. Replace the cover, cool in a desiccator and weigh. Multiply the loss in weight by 10, to obtain the percentage of moisture in the air-dry soil.

LOSS ON IGNITION.

After determining the moisture, remove the aluminium cover from the silica dish, containing the oven-dried soil, and transfer the dish and soil to a cool muffle furnace. Raise the temperature and ignite for 30-40 minutes at a bright red heat. Replace the aluminium cover, cool in a desiccator and weigh. The further loss in weight corresponds to the loss on ignition. Multiply by 10, to obtain the loss on ignition, expressed as a percentage of the air-dry soil.

LOSS ON ACID TREATMENT.

Transfer 10 g. of air-dry soil to a 250 ml. beaker, add 100 ml. of 0·2N hydrochloric acid, and allow the soil and acid to react for one hour, stirring at intervals. If the soil contains more than 2 per cent. of calcium carbonate add an additional 1 ml. of 2N hydrochloric acid for each per cent. of calcium carbonate. Then filter through an 11 cm. Whatman No. 44 filter paper, which has been previously dried overnight in a weighing bottle, in an oven at 105° C., and weighed. Wash the soil and filter paper with three portions of 40 ml., 20 ml., and 20 ml. respectively of the fifth normal acid. Then wash with distilled water until the filtrate is nearly neutral to litmus. After draining as completely as possible, transfer the soil and filter paper to the original weighing bottle, place in an oven at 105° C., dry for 24 hours, cool in a desiccator and weigh. The difference between the first weight (the weighing bottle containing the dried filter paper plus 10·000 g., the weight of soil taken) and the weight after the acid treatment corresponds to the moisture and loss on acid treatment. Deduct the weight of moisture, already determined in another sample, and multiply by 10, to obtain the loss on acid treatment, expressed as a percentage of the air-dry soil.

In the case of soils containing hard nodular calcium carbonate or gypsum, the determination of loss on acid treatment should be made comparable with the method used in the actual preparation of the sample for mechanical analysis.

LOSS ON SOLUTION (An alternative to loss on acid treatment).

Transfer 5 g. of air-dry soil to a 250 ml. beaker, add 20 ml. of 6 per cent. hydrogen peroxide and leave to stand overnight. Cover the beaker with a clock glass and place it on a boiling water bath for 15 minutes, watching it carefully to avoid the soil frothing over. Then immerse the beaker in the bath for 5 minutes, stirring the contents if necessary, to reduce frothing. Remove the beaker, add a further 10 ml. of hydrogen peroxide and replace it on the top of the bath for 10 minutes and in the bath for 5 minutes as before. Then rinse the cover and sides of the beaker with water, dilute to about 40-45 ml., and boil the contents gently for five minutes, over a burner or hot plate.

When the contents of the beaker are cold, dilute to about 50 ml., add 5 ml. of 2N hydrochloric acid and stir well. If the soil contains more than 2 per cent. of calcium carbonate, add an extra 0·5 ml. of 2N hydrochloric acid for each per cent. present. Allow the soil to react with the acid for one hour, stirring occasionally during this period.

After standing for one hour filter through an 11 cm. Whatman No. 44 filter paper, which has been previously dried overnight in a weighing bottle, in an oven at 105° C., and weighed. Wash the soil and filter paper with three portions of 40 ml., 20 ml. and 20 ml. respectively of 0·2N hydrochloric acid. Then wash with distilled water until the filtrate is nearly neutral to litmus. After draining as completely as possible transfer the soil and filter paper to the original weighing bottle, place in an oven at 105° C., dry for 24 hours, cool in a desiccator and weigh. The difference between the first weight (the weighing bottle containing the dried filter paper plus 5·000 g., the weight of soil taken) and the weight after the acid treatment corresponds to the moisture, loss on peroxide treatment and loss on acid treatment. Deduct the weight of moisture originally present in the air-dry soil, already determined in another sample, and multiply by 20 to obtain the loss on peroxide and acid treatment, expressed as a percentage of the air-dry soil. Report this value as loss on solution.

GRADING INTO FRACTIONS.

Equipment: The following equipment is necessary or desirable for the routine mechanical analysis of soils by the pipette method. If large numbers of analyses are being carried out it is convenient to work with batches of ten soils at a time. In such cases it is necessary to have ten sets of sieves, sedimentation cylinders, pipettes and beakers. The apparatus described below has been selected or specially designed to facilitate the operations involved and to reduce the manipulations required.

Shaking Machine. Any standard shaking machine can be used for this determination, but the most convenient type is an end-over-end shaking machine, designed to hold ten sedimentation cylinders and operating at about 15-16 revolutions per minute. Since the machine takes standard sedimentation cylinders, the transference of the suspension to a shaking bottle for shaking is avoided.

Motor Dispersion Unit. For this the usual laboratory type of stirring motor (1/50 h.p.) is suitable. It should be fitted with a propeller shaft and double-bladed propeller of stainless steel. The propeller is 57 mm. across and each blade is 13 mm. at its widest point. The base of the motor dispersion unit contains a circular recess, just large enough to hold a 600 ml. squat-shaped beaker. The beaker is held in the proper position, in relation to the propeller, by means of a clamping screw. A baffle grid, consisting of six vertical baffle plates 45 mm. high and 6 mm. wide, attached at the top and bottom to flat rings 82 mm. external diameter and 6 mm. wide, fits snugly into a 600 ml. squat-shaped beaker. The baffle grid is prevented from rotating by a vertical extension which projects over the top of the beaker and clamps against the lip. Under the influence of the rapidly rotating propeller the soil and water are continuously thrown against the vertical plates of the baffle grid. These tend to oppose the rotation of the liquid and ensure efficient mechanical dispersion of the soil aggregates.

Suction Regulating Device. The arrangement, shown diagrammatically in Fig. 10, enables a filter pump to be used for filling the pipettes, and ensures a steady gentle suction on the pipette, yet prevents this suction from becoming too strong. The filter pump is turned on sufficiently full to ensure that it will continue to operate in spite of variations in the water pressure in the main supply. It is connected to the top of tube A, 4-4·5 cm. diameter and 130 cm. long. This tube is filled with water to a depth of 30-35 cm. above the lower end of tube B. It thus acts as a valve, and, should the reduction of pressure exceed this predetermined amount (30-35 cm. of water), air flows in through tube B and maintains constant suction, by preventing the vacuum increasing beyond this. Under these conditions, the pipette, connected by a rubber tube to E, can be filled very gently, without disturbing the lower layers of the suspension. To obtain uniform conditions of filling, all pipettes used with this device should have approximately the same sized jets, that is, their time of delivery should be standardized within certain limits.

Sieves. For the accurate separation of the fine and coarse sands the sieves should be made from 70 mesh Institute of Mining and Metallurgy standard brass or copper gauze. This should be tightly stretched across the sieve frame, but the

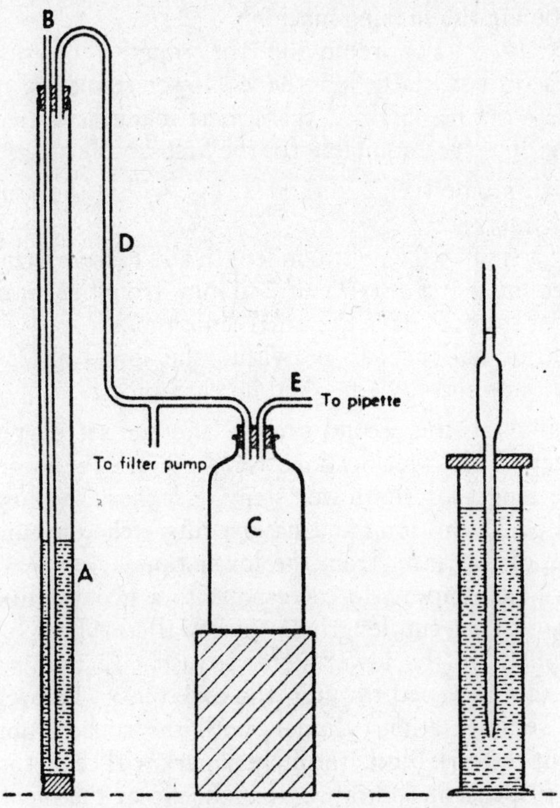

Fig. 10. A constant suction device for taking the pipette samples in mechanical analysis.

apertures must not be strained. The frames are 9 cm. in diameter at the top and at the screen. Below the screen they are funnel-shaped, tapering to 6 cm. diameter, so that they fit into the mouths of the sedimentation cylinders.

Sedimentation Cylinders. Gas cylinders, 40 cm. high and approximately 6·5 cm. internal diameter, are most convenient, provided that they are graduated to contain 1,250 ml.

and that this graduation mark is not less than 25 mm. nor more than 75 mm. below the top of the cylinder. The cylinders should have flanged tops and are closed by means of brass caps, fitted with sheet rubber gaskets, to ensure a watertight joint. The brass caps also enable the cylinders to be clamped in position in the shaking machine.

Pipettes. For removing the samples of suspension, pipettes with specially lengthened lower stems are necessary and these can be obtained from most manufacturers. Specifications of a pipette suitable for the first sample are as follows:

Pipette: to deliver 20 ml. at 20° C. within the usual limits of tolerance.

Lower stem: to be 39 cm. in length and to have a ring etched around it at a distance of 280 mm. from the lower tip.

Upper stem: to have the usual dimensions.

Time of delivery: to be within the limits of 25 and 30 seconds and to be marked on each pipette.

For removing the second sample another set of pipettes is necessary. The specifications for these are the same, except for the length of the lower stem. In these the lower stem should be 32 cm. long and have a ring etched around it at a distance of 220 mm. from the lower tip.

Mark each pipette to correspond to a given cylinder and, by means of a 1 cm. length of thick-walled rubber tubing fitting tightly on the lower stem, adjust it so that, when the lower stem is passed through the cork block (below), placed on the sedimentation cylinder, until the rubber tubing rests on top of the cork block, the etched mark will be on the surface of the suspension, indicating that the tip of the pipette is the correct distance below the surface. Note that, for the second pipette sample, the surface of the liquid will be lower than the graduation mark on the cylinder, owing to the withdrawal of 20 ml. of suspension for the first sample. Make allowance for this when adjusting the pipettes. If the rubber tubing fits closely the pipettes will remain correctly adjusted, provided that they are always used with the proper cylinder and the same cork block.

Cork Block. Cut a piece of compressed cork sheet, about 2 cm. thick, to a 10 cm. circle or square and drill a hole through the centre, 6 mm. in diameter, or just large enough to take the lower stem of the pipettes freely.

Beakers. For the fine sand decantation 500 ml. tall shaped beakers are most convenient. These should be permanently marked on the outside by means of a thin line of glass writing ink, to indicate a height of 100 mm. above the inside bottom.

Camel Hair Mop. Shorten the hairs of a camel hair mop, fitted with a wooden handle, to about 12 mm. in length and dip it in molten paraffin wax, to impregnate the base of the hairs. When cold, free the hairs from excess paraffin by working between the fingers.

Rubber Pestles. Take a rubber stopper about 25 mm. long and 20 mm. diameter at its base and drill it about three-quarter way through from its narrower end. Stretch out the core, clip it off and insert a stout glass rod about 22 cm. long into the hole.

Stirring Paddle. To one end of a spindle, 46-48 cm. long, attach a circular piece of sheet brass 6 cm. in diameter and 3 mm. thick. Drill 6-8 holes in it, each about 6 mm. diameter. This paddle produces a vigorous stirring action in the cylinders by acting in the same manner as the piston of a pump.

Reagents:

2N Hydrochloric Acid. Dilute 175 ml. of concentrated hydrochloric acid to 1 litre.

0·2N Hydrochloric Acid. Dilute 175 ml. of concentrated hydrochloric acid to 10 litres.

N Sodium Hydroxide. Dissolve 40 g. of sodium hydroxide in water and dilute to 1 litre.

None of the above reagents require standardization.

Treatment with Hydrogen Peroxide:

Transfer 25 g. of the air-dry soil to an 800 ml. tall shaped beaker, add about 50-60 ml. of 6 per cent. hydrogen peroxide and allow the reaction to proceed in the cold, preferably overnight. Then stand the covered beaker on the top of a boiling water bath, watching it carefully and removing it when necessary, to prevent the soil frothing over.

If the soil contains much organic matter, add another 30 ml. of hydrogen peroxide, after the first vigorous reaction on the water bath has ceased, say after 5-10 minutes, and leave the beaker on the top of the bath for a further 10 minutes. If the soil is particularly rich in organic matter, add a third lot of 30 ml. of hydrogen peroxide, again heating for 10 minutes on the top of the water bath. This third treatment is, however, seldom necessary. After the reaction on the top of the bath has slowed down, immerse the beaker in the boiling water of the bath for 5 minutes. Then remove it from the bath, add 25–40 ml. of hydrogen peroxide, and replace the beaker on the top of the bath for 10 minutes and in the bath for 5 minutes as before.

If the soil contains only small to moderate amounts of organic matter (e.g. most subsoils and a few surface soils) it does not require so much treatment with hydrogen peroxide. After heating such soils for about 15 minutes with the initial amount of hydrogen peroxide, on the top of the bath, immerse the beaker in the bath for 5 minutes. Then remove, add a further 25-40 ml. of hydrogen peroxide and, after a minute or two, replace the beaker on the top of the bath for 10 minutes and in the bath for 5 minutes as before. In each case, when the treatment with hydrogen peroxide on the water bath is finished, rinse the cover and sides of the beaker with water and dilute to about 150 ml. Bring the contents to the boil over a burner or hot plate and keep gently boiling for 5 minutes, watching carefully to avoid frothing over. Place aside to cool.

If the soil contains manganese dioxide, before commencing the hydrogen peroxide treatment add 100 ml. of water and 0·5-2·0 g. of sodium hydrogen sulphite and boil until the volume is reduced to about one-half. Then add hydrogen peroxide and continue with the treatment as described above.

Acid Treatment and Filtration:

When the contents of the beaker are cold clean the sides with a rubber pestle and add 25 ml. of 2N hydrochloric acid. If the soil contains more than 2 per cent. of calcium carbonate add an extra 2·5 ml. of 2N acid for each per cent. present. Then dilute until the volume is approximately 250 ml. and thoroughly rub the soil with a rubber pestle. Commence the above operation early in the morning so as to give sufficient time for the subsequent filtration and washing, which may require up to 7 hours for medium to heavy soils.

Allow the acid and soil to react for one hour, rubbing well at intervals. Then test the solution with a piece of blue litmus paper, to ensure that an excess of acid is present, and filter through a Buchner funnel, carefully fitted with an 11 cm. Whatman No. 50 hardened filter paper. Wash the soil with four successive portions, each of 50 ml., of 0·2N hydrochloric acid, draining the filter completely between each addition. Then wash thoroughly with water, adding it in small portions at a time and draining completely between each addition. Continue the leaching with water until, on testing a few drops of the filtrate, it is nearly neutral to litmus. It is usually necessary to continue the filtration until the filtrate amounts to about 800-900 ml.

Acid Treatment and Filtration (Procedure for Soils Containing Gypsum):

If the soil contains more than 2-3 per cent. of gypsum the foregoing procedure may not dissolve it completely and the gypsum remaining in the soil will interfere with the proper dispersion of the clay. After the peroxide treatment of such soils, transfer the soil and water to a shaking cylinder, dilute it to about 650 ml. and add 115 ml. of 2N hydrochloric acid. If the soil contains more than 10 per cent. of calcium carbonate, add an extra 2·5 ml. of hydrochloric acid for each per cent. present. After allowing any carbon dioxide evolved to escape, close the cylinder and shake for 8-16 hours in the shaking machine, to dissolve all the gypsum present. After shaking, filter through a Buchner funnel, fitted with an 11 cm. Whatman No. 50 filter paper. Wash the soil with six portions, each of 50 ml. of 0·2N hydrochloric acid and then with water, as before, until the filtrate is nearly neutral to litmus.

Dispersion:

After the soil has been washed nearly free from acid and drained, transfer it from the Buchner funnel to a 600 ml.

squat shaped beaker, using a spatula and two large clock glasses to assist in the transfer. After removing most of the soil spread the filter paper on one of the clock glasses and clean the last traces of soil from it by means of a camel hair mop freely moistened with water. Use the same camel hair mop to clean the Buchner funnel and the second clock glass, rinsing each with a jet of water from the wash bottle. Clean the camel hair mop by rinsing it in two or three portions of water contained in a small beaker. When the whole of the soil has been transferred to the 600 ml. beaker the volume of water should not exceed 200-250 ml. Now add 10 ml. of N sodium hydroxide, place the baffle grid in position in the beaker and clamp it rigidly in position. Place the beaker beneath the propeller of the motor dispersion unit, lower the propeller into position and set the latter in motion, controlling its speed by means of a rheostat. Adjust the speed so that vigorous stirring is effected, avoiding loss from splashing. Continue the mechanical dispersion for 10–15 minutes, then raise the motor and propeller and rinse the propeller blades into the beaker. Remove and rinse the baffle grid similarly.

Place a 70 mesh sieve in the mouth of a sedimentation cylinder. Pour the suspension on to the sieve and, with a stream of water from the wash bottle, wash as much material as possible through the sieve, until no more clay and silt remain on the sieve and the cylinder is about one-half full. Transfer the sieve to a small aluminium tray, dry and keep for the sand separation.

Place the brass cap on the cylinder and clamp it in position in the shaking machine. Shake for 12-16 hours to ensure complete dispersion. Shaking is most efficient if the cylinder is not more than one-half to two-thirds full at the time of shaking.

Dispersion (Alternative Method):

If a motor dispersion unit is not available, use the following method of hand dispersion.

Transfer the soil from the Buchner funnel to a porcelain basin of about 100 ml. capacity, using a spatula and two large clock glasses to aid in the transference. Leave the filter paper on one of the clock glasses. Into a second basin wash the funnel and clock glass, using a stream of water from the wash bottle and a camel hair mop, to loosen any adhering soil. Then remove the last traces of soil remaining on the filter paper, rubbing it with the brush and collecting the suspension in the second basin.

Then using the camel hair mop or a rubber pestle, work the soil in the first basin into a thick paste or cream, adding just sufficient water for this purpose. Rub the paste thoroughly to bring about complete mechanical dispersion of all the soil particles. This rubbing is essential for good dispersion in many soils. Continue working the soil, adding the water gradually (10-20 ml. portions at a time) until the basin is about three-quarters full. Set aside to stand for a minute or two, then decant the suspension through a 70 mesh sieve into a sedimentation cylinder. Again rub the residue with a little more water, allow to stand and decant as before. Repeat this twice more or until nearly all the clay has been separated from the sandy residue. When most of the clay has been removed the rubber pestle will be found more suitable than the brush for working the soil. Now pour the liquid from the second basin on to the sieve, puddle any sediment in the basin and transfer it completely to the sieve.

Transfer the sandy residue from the first basin to the sieve, rubbing it lightly with the brush and washing as much through as will readily go. Finally clean the brush in about six small portions of water, in one of the basins, washing the residue on to the sieve each time. Rinse the lower rim of the sieve, place it on an aluminium tray, dry, and set aside for the sand separation.

Add 10 ml. of N sodium hydroxide to the suspension in the sedimentation cylinder. The latter should not be more than one-half full. Close with the brass cap and clamp it in position in the shaking machine. Shake for 12-16 hours to complete dispersion.

Silt and Clay:

After shaking, stand the sedimentation cylinder on the bench, rinse the brass cap into it and dilute the suspension to the 1,250 ml. graduation mark. Determine the temperature of the suspension and, from Table 2, the proper time of sedimentation, corresponding to this temperature, for the first pipette sample.

TABLE 2.

The Time of Sedimentation at Different Temperatures. International System.

Temperature.	Fine Sand Decantation. Depth 10 cm.		First Pipette Sample. Depth 28 cm.	Second Pipette Sample. Depth 22 cm.	Second Pipette Sample. Depth 28 cm.
Deg. C.	Mins.	Secs.	Mins.	Hrs.	Hrs.
8	6	40	18¾	24½	—
9	6	30	18	23¾	—
10	6	20	17½	23	—
11	6	10	17	22½	—
12	6	0	16½	21¾	—
13	5	50	16¼	21¼	—
14	5	40	15¾	20½	—
15	5	30	15¼	20	—
16	5	20	15	19½	24¾
17	5	10	14½	19	24¼
18	5	0	14¼	18½	23½
19	5	0	13¾	18	23
20	4	48	13½	17½	22½
21	4	40	13¼	17¼	22
22	4	30	13	16¾	21½
23	4	30	12½	—	21
24	4	20	12¼	—	20½
25	4	15	12	—	20
26	4	10	11¾	—	19½
27	4	5	11½	—	19
28	4	0	11¼	—	18½
29	3	55	11	—	18¼
30	3	50	10¾	—	17¾
31	3	45	10½	—	17½
32	3	40	10¼	—	17
33	3	35	10	—	16¾

At a convenient time stir the suspension, by steady but vigorous up and down strokes of the stirring paddle, until all sediment is removed from the bottom of the cylinder, usually 30-45 seconds. Remove the paddle from the suspension and note the time of commencement of sedimentation.

When the proper time of sedimentation has nearly elapsed select the 39 cm.-stemmed pipette belonging to the cylinder, pass its lower stem through the hole in the cork block and check the adjustment for depth of sampling against the outside of the cylinder. Ten seconds before the correct time of sampling close the upper stem of the pipette with the forefinger, introduce the pipette into the suspension and lower the cork block and pipette until the cork block rests on the top of the cylinder, indicating that the tip of the pipette is 28 cm. below the surface of the suspension. Then fill the pipette by applying continuous gentle suction to it, from the device shown in Fig. 10. Remove the pipette from the suspension, adjust the liquid in it to the graduation mark and deliver the sample into a weighed silica capsule (64 mm. diameter and 19 mm. deep). Evaporate the sample to dryness on a water bath, transfer the capsule to an oven at 105° C., dry overnight, cool in a desiccator and weigh. Record the increase in weight as oven-dry silt and clay. After weighing transfer the capsule to a cool muffle furnace, raise the temperature and ignite for 15 minutes at a bright red heat. Again cool in a desiccator and weigh, to obtain the corresponding value for ignited silt + clay.

Leave the sedimentation cylinder to stand, away from heat and direct sunlight, until the proper time to withdraw the second pipette sample, determined from Table 2 and the average air temperature during sedimentation. For this sample use the 32 cm.-stemmed pipette, withdrawing the sample from a depth of 22 cm. below the new surface level of the suspension. Withdraw the sample, evaporate it in a silica capsule, dry and weigh exactly as in the case of the first sample. Record the weight as oven-dry clay. If desired, determine the corresponding ignited value.

The percentage of silt in the original soil then corresponds to

$$(\text{Weight of first pipette sample—Weight of second sample}) \times \frac{1{,}250}{20} \times \frac{100}{25}$$

$$= (\text{Weight of first pipette sample—Weight of second sample}) \times 250.$$

The percentage of clay corresponds to

$$\text{Weight of second pipette sample} \times 250.$$

but this latter value must be corrected for the amount of sodium hydroxide used for dispersing the suspension, since most of this remains in solution. This correction can be made with sufficient accuracy by deducting 1·6 from the percentage of clay found above.

The second pipette sample is always withdrawn after an overnight period of sedimentation. When the temperature exceeds 19-21° C. the time of sedimentation becomes inconveniently short and, for temperatures higher than this, it is more convenient to take the second sample at a greater depth, giving a correspondingly increased time of sedimentation. The times of sedimentation for a depth of 28 cm. are shown in the last column of Table 2. When using this depth of sampling for the second pipette sample it is desirable to stir the suspension with the stirring paddle after removing the first sample and to take the commencement of sedimentation from this second stirring. It will be noted that the surface of the suspension in the cylinder will be lower, due to the removal of the first pipette sample, so that the 39 cm.-stemmed pipette used for the second sample will require a different adjustment, in relation to the cork block, to that used in the first sample.

Coarse Sand:

When the sandy residue on the 70 mesh sieve is dry, rub it lightly with the forefinger and sieve until no more fine material passes through the screen. This can be done by hand but more concordant results are obtained if it be done in an automatic sieving machine, sieving for 4 periods each of three minutes, systematically changing the positions of the sieve in the machine each time. When sieving is complete, transfer the coarse sand remaining on the 70 mesh screen to a weighed crucible, dry overnight in an oven at 105° C., and weigh. To obtain the percentage of coarse sand multiply the weight by 4.

Fine Sand:

After withdrawing the second pipette sample from the sedimentation cylinder, remove the suspension to within 3-4 cm. of the bottom, by means of a siphon tube connected to a filter pump. The lower end of the siphon tube should be bent upwards to prevent disturbing the sediment on the bottom of the cylinder. Then bring this sediment into suspension and, by means of a stream of water from the wash bottle, transfer it to a 500 ml. tall shaped beaker, marked at a depth of 10 cm. above the bottom. To this beaker also add any fine sand, which has been separated from the coarse sand by sieving.

Place the beaker containing the suspension aside and, after 10-15 minutes, carefully pour off about three-quarters of the supernatant liquid so as to disturb the sediment on the bottom as little as possible. Fill with water to the 10 cm. mark, stirring the sediment on the bottom, and decant as before, after 10–15 minutes. Repeat this operation once more, by which time the turbidity will have been considerably reduced and the proper times of decantation can be used.

Determine the temperature and, from the second column of Table 2, the corresponding time of decantation for the fine sand. Now fill the beaker rapidly to the 10 cm. mark, so ensuring a thorough mixing of all sediment on the bottom, and set aside to sediment. After the pre-determined time interval, pour off the supernatant suspension, as completely as possible, without losing any material which deposits on the bottom of the beaker. Refill the beaker to the 10 cm. mark and repeat the decantations, until a negligible amount of material remains in suspension at the end of the time of sedimentation. Towards the end of the decantations rub the sandy residue twice with a rubber pestle, to remove the last of the finer fractions. Time the last few decantations so that the correct time interval occurs in the middle of the pouring off.

When decantation is complete, transfer the fine sand to a weighed silica basin, decant the excess water from it, evaporate to dryness on a water bath, dry overnight in an oven at 105° C., cool in a desiccator and weigh. To obtain the percentage of fine sand, multiply its weight by 4.

Statement of Analysis:

Tabulate the analysis showing the oven-dry percentage of each fraction in the original air-dry soil, as follows:

Coarse Sand
Fine Sand
Silt
Clay
Loss on Acid Treatment (or Loss on Solution)
Moisture
Total

Report these values, or recalculate and report the values on a moisture-free basis. If the Loss on Acid Treatment value is used instead of the Loss on Solution the total will be less than 100, since the amount of organic matter destroyed by hydrogen peroxide is not taken into account in such cases.

In order to check such analyses against loss, or possible errors in weighing, tabulate and summate the following values:

Coarse Sand (oven-dry)
Fine Sand (oven-dry)
Silt (ignited)
Clay (ignited)
Loss on Acid Treatment
Moisture
Loss on Ignition

Oven-dry values for coarse and fine sands are used since these fractions usually show little loss on ignition. The summation should approximate to 100 and, if the total is less than 98·5 or more than 102, the analysis should be regarded with suspicion. Before calculating the total for soils containing calcium carbonate or gypsum, deduct 0·44 times the percentage of calcium carbonate or 0·26 times the percentage of gypsum, corresponding to the amounts of carbon dioxide and water of crystallization respectively, since these are both included in the loss on acid treatment and the loss on ignition. Instead of determining the values for ignited silt and clay separately the ignited value for the first pipette sample may be used, provided that it is corrected by deducting 1·6 for the amount of dispersing agent in solution.

Mechanical Analysis: Beaker Method.

Transfer 10 g. of soil to a 500 ml. tall shaped beaker and treat it with hydrogen peroxide, exactly as described on p. 65 but using proportionately smaller amounts of peroxide. When cold, dilute to about 100 ml. and add 10 ml. of 2N hydrochloric acid or, if the soil contains calcium carbonate, an additional millilitre for each per cent. of carbonate. Stir well and allow to stand for one hour. Then dilute with water until the depth of liquid is about 10 cm., stir well and remove the stirring rod. Allow to stand for a few minutes, then decant the clear supernatant liquid as completely as possible and discard it. Refill with water, leave to stand until clear and again decant as completely as possible. Thoroughly rub the residue on the bottom of the beaker, with a camel hair mop or rubber pestle (p. 65), to puddle it and give good mechanical dispersion. Dilute the paste with about 100 ml. of water and wash it through a 70 mesh I.M.M. standard wire screen, collecting the turbid suspension in another 500 ml. tall shaped beaker, which is marked on the outside at a depth of 100 mm. above the bottom. Lightly rub the material on the sieve and wash thoroughly until no more clay and silt remain. Then rinse the lower rim of the sieve, place it on a small aluminium tray and dry. When dry, rub the sandy residue on the screen with the forefinger and sieve by hand until no more fine sand passes the screen. Add the fine sand to the suspension in the beaker. The coarse sand remains on the screen. Transfer it to a weighed crucible, dry overnight in an oven at 105° C., cool in a desiccator and weigh. To obtain the percentage of coarse sand in the original air-dry soil multiply the weight by 10.

Add 4 ml. of N sodium hydroxide to the suspension in the 500 ml. beaker, dilute it to the 10 cm. mark, stir well and place aside to sediment. If it still flocculates decant and discard the clear supernatant liquid and refill to the 10 cm. mark. If the suspension is turbid leave it to sediment for the time shown in the second column of Table 3, corresponding to the temperature of sedimentation. After the lapse of this time, decant the turbid suspension, without disturbing the sediment on the bottom, and collect it in a large bottle. With the camel hair mop or rubber pestle, rub the sediment on the bottom to assist dispersion, add about 0·5 ml. of concentrated ammonia and again fill to the 10 cm. mark, stirring well. Between decantations keep the camel hair mop or rubber pestle in a small beaker alongside the suspension. Continue the decantations, rubbing the residue and refilling the beaker with ammoniacal water until only negligible amounts of clay remain in suspension at the end of the period of sedimentation. This may require from 12 decantations for sandy soils to 20-30 decantations for some heavy clays. The first two or three sedimentations should be allowed to stand somewhat longer than usual for, when the suspension is very turbid, it is not easy to see that the sediment on the bottom is not disturbed during the decantation.

TABLE 3.

The Times of Sedimentation for the Clay and Silt Fractions. Depth of Sedimentation 10 cm.

Temperature. Deg. C.	Clay Decantation. Hrs. Mins.	Silt Decantation. Mins. Secs.	Temperature. Deg. C.	Clay Decantation. Hrs. Mins.	Silt Decantation. Mins. Secs.
8	11 0	6 40	21	7 50	4 40
9	10 40	6 30	22	7 40	4 30
10	10 25	6 20	23	7 25	4 30
11	10 10	6 10	24	7 15	4 20
12	9 50	6 0	25	7 5	4 15
13	9 35	5 50	26	6 55	4 10
14	9 20	5 40	27	6 45	4 5
15	9 5	5 30	28	6 40	4 0
16	8 50	5 20	29	6 30	3 55
17	8 35	5 10	30	6 20	3 50
18	8 25	5 0	31	6 15	3 45
19	8 10	5 0	32	6 5	3 40
20	8 0	4 48	33	5 55	3 35

When the whole of the clay suspension has been collected make it acid with a small excess, about 3-4 ml., of acetic acid and set aside to flocculate. When flocculation has occurred siphon off the clear supernatant liquid, as completely as possible, and wash the residue into a tall beaker. Again leave until the supernatant liquid is clear and siphon it off. Transfer the clay residue to a weighed evaporating basin, add one or two millilitres of ammonia and evaporate it to dryness on a water bath. Dry in an oven at 105° C. for 24 hours, cool in a desiccator and weigh. To obtain the percentage of clay multiply by 10.

To separate the silt from the residue remaining in the decantation beaker add water, stir well and fill to the 10 cm. mark. Note the temperature and set aside to sediment for the time indicated in the third column of Table 3. After the proper time decant the turbid suspension and collect it in a large bottle. Repeat the sedimentations and decantations, until negligible amounts of silt remain in suspension.

When the whole of the silt suspension has been obtained allow it to stand for 2-3 days so that the silt settles to the bottom of the bottle. Remove the supernatant liquid by siphoning, concentrate the silt further by leaving the remaining suspension to settle in a tall shaped beaker and then rinse it into a weighed evaporating basin. Evaporate the suspension to dryness on the water bath, dry overnight in an oven at 105° C., cool in a desiccator and weigh. To obtain the percentage of silt multiply by 10.

Transfer the fine sand remaining in the decantation beaker to a small evaporating basin, decant the surplus water, evaporate to dryness, dry overnight in an oven at 105° C., cool in a desiccator and weigh. Multiply the weight by 10 to obtain the percentage of fine sand.

Determine Moisture and Loss on Acid Treatment by the methods given on pp. 59–61.

Mechanical Analysis: Hydrometer Method.

This method, although more rapid than the pipette method, gives approximate values only, for silt and clay. It should not be used for saline or organic soils nor for soils which are known to be difficult to disperse. It also gives anomalous results with calcareous soils since there is no pre-treatment with hydrochloric acid to remove the calcium carbonate. Details of the method used by T. J. Marshall in these laboratories are as follows:

Equipment:

Hydrometer. The ordinary Bouyoucos hydrometer is used. The stem of this hydrometer is calibrated to read directly in percentages of soil remaining in suspension. It is adjusted for a temperature of 67° F. (19·4° C.). The overall length of the hydrometer is 285 mm., the stem being 150 mm. long.

Motor Dispersion Unit. See p. 62.

Stirring Paddle. See p. 65.

Method:

Transfer a weighed quantity of air-dry soil, equivalent to 50 g. of oven-dry soil for medium to heavy textured soils or 100 g. for sandy soils, to a 600 ml. squat shaped beaker, add about 200 ml. of water and 15 ml. of 0·5N sodium oxalate. Place the baffle grid in the beaker and disperse in the motor dispersion unit for about 20 minutes. Then wash the soil into a suitable sedimentation cylinder and dilute the suspension to 1 litre. Thoroughly stir the suspension, working the stirring paddle steadily up and down in it. Then remove the paddle and note the time of commencement of sedimentation.

Determine the percentage of silt and clay in suspension, by noting the hydrometer reading 5 minutes after the commencement of sedimentation, and the percentage of clay, from the hydrometer reading after 5 hours' sedimentation. To make these readings, carefully introduce the hydrometer into the suspension 20-30 seconds before the pre-determined time and note the reading at this time. If frothing at the surface of the suspension makes it difficult to read the scale, add one or two drops of amyl or capryl alcohol to break the froth.

If the temperature of the suspension differs markedly from 19-20° C., make a correction to the scale reading by adding 0·3 units for every degree above 19·4° C. or subtracting the same amount for each degree below 19·4° C. The values so obtained correspond directly to the percentages of silt + clay and clay in the oven-dry soil if a 100 g. sample was taken. If, however, only the equivalent of 50 g. of oven-dry soil was used multiply the corrected scale readings by 2.

If the values for fine and coarse sand are required these should be determined by sieving and decantation, exactly as described for the pipette method (pp. 72–73).

REFERENCES

(1) Bouyoucos, G. J.: *Soil Sci.*, 1927, *23*, 319–330 and 343–353.

(2) First Commission I.S.S.S.: *Trans. 1st Comm. Int. Soc. Soil Sci. Versailles*, 1934, 303 and 305.

(3) Hosking, J. S.: *J. Agric. Sci.*, 1932, *22*, 92–100.

(4) Prescott, J. A., Taylor, J. K. and Marshall, T. J.: *Trans. 1st Comm. Int. Soc. Soil Sci. Versailles*, 1934, 142–153.

(5) Robinson, G. W.: *J. Agric. Sci.*, 1922, *12*, 287–291.

(6) Robinson, G. W.: *J. Agric. Sci.*, 1922, *12*, 306–321.

(7) Robinson, G. W.: *Imp. Bur. Soil Sci.*, 1933, Tech. Comm. 26.

(8) Thoreen, R. C.: *Public Roads*, 1933, *14*, 93–105.

(9) Troell, E.: *J. Agric. Sci.*, 1931, *21*, 476–483.

(10) Walkley, A.: *J. Counc. Sci. Ind. Res. Australia*, 1936, *9*, 321–322.

CHAPTER IV

SINGLE VALUE PHYSICAL CONSTANTS

From time to time many different single value constants have been proposed in attempts to characterize a soil by a single number instead of a group of figures, such as is obtained, for example, in a mechanical analysis. Most of these methods are highly empirical but the determinations are generally simple and rapid and enable large numbers of soils to be examined. Many of these single value determinations measure related soil properties. For instance, in many soils sticky point can be correlated with loss on ignition or with total cation exchange capacity, both of these latter values being related to the colloidal content of the soil. Moisture equivalent, sticky point and heat of wetting also frequently serve as good indexes of soil texture. The determination of any one of these constants can be carried out in a fraction of the time required for a complete mechanical analysis. The correlations with texture are not perfect, due to the varying contributions of the organic and inorganic fractions of the colloid to the total. However, provided that they are restricted to soils within closely related groups and to soils within these groups having similar amounts of organic matter, the correlations are sufficiently good to warrant the use of these single value constants as indexes of texture in much preliminary survey work.

Many of the single value constants measure the water relationships of the soil and the introduction by Schofield (13, 14) of the pF scale, for expressing these water relationships in terms of free energy, has done much to bring out the relationship of several of these constants to each other. A soil will absorb or hold water against a pressure difference until equilibrium is reached and this pressure difference is a measure of the force with which water is bound to the soil. It has been spoken of as a suction force and varies from a very small value in a saturated soil to that equivalent to the suction of a column of water 10 kilometres high in a soil in equilibrium with an atmosphere of 50 per cent. relative humidity, or nearly 10 times this value in the case of an oven-dry soil. Buckingham originally introduced the expression "capillary potential" to express this tension, on the energy scale, with which water is bound to the soil. Schofield's expression pF corresponds to the logarithm of the capillary tension and its use gives a scale analagous to the pH scale. The term pF is defined as the logarithm of the height of a column of water equivalent to the suction force in equilibrium with a soil at the given moisture content. The use of the pF scale enables the expression of these widely differing tensions on a simple scale of 0–7.

Soils behave differently depending on whether they are drying out or wetting and the curve connecting pF and moisture content for a soil drying out differs from that for a soil being wetted. When the whole curve connecting pF and moisture content is plotted the positions of the various single value moisture constants can be marked on it. The pF values generally assigned to some of these constants, on the drying curve of the soil, are as follows:

Field capacity	pF 3·2
Moisture equivalent	pF 2·7 - 2·9
Permanent wilting percentage	pF 4·2
Moisture at 50 % relative humidity . .	pF 6·0
Oven-dry soil	pF 6·9

When it is recognized that the slopes of the moisture curves differ from soil to soil, it is clearly seen why no general correlation exists between permanent wilting percentage and moisture equivalent. The value of this ratio is a characteristic of the soil, being a measure of the slope of the curve between pF 2·9 and pF 4·2.

Over the important range of soil moistures (pF 3·0-4·4) the pF value is most readily determined by the freezing point depression method since pF = 4·1 + log t, where t is the freezing point depression in degrees C. For the range pF 4·15-7 vapour pressure determinations can be used.

In addition to the single value constants described below, engineers find other physical constants, such as the Atterberg plastic tests, soil shrinkage, and the Proctor value, or the moisture content at the point of maximum density, of considerable importance. For further information on these and other values the reader should consult one of the standard works on the subject, such as *Engineering Properties of the Soil*, by C. A. Hogentogler.

WATER HOLDING CAPACITY

The water holding capacity, or the amount of water taken up by unit weight of dry soil when immersed in water under standardized conditions, is most conveniently determined by the Keen-Raczkowski box experiment using the circular shaped boxes described by Coutts (4). The dimensions of the boxes used in these laboratories differ slightly from those described by Coutts. Their height is the same but the internal diameter of 5·6 cm. was chosen to permit the use of a standard filter paper, 5·5 cm. in diameter. In their original paper, Keen and Raczkowski (10) showed that by extending Hilgard's experiment several other values, such as apparent specific gravity, pore space, true specific gravity and volume expansion, could be determined, in addition to the water holding capacity of the soil. The determination of the latter value is the most generally useful and details of the method for its determination are given below. Reference should be made to the original paper if the other soil constants are required.

Knowledge of the water holding capacity of a soil is desirable in pot culture technique, since it provides a simple means of determining useful moisture levels, at which to maintain soils for good plant growth. With medium textured soils in pots, good growth is obtained at moisture levels corresponding to 50-70 per cent. of the total water holding capacity of the soil.

For the determination of water holding capacity the air-dry soil is crushed, so that as much as possible will pass through a fine sieve. For most purposes a sieve with round holes, 0·5 mm. in diameter, is sufficiently fine. After mixing the coarse material back into the sample, the soil is packed into the container, by adding small quantities at a time, with gentle but repeated tapping of the container, on a block of cork or wood, between each addition. When the box is full it is placed in water so that the soil is wetted through the perforated bottom. After standing overnight the soil is fully saturated and the amount of water retained is determined and expressed as a percentage of the oven-dry soil.

Provided that the prescribed conditions are rigidly followed, closely reproducible values can be obtained by different observers, or by the same observer on different occasions. It is essential that the air-dry soil be carefully and uniformly packed into the box. The preliminary fine crushing of the soil and the addition of small quantities at a time, with repeated tapping of the box between additions, help to secure uniformity in the packing. Lack of uniformity leads to inconsistent results. Insufficient packing is the most frequent source of error and gives unduly high values for water holding capacity. The depth of water surrounding the boxes during soaking should also be standardized. A depth of one-quarter of an inch should be maintained, more water being added to restore this depth if much is absorbed by the soil.

The Determination of Water Holding Capacity.

Equipment:

Circular Brass Boxes. These are spun from 20 gauge brass and have an internal diameter of 5·6 cm. and a height of 1·6 cm. The bottom is perforated with numerous holes, 0·75 mm. in diameter, spaced at 4 mm. centres. Each box is supplied with a split brass ring, made from spring wire, which serves to hold the filter paper in position over the perforated bottom.

Method:

Crush the air-dry soil in a porcelain mortar and sieve through a small sieve having round holes, 0·5 mm. in diameter. Continue crushing the coarse residue, so as to disintegrate clay aggregates yet avoiding the actual grinding of any sand particles. When crushing is as complete as possible return the coarser particles, remaining on the sieve, to the finer fraction and incorporate them with it by thorough mixing. Crush sufficient soil to enable three determinations to be made.

Place a thin 5·5 cm. filter paper (Whatman No. 1 or No. 44) on the bottom of one of the circular brass boxes and fix it in position by means of the split brass spring. Weigh the box and filter paper and record it as "weight of box unfilled" *(a)*. By means of a spatula transfer several small portions of soil, from different parts of the heap of crushed soil, to the box until the latter is about one-quarter to one-third full. Pack the soil, by tapping the box smartly but gently, 20-30 times, on a sheet of cork or wooden bench top. Add further small portions of soil, not more than 5-6 g. at a time, and tap the box as before, between each addition, so as to secure uniform packing. Continue the addition of soil, tapping the box systematically between each addition, until it is nearly full. Then add sufficient soil to fill the box and strike off the surplus until level with the top of the box, using the straight edge of the spatula. Smartly tap the upper edge of the box several times with the edge of the spatula, add a little more soil and again strike off the surplus as before. Avoid the loss of the coarser grained particles, which tend to work their way on to the upper surface. For the determination of water holding capacity it is not necessary to continue packing the soil until further settlement ceases.

Having properly packed the soil into the box, place it in a 9-11 cm. Petri dish and add water to the dish until the depth of water is one-quarter of an inch. After a time add a little more water, if necessary, to restore this depth. Place the dish aside and leave overnight. During the absorption of water considerable movement of the soil may take place. If there is much expansion, crumbs of soil may fall from the top but this will not affect the determination of water holding capacity.

After standing in water for 12–16 hours equilibrium will have been reached and the soil will be fully saturated. Remove the box containing the saturated soil, carefully wipe it dry on the outside and weigh it, recording the weight as "weight of box + saturated soil" *(b)*. After weighing place it in an oven at 105° C., supporting it on a triangle so as to allow free access of air to the perforated bottom. Dry for 24 hours, or until constant weight is attained, cool in a desiccator and weigh again, recording the weight as "weight of box + oven-dry soil" (c).

Determine the amount of water absorbed by the filter paper as follows. Weigh five filter papers together, saturate them with water, place them on a flat glass plate and squeeze gently by rolling with a glass rod. Weigh again to determine the amount of water retained. From this calculate the average amount (d) retained by one paper. The amount of moisture present in the air-dry filter paper can be neglected. Calculate the water holding capacity, as a percentage of the oven-dry soil, from the expression

$$\frac{b - c - d}{c - a} \times 100$$

a, b, c, d having the values noted above.

Since the box containing the saturated soil should not be placed directly on the scale pan, owing to moisture seeping from the box, it is preferable to place a watch glass on the scale pan and to stand the box on this watch glass during all the weighings. Its weight will not then enter into the calculations.

MOISTURE EQUIVALENT

The expression "moisture equivalent" was originally introduced by Briggs and McLane (2) to represent the moisture held in the smaller capillary spaces of the soil and was later more precisely defined by them as the amount of water retained by a soil when a layer of that soil, 1 cm. deep, is centrifuged for 40 minutes in a gravitational field equal to 1,000 times that of gravity. Although they clearly described the procedure to be used in making moisture equivalent determinations, significant variations in technique have developed in different laboratories. Viehmeyer and his colleagues (19, 20) have studied the influence of many factors on the determination and have worked out a technique, leading to more concordant values in the hands of different operators. Their method differs essentially from the original in that the time of centrifuging is reduced to 30 minutes, while no attempt is made to use an amount of soil, which will give a centrifuged layer, 1 cm. in depth. They found that the most concordant values were obtained when a constant weight of soil was taken for each determination.

Moisture equivalent is an empirical value and the conditions for its determination must be closely standardized. Viehmeyer's technique has been adopted in so many different laboratories that the commonly accepted definition of moisture equivalent may now be stated as the amount of water, expressed as a percentage of the oven-dry soil, retained by a soil when approximately 30 g. of it are centrifuged for 30 minutes in a gravitational field of 1,000 *g*.

The reduction in the time of centrifuging, from 40 to 30 minutes, probably alters the value very little, since most soils rapidly lose their excess water during the first 10 minutes in the centrifuge. If loams and lighter textured soils are centrifuged for longer than 30 minutes, an undue amount of cracking occurs and this increases the losses by evaporation during centrifuging, so introducing greater errors from this source. With these light textured soils equilibrium is rapidly reached and it is undesirable to use a period of centrifuging longer than 30 minutes. Since this period is also sufficiently long for most heavy soils to reach equilibrium its adoption for all soils is warranted.

While the depth of the centrifuged soil layer is of considerable importance from theoretical considerations, Viehmeyer considered that the depth of soil need not be rigorously controlled. It is not possible to choose a suitable weight of soil, which, on centrifuging, will give a final layer 1 cm. thick, without making preliminary trials for every sample. Practical considerations, therefore, make it much more convenient to take a constant weight of soil for every determination, rather than attempt to choose an amount to give a final depth of 1 cm. By the use of a fixed weight of soil, concordant values can be more readily obtained.

The method given on p. 89 is essentially the technique of Viehmeyer. It differs only in the regular use of 30 g. of air-dry soil, instead of 30·5–31·5 g., an amount chosen to represent 30 g. of oven-dry soil. This difference leads to negligible differences in the values obtained. Moisture equivalent determinations should always be carried out on air-dry soils and the preliminary crushing to pass the 2 mm. screen should be as light as possible.

Soils, in which exchangeable sodium represents an appreciable proportion of the total exchangeable cations present, frequently fail to reach equilibrium in the moisture equivalent centrifuge and this is indicated, either by free water on the surface, or by the upper layers of the centrifuged block being wetter and softer to the touch than the lower layers. Owing to the impermeability of these sodium saturated soils, the water cannot filter through the block, even in the high gravitational field, to reach equilibrium in 30 minutes. Some soils remain so fluid that a separation of clay and sand occurs during centrifuging. The moisture equivalent value for any soil, the upper surface of which feels wetter to the touch than the lower surface, should be rejected. However, since moisture equivalent behaves as an additive function a dilution method can be used for the determination of the probable moisture equivalent of these soils. By mixing the soil with 30, 50, and 70 per cent. of pure sand and determining the moisture equivalent of each mixture a useful approximation can be obtained by extrapolation (12).

Coutts (5) has described a modification of the ordinary method, by which moisture equivalent determinations can be carried out on about 4 g. of soil. The soil is placed in a hole, ⅝ inch in diameter, drilled in the centre of a paraffined wooden block, which just fits the moisture equivalent boxes. As the result of trials with several soils, he found that, when 3·8 g. were centrifuged in this block, the results were the same as those given by the standard method, using 30 g. of soil. Before using this modification the relationship should be checked, since the amount of soil taken represents a considerably greater depth of soil than in the standard procedure.

Bouyoucos (1) has shown that values closely approximating to the moisture equivalent can be obtained by a simple suction method, using a Buchner funnel and filter pump. In this method the air-dry soil is placed on a filter paper in a small Buchner funnel and carefully compacted. After soaking in water for 24 hours the funnel is connected to a good vacuum pump and suction applied until the whole of the excess of water has drained from the top of the soil. Suction is continued for a further 15 minutes, during which time the funnel is covered with a moist cloth to prevent evaporation. The soil is then removed from the funnel and its moisture content determined. In general this method gives higher values than the true moisture equivalent but may be of some value in those laboratories in which the few determinations required do not warrant the purchase of the expensive centrifuge equipment. Values obtained by the Bouyoucos procedure should be clearly designated as such.

Moisture equivalent determinations are of value in giving a rapid index of soil texture. The correlation between moisture equivalent and mechanical analysis, although imperfect, is valuable. The contribution of soil organic matter to the moisture equivalent is about 130 per cent., i.e. the moisture equivalent of organic matter is 130 (12). This value is the mean of several experimental determinations. The contribution of clay is much more variable and is probably of the order of 40-70 per cent. In using moisture equivalent as an index of texture, comparisons should be restricted to soils of the same family and, within the family, the soils should be grouped according to their character and position in the profile. Values for heavy clay subsoils should not be compared with those for the lighter and more organic surface horizons. By restricting the comparisons in this way, Taylor (16) has used moisture equivalent as a rapid means of determining the approximate clay content of soils from certain irrigation areas, the texture of which proved difficult to assess in the field.

Moisture equivalent is more frequently used in the study of the water relationships of irrigation soils. Viehmeyer and Hendrickson (18) have found that, for fine textured soils, the moisture equivalent gives a good measure of field capacity, and moisture equivalent is frequently used to enable the laboratory determination of this value. The relationship between moisture equivalent and field capacity does not hold for sandy soils nor for very heavy clays.

The Determination of Moisture Equivalent.

Equipment:

Briggs-McLane Moisture Equivalent Centrifuge. A centrifuge fitted with a special drum head is necessary for the determination of moisture equivalent. The head is constructed to hold 16 brass cups or boxes, with gauze bottoms and sheet metal lids. Each box is 2 in. square and 1 in. deep and the bottoms are curved in the arc of a circle, so as to fit closely against the circumference of the revolving drum head. The boxes and lids are numbered and the weight of each pair is balanced, to ensure smooth running of the centrifuge. On no account should the centrifuge be run with fewer than the full number of boxes. As usually supplied, the centrifuge is fitted with a speed governor, to control its speed at 2,440 r.p.m., corresponding to a centrifugal force 1,000 times that of gravity. Better control of the speed can, however, be obtained by fitting the machine with a speed indicator and controlling the rheostat manually.

Aluminium Dishes. Numbered aluminium dishes, with slip over lids, 8 cm. in diameter and similar in shape to Petri dishes are very suitable for holding the centrifuged soil during drying and weighing.

Method:

Carry out this determination in duplicate, using a balanced pair of boxes and placing the duplicate samples diametrically opposite each other in the head of the centrifuge, so as to obtain a well balanced load during centrifuging. If there is any doubt about the constancy of the speed control on the centrifuge, it is desirable to include a standard soil in each series of determinations. For this purpose a large bulk sample of a suitable soil should be prepared. Lack of reproducibility of the moisture equivalent for this standard soil will then indicate faulty speed control. For a soil with a moisture equivalent of 20, values should be reproducible within $\pm$ 0·3 units.

Cover the gauze bottoms of the centrifuge boxes with closely fitting squares of Whatman No. 2 filter paper and to each box add 30 g. ($\pm$ 0·1 g.) of air-dry soil, which has been screened through a 2 mm. screen. Tap the box lightly, to level the surface of the soil. When all the centrifuge boxes have been filled stand them in a metal tray and pour water into the tray until it surrounds the boxes to a depth corresponding to that of the soil in the boxes. Leave to stand overnight, then remove the boxes and stand them on a wet towel. Cover them with another damp cloth, to prevent evaporation from the soil surface, and leave them to stand for 30 minutes, so that excess water drains from the soil.

After draining, close each box with its correspondingly numbered lid, securely clipping it in place. Then load the boxes into the drum head of the centrifuge, placing corresponding boxes containing duplicate determinations directly opposite each other, to maintain good balance during centrifuging. When all 16 boxes have been placed in position close the drum head with its cover plate and screw down securely. Close the protecting cover of the centrifuge and turn on the power. Then, by moving the rheostat arm slowly and evenly, bring the centrifuge up to its proper speed of 2,440 r.p.m., taking not less than 3 minutes nor more than 5 minutes for this operation. It can be done conveniently in 3 minutes by keeping the rheostat on each stud for 20 seconds; speeding up at a rate greater than this puts an excessive strain on the electrical circuit and leads to trouble. Maintain the speed at 2,440 r.p.m. for 30 minutes, either by means of a governor or by manual control. After 30 minutes shut off the power and bring the machine to rest within 3 minutes, by braking for 20 seconds on each rheostat position.

Remove the boxes from the centrifuge head and quickly transfer the soil to covered aluminium dishes, previously weighed (weight = a). Carry out this transference without delay, so as to avoid errors arising from evaporation losses at this stage. Remove the filter paper from each soil block and discard it. Feel the top and bottom surfaces of the soil block so as to detect and reject any soils (e.g. sodium saturated soils) which have not reached equilibrium. Immediately cover the dish with its lid, and, as soon as all samples have been transferred, weigh each dish, to obtain the weight of soil and moisture at the moisture equivalent (b). Remove the lid, place the dish and moist soil in a well ventilated oven at 105° C., dry for 16–24 hours, cover the dish with its lid, cool in a desiccator and weigh as oven-dry soil (c). An accuracy of 0·01 g. is sufficient in all three weighings.

The moisture equivalent of the soil then corresponds to

$$\frac{b-c}{c-a} \times 100$$

If it is desired to carry out more than 16 moisture equivalent determinations per day, without going to the expense of additional sets of centrifuge boxes, this can be done by using filter papers folded to fit the inside of the metal boxes. Over a wooden former, fold a 4-inch square of filter paper so that it will fit snugly into the centrifuge box. Place the folded filter paper in a bottomless brass frame, 2 in. square and 1 in. deep, and transfer 30 g. of soil to it. Soak the soil in water, as previously described, then lift the filter paper and soil from the brass frame and place it in the centrifuge box. Drain for 30 minutes as before and continue with the determination in the standard manner.

FIELD CAPACITY

Field capacity is defined as the amount of water, expressed as a percentage of the oven-dry soil, held in a soil after the excess of gravitational water has drained away, and after the rate of downward movement of water has materially decreased (18). From this definition it is seen that field capacity is essentially a measurement to be made in the field, on an undisturbed soil. There must be free drainage through the profile, and there must be no water table near the horizons investigated.

It is generally assumed that the downward movement of water ceases within 48-72 hours after the soil has been fully wetted, either by rainfall or irrigation. During this time the surface of the soil must be protected from further rain or from loss of moisture by evaporation.

The value for field capacity is less than that for water holding capacity since the latter corresponds to the moisture present in a fully saturated soil resting on a water table. Under these conditions, all the soil pores are completely filled with water. The field capacity of sands and other light textured soils is generally higher than the moisture equivalent. For the finer textured soils, with the exception of very heavy clays, field capacity corresponds very closely to moisture equivalent. For soils with a moisture equivalent of about 14 to 30 or 35, the laboratory determination of moisture equivalent may replace the determination of field capacity.

The Determination of Field Capacity.

After rainfall or an artificial irrigation, sufficiently heavy to saturate the profile, cover the soil surface to prevent evaporation or further wetting. After 2-3 days, by which time equilibrium will have been established, sample the soil at the various depths desired and determine moisture in each sample. If a pit is dug to facilitate sampling, remove the face of the soil to a depth of 2-3 inches, immediately before collecting the sample at each depth. Without undue delay transfer the sample to a tin with a closely fitting lid and transport it to the laboratory, protecting the tin from the direct heat of the sun's rays. In the laboratory crumble or break the soil into small pieces, to facilitate subsampling, and determine moisture by drying representative portions, usually about 50 g., in an oven at 105° C. until constant in weight.

To obtain the field capacity, express the moisture lost on drying as a percentage of the weight of oven-dry soil remaining.

PERMANENT WILTING POINT

Briggs and Shantz (3) defined the wilting coefficient as the moisture content of the soil, expressed as a percentage of the dry weight, at the time when the leaves of a plant growing in that soil first undergo a permanent reduction in their moisture content, as a result of the deficiency in the soil-moisture supply. At this moisture content the leaves of the plant wilt and cannot recover their turgor in an approximately saturated atmosphere, without the addition of water to the soil. This moisture content is often referred to as the permanent wilting percentage; it corresponds to the point at which the soil can no longer supply moisture to a plant growing in it at a sufficient rate to maintain turgor and the plant wilts permanently.

Most plants reduce the moisture content of a given soil to approximately the same level before wilting occurs. This permanent wilting point is a characteristic of the soil. Since it corresponds to the lower limit of moisture available for plant growth, a determination of its value, together with that of field capacity, defines the total amount of water that can be held by a given soil in a form available for plant growth.

The permanent wilting percentages may be determined either directly or indirectly. Its direct determination is desirable whenever possible. The method is simple although it is tedious and slow. It is necessary to make four or five replicate determinations in order to obtain statistically reliable values, while the plants take about six weeks to reach a suitable stage of development. Since most plants reduce the moisture content of a given soil to the same level, the choice of plant is not important. However, some plants show signs of wilting more clearly than others and are, therefore, more suited for use in the determination. The dwarf sunflower gives excellent indications of the onset of wilting and is very convenient to handle experimentally. As it has been used by numerous investigators its choice is preferable, whenever possible. Instead of growing the plant in soil contained in an impermeable pot and sealing the surface of the soil with wax, a closed tin may be used. The small opening in the lid, through which the stem of the plant protrudes, is plugged with cotton wool to prevent evaporation from the soil surface. During the growth of the plant the tin should be protected from the direct rays of the sun, to prevent big fluctuations in temperature.

In order to avoid the tedium and delay associated with the direct determination, Briggs and Shantz suggested that the wilting coefficient could be obtained from the moisture equivalent by dividing the latter by 1·84, a value derived from direct experimental determinations on several different soils. Unfortunately this relationship is far from constant (p. 81). With different soil types values ranging from 1·4-3·8 have been recorded for the ratio $\frac{\text{moisture equivalent}}{\text{permanent wilting percentage}}$.

The general use of a constant factor for this ratio is fundamentally unsound and leads to entirely erroneous values. It is, however, permissible to use the relationship between moisture equivalent and permanent wilting percentage to obtain a quick estimate of the latter value, provided that the correlation is restricted to soils in a closely related group and that the ratio is established experimentally for a few typical soils of the group.

Of the indirect methods for the determination of permanent wilting percentage, the determination of the moisture content of the soil, corresponding to pF 4·2, by the depression in the freezing point method (p. 107) is based on the soundest theoretical approach. The method is convenient and avoids the delay inherent in the direct determination. As the pF curve is nearly straight in the neighbourhood of the permanent wilting point, freezing point determinations on a soil at two different moisture contents, chosen to give depressions of 1-2° C., are sufficient. From these determinations the moisture content corresponding to pF 4·2 (or a freezing point depression of 1·26° C.) is obtained by interpolation or extrapolation, and this corresponds to the permanent wilting percentage. Procedure B for the determination of the freezing point depression is sufficiently accurate, but, if this procedure is used, allowance should be made for the difference of about pF 0·05 in the two methods and the moisture content corresponding to pF 4·25 should be taken. These values have been chosen by da Costa (6) as the mean experimental values corresponding to the permanent wilting point of many widely differing soils.

The indirect freezing point method is not applicable to the determination of the permanent wilting point in soils containing more than 0·05 per cent. of total soluble salts, since soluble salts affect the pF value corresponding to the permanent wilting point. Soluble salts do not affect the direct method for the determination of permanent wilting point unless they are present in quantity sufficient to induce marked pathological symptoms in the plant used.

The Determination of the Permanent Wilting Point.

The technique used in these laboratories has been described by Marshall and Williams (11) and is as follows:

Equipment:

Lever Lid Tins. Commercially obtainable 1 lb. honey tins, with a lever lid, serve for this determination. Number the tins and lids and punch a hole about 10 mm. in diameter in the centre of each lid. For convenience, adjust the weight of all tins used to the same value, by means of small pieces of lead. An accuracy of ± 0·5 g. is sufficient.

Method:

To obtain sufficient accuracy in this determination 4-5 replications are desirable. The soil should be air-dry; it may be sieved through a 2 mm. screen or used unsieved. Weigh out 600 g. portions of the air-dry soil and transfer this amount to each honey tin, tamping the soil lightly to pack it into the tin.

In order to find out how much water to add to moisten the soil sufficiently, determine the hygroscopic moisture present in the soil at the time of filling the tins, by drying duplicate 10 g. portions of the original sample (p. 59). Also determine the moisture equivalent on further subsamples of the original soil (p. 89).

In each tin plant three dwarf sunflower seeds. Then add sufficient water to bring the moisture content of the soil to that of the moisture equivalent, making allowance for the hygroscopic water already present. This ensures thorough wetting of the soil without waterlogging it. Determine the weight of the tin and moist soil at this stage. Place aside in the glasshouse, leaving the lid off until the seeds germinate.

After the seeds have germinated, select the best seedling and lead it through the hole in the lid. Remove the other seedlings and plug the space between the stem of the selected plant and the edge of the hole with cotton wool. Protect the tin from the direct heat of the sun in the glasshouse, by surrounding it with damp sawdust. At intervals during the growth of the plant weigh the tin and bring it back to its original weight by the addition of water, so as to replace transpiration losses from the soil. To do this remove the lid sufficiently to enable the water to be poured on to the soil.

Allow growth to proceed until four pairs of leaves have

been developed, in addition to the original cotyledons. At this stage weigh the tin and bring the soil to its original moisture content for the last time. Then leave the plants to grow until the first definite signs of wilting appear and all the leaves tend to lose turgor. To determine whether this is permanent wilting, place the tin and plant in a closed bin, which contains a few inches of water on the bottom, so that a humid atmosphere is obtained around the plant. Leave overnight. If the leaves then become turgid the permanent wilting point has not been reached. Return the tin and plant to the open glasshouse and leave for a few hours until further wilting is noticed. Again check by returning it to the humid atmosphere of the bin. Repeat this procedure until complete recovery no longer occurs, indicating that the permanent wilting point has been reached.

When the plant remains permanently wilted, cut off the stem and leaves and determine their fresh weight. Then weigh the tin and soil, place it in an oven at 105° C., dry for several days, or until constant weight is attained, close the tin, cool and weigh again. The difference between these two weighings corresponds to the amount of moisture present in the soil and roots at the permanent wilting point, while the weight of oven-dry soil is obtained by deducting the weight of the tin from the final weighing. A correction for the oven-dry weight of the roots is unnecessary.

To obtain the amount of water present in the soil at the permanent wilting point it is desirable to make a correction for the amount of water in the roots and this is done by assuming that the roots correspond to one-half of the fresh weight of the stem and leaves and that they contain 80 per cent. of water. The weight of water in the roots thus corresponds to

$$0{\cdot}4 \times \text{weight of the stem and leaves.}$$

This empirical factor represents the average of many experimental determinations and is sufficiently accurate for the purpose required.

The permanent wilting percentage is then given by the expression:

$$\frac{\text{Weight of water in the soil and roots} - \text{Weight of water in the roots}}{\text{Weight of oven-dry soil}} \times 100.$$

STICKY POINT

The "point of stickiness" or "sticky point," as it is now more commonly called, was originally proposed by Hardy (7) as a useful single value constant for characterizing soils. It is defined as the moisture content, expressed as a percentage of the oven-dry soil, at which kneaded moist soil just ceases to adhere to external objects (7) or at which a thoroughly kneaded plastic mass of the soil is just about to stick to the fingers or to a knife (9). The sticky point differs slightly from the point of maximum plasticity or that point at which the kneaded soil can be moulded with the greatest facility. At the sticky point the colloidal components of the soil are saturated with water whereas, at the point of maximum plasticity, they are not quite saturated.

The sticky point is determined by kneading a small quantity of soil, with successive additions of water, until the proper state of stickiness is reached. Moisture is then determined by drying a portion of the wet soil. All determinations should be made on air-dry soil. Values obtained on soil which has been oven-dried are slightly lower.

The determination is empirical and a certain amount of personal judgment is involved in the recognition of the correct sticky point. However, after a little experience, the sticky point is easily recognized and satisfactory agreement can be obtained by different workers, provided that reasonable care is taken. The method is particularly valuable since the determination is rapid and requires no apparatus other than a drying oven and simple balance.

Sticky point ceases to possess significance in light sandy soils, for in these soils the wet sample is not plastic and there is a considerable amount of free interstitial water before exterior adhesion is pronounced. Sandy soils do not "stick" like clay soils and sticky point can only have a conventional meaning extrapolated from the known behaviour of heavier soils. Prescott and Poole (12) found that as the amount of clay in a soil increases the value for sticky point falls at first but, when the amount of colloidal material reaches the equivalent of about 10 per cent. of clay it begins to rise in proportion to the colloidal content. The initial decline in the value for the sticky point appears to be associated with the filling up of the pore space in the sandy framework by clay.

Values for sticky point and moisture equivalent are closely related in most mineral soils with more than 12-15 per cent. of clay. The value for sticky point is, in general, somewhat higher than that for moisture equivalent. The sticky point values of normal clays are probably very close to the moisture equivalents of the same material. However, organic matter contributes proportionately much more to the sticky point than it does to the moisture equivalent so that the relationship between sticky point and moisture equivalent does not hold for soils rich in organic matter. The value for the sticky point of soil organic matter is probably 350-440 while that for clay is of the order of 50-60.

In soils containing gypsum the determination of sticky point is unsatisfactory, since the colloidal properties of the clay are affected by the flocculating action of the gypsum, and the sticky point cannot be readily recognized. However, the presence of calcium carbonate does not affect the determination.

Apart from its value as a quick index of texture, the determination of sticky point is very useful in many field studies on the movements of moisture in a soil. Its use enables the soil moisture figures to be corrected for local variations in texture due to differences in the sites of successive samplings. By referring all moisture determinations to the sticky point of the same sample West (21) obtained a big increase in the precision of the measurement of field moisture values. In some cases the gain in precision was equivalent to a fifteen-fold increase in the number of samples. Moisture equivalent could be used for the same purpose, but the simplicity, both of the technique and apparatus required, makes the determination of the sticky point particularly valuable.

The Determination of Sticky Point (12).

Take about 20-30 g. of air-dry soil, moisten it with small portions of water and mix it with a flexible steel spatula on a glass plate until the mixture is definitely wet and sticky. At this stage it should have the consistency of a thick paste. Scrape the soil from the glass plate and knead thoroughly between the fingers, until the soil reaches the stage at which it no longer sticks to the fingers or to the spatula. In this condition it should be possible to cut cleanly through the moulded soil; it should not stick to the knife. When this state is reached the soil is at the sticky point.

When it is considered that the sticky point has been reached, place the soil block in a covered aluminium dish, the weight (a) of which has been previously noted. Weigh without undue delay, to obtain the weight (b) of dish and moist soil at the sticky point. Then dry in an oven at 105° C. for 16 hours, cool in a desiccator and weigh again (c). From the weight of oven-dry soil obtained ($c - a$) and the weight of water lost ($b - c$) calculate the moisture content, per 100 g. of soil, at the sticky point. The necessary calculation is:

$$\text{Sticky point} = \frac{(b - c)}{(c - a)} \times 100.$$

With medium to heavy soils the best guide to the recognition of the sticky point is the ability to handle the moulded soil with gentle squeezing or pressing in the palm of the hand, with little or no adherence to the skin. With very heavy soils it is not always possible to detect the sticky point readily; the knife test is the most satisfactory criterion with these soils. Highly organic soils are the most difficult to handle. With such soils, in addition to applying the knife test, the soil should be moulded into a block and broken across. When at the sticky point the broken surface just begins to glisten with water.

With sandy soils the sticky point loses significance, but a value can be obtained and the end point recognized, by moulding the kneaded soil into a block, breaking it and examining the broken surface. At the sticky point the broken surface just glistens with water but does not show any free water when the block is tapped with the finger.

Make all sticky point determinations at least in duplicate, preferably allowing an interval of time to separate the duplicate determinations. This reduces the personal error associated with the recognition of the end point. Duplicate determinations should not differ by more than 1 per cent.

HEAT OF WETTING

The heat of wetting is the heat evolved when dry soil is wetted. It is expressed in calories per gram of dry soil. The amount of heat to be measured is usually very small and, in order to secure accurate values, many precautions must be taken. Janert (8) has simplified the apparatus required for the determination, so enabling a large number of observations to be carried out in the course of a day. A known amount of dry soil is added to water, contained in a calorimeter, and the rise in temperature is noted. The method has been used extensively by Tisdall (17) in studies of the irrigation soils at Merbein and his technique is given below.

In order to obtain reproducible values, the determination should be carried out in a constant temperature room, or in a room in which the temperature change is less than 0·02° C. per minute. With changes exceeding 0·04° C. per minute it is not possible to obtain agreement between duplicate determinations to within 5 per cent., the maximum variation permissible on any one sample.

Particular attention must be paid to the oven-drying of the sample before the determination, since the heat of wetting is very sensitive to the times and temperatures employed in drying the soil. The conditions of drying should be closely standardized and the weighing bottles, containing the dried soil, should be tightly stoppered as soon as they are removed from the oven. Drying overnight at 110° C. is usually adopted.

The heat of wetting is a very useful single value constant for estimating soil texture. In this connexion it has been extensively used in determining suitable depths for tile drains in irrigated soils. The heat of wetting depends largely upon the relative proportions of organic matter, clay and silt. Organic matter possesses a higher heat of wetting, weight for weight, than does the inorganic fraction of the soil. The heat of wetting of a soil also depends upon the nature of the exchangeable cations present. However, provided that correlations are restricted to soils within any one group, the variations in the proportion of the different exchangeable cations is usually insufficient to affect seriously the estimates of texture, derived from the heat of wetting. Owing to the high heat of wetting of organic matter, correlations with texture should be restricted to soils of similar organic contents. Cognizance should be taken of the accumulation of organic matter in surface soils.

Tisdall (17) found that, for the soils of the Murray Valley, the heat of wetting could be expressed by the following regression equation:

$$\text{Heat of Wetting} = 0{\cdot}2127 + 0{\cdot}1076 \text{ clay} + 0{\cdot}0783 \text{ silt} + 1{\cdot}0809 \text{ organic carbon.}$$

For these soils, in which silt and organic matter are usually low, heat of wetting is particularly valuable, since there is also a very high correlation coefficient between it and the clay content alone. For these soils the linear relationship between heat of wetting and clay holds over a wide range of clay contents. It holds for soils containing as little as 4 per cent. of clay.

Tisdall also found that the presence of calcium carbonate, in amounts up to 30 per cent., did not affect the value for heat of wetting while gypsum, in amounts up to 4 per cent., was also without appreciable effect on the value.

The Determination of the Heat of Wetting.

Janert's method, as used by A. L. Tisdall at the Commonwealth Research Station, Merbein, is as follows:

Equipment:

Beckmann Thermometer. This should be fitted with a reading lens to enable estimates of temperature to be made to the nearest one-thousandth of a degree C.

Calorimeter. A straight walled Dewar flask, 5 cm. internal diameter and 10 cm. internal depth, is most suitable. The external dimensions corresponding to these measurements will be about 7 cm. diameter and 15 cm. overall length. The flask is supported in a stand of porcelain, bakelite or wood, 7·75 cm. internal diameter. A hole 2·5 cm. in diameter in the base takes a cork, drilled to receive the sealed-off tip of the flask. The flask is further held in position in the supporting stand by a rubber band slipped over the outside of the flask, near its top. The stand has a lip, 1·5 cm. deep and 0·3 cm. wide, to take the lid. The latter is 1 cm. deep and 8·25 cm. in diameter, bored with three holes 3·7 cm., 1·2 cm. and 0·6 cm. in diameter, to take the cap, the thermometer and the stirrer, respectively. The holes are arranged within the inside diameter of the Dewar flask, when the lid is in position. The cap, 4 cm. in diameter and 4 cm. high, is hollowed out on the inside to reduce its weight. It is cut to fit into its appropriate hole in the lid as snugly as possible, consistent with ease of removal and insertion. The spoon-shaped, perforated metal stirrer is 4·25 cm. in diameter and is attached to a glass handle 0·5 cm. in diameter and 20 cm. long. The stirrer is cut to fit around the thermometer, so that the whole of the contents of the Dewar flask can be effectively stirred. The thermometer is permanently fitted in place in the lid by means of de Khotinsky cement. The joint between the lid and the lip of the stand is made as snug as possible, by pasting a strip of linen to the edge of the lid.

Weighing Bottles. These should be 5 cm. high and 2·5 cm. diameter. They should be fitted with rubber stoppers and the bottles and stoppers numbered.

Method:

Carry out this determination on an air-dry sample that has been ground further, so that it will pass a 1 mm. screen. Use a quantity of soil sufficient to give a rise in temperature of about 0·5° C. This may require from 30-40 g. for very light soils to 5-10 g. for very heavy clays. Usually an amount of 15–25 g. is satisfactory.

Transfer a suitable amount of the air-dry soil to a weighing bottle, which has been previously weighed, dry in an oven for 16 hours at 110° C., then close the bottle tightly with its rubber stopper, cool in a desiccator and weigh. After weighing, tightly seal the junction of the weighing bottle and stopper by painting it with molten Sira wax.

From a burette, run sufficient water into the Dewar flask so that the volume of soil and water will be exactly 100 ml. For this purpose make the assumption that the specific gravity of the soil is 2·65 and calculate the volume occupied by it, by multiplying the weight taken by 0·38. Now place the sealed bottle, containing the dried soil, in the calorimeter and leave to stand until the temperature changes approach a constant amount. If the whole of the operations are carried out in one room and the distilled water is also at room temperature, this period of waiting is about 15 minutes. By the use of two calorimeters, one can be reaching equilibrium while a determination is in progress in the other.

After a suitable period of standing, move the stirrer up and down, with a stroke of 1-2 cm., for 3-4 minutes, stopping the stirring only to note the temperature at regular half-minute intervals. Before making each reading, tap the thermometer lightly with a pencil, covered with a piece of rubber tubing, to avoid errors due to the sticking of the mercury column.

Record each thermometer reading. When the differences between successive readings reach a constant amount, start the determination proper. Note this last thermometer reading as t_1; also note the time corresponding to it. Immediately after reading the thermometer, remove the lid, lift the weighing bottle out by means of the stirrer, grasp it lightly in the fingers, break the seal and tip the soil into the water. Tap the bottle with the finger to remove the soil as completely as possible. Immediately replace the cap and commence re-stirring. The whole operation should not require more than 10 seconds. Continue stirring as before, noting the temperatures at each half-minute interval until the differences between successive readings again reach a constant value. Designate the first of the readings showing a constant difference as t_2.

Since there is a definite trend in temperature, due to heat-transfer between the calorimeter and the surroundings, the observed temperatures must be adjusted to eliminate this trend and to ensure that the initial and final temperatures correspond to the same instant of time. Let x be the mean or constant difference between the successive half minute readings, immediately prior to introducing the soil, and y the number of half minutes that elapse between noting the reading t_1, and the time of making the final reading t_2. Then the corrected value for t_1, adjusted to correspond to the time of making the reading of t_2, is given by $t_1 + xy$. In other words, t_1, corrected, corresponds to the temperature that would have been reached at the time of the reading t_2 if the normal temperature trend had not been affected by the heat of wetting of the soil. Note that the value for x may be either positive, corresponding to an increasing temperature trend, or negative, corresponding to a decreasing trend.

The temperature generally reaches its "peak" within a minute or so of adding the soil. In the case of gypseous soils the temperature usually reaches a maximum within about one minute after mixing and then falls slightly. For these soils the highest reading should be taken for t_2.

Before calculating the heat of wetting it is necessary to know the water equivalent of the calorimeter. Determine this for each instrument used, by dissolving 2·583 g. of sodium thiosulphate ($Na_2S_2O_3.5H_2O$) in 75 ml. of water in the calorimeter, using the same technique as that described above for determining the heat of wetting of a soil. Sodium thiosulphate has a heat of solution of −11,370 calories per gram-molecule at this dilution.

Since the heat gained by the system equals that lost by the salt then

$$\frac{H\,w}{m} = (t_2 - t_1)\,[(V + w)S + W.E.]$$

where H = the heat of solution of the salt used, in calories per gram
w = the weight of the salt taken
m = the molecular weight of the salt
t_1 = the initial temperature
t_2 = the final temperature
V = the volume of water in the calorimeter
S = the specific heat of the resultant solution
and W.E. = the water equivalent of the calorimeter.

From the above equation it follows that

$$W.E. = \frac{H\,w}{m(t_2 - t_1)} - (V + w)$$

or using 2·583 g. of sodium thiosulphate and 75 ml. of water

$$W.E. = \frac{11{,}830}{t_1 - t_2} + 77{\cdot}58 \text{ calories.}$$

Calculate the heat of wetting of the soil, in calories per gram from the equation

$$\text{Heat of Wetting} = \left(\frac{0\cdot 2W + K + M}{W}\right)(t_2 - t_1)$$

where W = the weight of soil taken
0·2 = the specific heat of the soil
K = the water equivalent of the calorimeter, as determined above
M = the weight of water in the calorimeter
t_1 = the initial temperature (corrected)
and t_2 = the final temperature.

Duplicate determinations should agree to within 5 per cent.

DEPRESSION IN THE FREEZING POINT

The depression noted in the freezing point when a moist soil is frozen depends upon the free energy of the water in the soil. This in turn is chiefly dependent upon the moisture level of the soil. The method has been applied by Schofield for the determination of the pF value of soils over a range of moisture contents extending from just below the permanent wilting percentage to the moisture equivalent. Outside this range it is of theoretical interest only; experimental difficulties preclude its use with drier or moister soils.

If the depression in the freezing point is plotted against the *water* content of the soil smooth curves are obtained. In making the determination it must be remembered that freezing dries the soil and that the value for water present, at the freezing point, is not that of the moisture present in the original soil, or in the soil after thawing, but corresponds to this value less the amount of ice formed at the time of freezing. In the determination of the depression of the freezing point it is therefore necessary to make accurate determinations both of the temperature at the freezing point and the *water* content, to which it corresponds.

The determination of the freezing point of a moist soil is made possible by the fact that freezing can be induced, by jarring, in a soil that has been super-cooled. Schofield and da Costa (15) have investigated the method originally used by

Bouyoucos and McCool and have defined the precautions necessary to obtain accurate values. They used the same apparatus as the earlier investigators, with the exception that they recommend wide-mouthed vacuum flasks to hold the freezing mixtures. Their method differs, however, in two important details from that of Bouyoucos and McCool. Before freezing is started, the temperature of the supercooled soil is accurately observed and shortly after freezing has started, the tube, containing the soil and thermometer enclosed in an air jacket, is placed in a freezing mixture, the temperature of which is close to that finally attained by the frozen soil. The first modification is necessary to enable the calculation of the amount of ice formed during freezing, so as to obtain a corrected value for the water content of the soil at the time of freezing. The second modification increases the accuracy of the determination, for the amount of heat lost after the commencement of freezing is reduced and the observed maximum temperature corresponds more closely to the true temperature at the freezing point.

Schofield and da Costa have outlined two procedures for the determination of the depression of the freezing point and these are given below. Procedure A includes all the precautions desirable to secure the greatest accuracy. It is elaborate and time-consuming, but is recommended for all fundamental determinations. Procedure B is designed to give results sufficiently accurate for such purposes as the indirect determination of the permanent wilting point and other routine determinations. In procedure A three freezing mixtures are used, whereas only one is used in procedure B. The use of three freezing mixtures gives more delicate control of the degree of supercooling and also reduces the heat losses, due to the surroundings, while freezing is in progress.

For reproducible values it is necessary to standardize the method used for wetting the soil and packing it into the tube. The wet soil should always be left for 48 hours before making the freezing point determination.

The automatic tapper used by Schofield and da Costa eliminates errors arising from the sticking of the mercury column in the Beckmann thermometer. A thermocouple may, with advantage, replace the Beckmann thermometer for the measurement of the freezing point. Its use avoids errors due to pressure exerted by the freezing soil on the bulb of the thermometer.

The Determination of the Freezing Point: Schofield and da Costa's Method (15).

Equipment:

Beckmann Thermometer. When not in use the thermometer should be kept with its bulb in ice and water. When so kept a second determination of the freezing point of water generally checks with the first determination within two or three thousandths of a degree. It is necessary to redetermine the zero each morning and afternoon before use. If the thermometer has been kept at room temperature it will give successively higher readings for the freezing point; reproducible values will only be obtained after it has been used for six or more times.

Vacuum Flasks. Wide mouthed vacuum flasks are necessary to hold the freezing mixtures.

Automatic Tapper. This is made from an electric bell and operates on the stem of the Beckmann thermometer.

Procedure A:

Mix several portions of air-dry soil with water so that a series of samples is obtained with a suitable range of moisture contents. Pack duplicate portions of each moistened sample, corresponding to 20 g. of dry soil, into boiling tubes, about 22 cm. x 2·5 cm., stopper and leave to stand for 48 hours. Adopt a standardized procedure for moistening the soil and packing it into the tubes. After the soil has been kept in the moist condition for 48 hours, place the tubes containing it in a pail of ice water for the preliminary cooling. Keep them in this until the time for making each determination.

Insert the Beckmann thermometer into one of the tubes so that the soil is packed snugly around the bulb. Then dip the tube into a cooling mixture at about -3° C. until the temperature has been depressed below 0° C. by about half the amount desired. Remove the tube, rapidly wipe it on the outside and insert it in a wider tube, about 15 cm. x 4 cm., supporting it by a loosely fitting stopper, so that it is entirely surrounded by an air jacket. Now continue the cooling in a second cooling mixture, contained in another vacuum flask, and adjusted to the temperature required for supercooling, usually about 1° C. below the expected freezing point. The air jacket partially insulates the soil tube and so reduces the rate of cooling. When no change of temperature occurs over a period of several minutes, record this temperature and start the freezing by giving a sharp twist to the Beckmann thermometer. When the temperature has risen to within about half a degree of the expected freezing point, remove the soil tube, still surrounded by the second tube, and rapidly transfer it to a third freezing mixture at this temperature. Then set the automatic tapper going and record the maximum temperatures shown by the thermometer. After the observation determine the amount of moisture in the soil used and correct it for the amount of ice formed, as directed below.

In this procedure the initial freezing mixture at −3° C. is only used to save time. The initial cooling can be carried nearly to the point of supercooling desired, but in such a case great care is needed in wiping the soil tube and inserting it in the wider tube, as a slight jolt may start freezing and spoil the determination.

To obtain a minimum estimate of the amount of ice formed the water equivalent of the tube containing the soil and thermometer must be known. To determine this, cool it to 0° C., then immerse it in a known weight of water at room temperature, contained in a vacuum flask, and note the drop in temperature. The water equivalent corresponds to $\frac{(t_1 - t_2)\,W}{t_2}$ where W = the weight of water in the vacuum flask and t_1 and t_2 its temperature before and after immersing the soil tube. The value obtained is the water equivalent of the tube, thermometer and moist soil. Deduct the weight of water contained in the soil used to obtain the water equivalent of the solids (tube, thermometer and dry soil).

The heat derived from the formation of ice per 1° C. rise in temperature during freezing is $\left(W + \frac{MS}{100}\right)$ calories,

where W = the water equivalent of the solids, expressed in g.
S = the weight of dry soil taken
and M = the weight of water present per 100 g. of dry soil.

Taking 80 calories per gram as the latent heat, the ice formed is $\frac{1}{80}\left(W+\frac{MS}{100}\right)$g. Expressed as a percentage weight of the dry soil this is $\frac{1}{80}\left(W+\frac{MS}{100}\right)\frac{100}{S}$ or $\frac{W}{80} \cdot \frac{100}{S} + \frac{M}{80}$. The amount of ice formed during freezing is, therefore, $\left(\frac{W}{80} \cdot \frac{100}{S} + \frac{M}{80}\right)$ C, where C is the rise in temperature during freezing. By subtracting this value from the amount of moisture present calculated as a percentage of the dry soil, the amount of *water* present after freezing is obtained. Provided that a constant weight of dry soil is used each time, a mean value can be determined for the water equivalent of the tube, thermometer and dry soil.

Having calculated the amount of ice formed, determine the pF value of the soil, corresponding to the given water content, from the equation

$$pF = 4 \cdot 1 + \log t$$

where t is the observed depression in the freezing point. From a series of observations at different moisture contents plot the values obtained for pF against the corresponding corrected moisture contents. The moisture content of the soil, corresponding to any desired pF value, can then be obtained by interpolation.

Procedure B:

This procedure is less elaborate but is sufficiently accurate for such purposes as the indirect determination of permanent wilting percentage.

Proceed as in the previous method as far as cooling the soil tube to about half the desired amount, wiping it and inserting it in the wider tube. From this point continue the cooling further, by replacing the jacketed tube in the freezing mixture (−3° C.). Allow the slow cooling to continue until the desired amount of supercooling is obtained, usually corresponding to about 1° C. below the expected freezing point.

When the desired temperature has been reached, record the reading of the thermometer, start the freezing by giving the thermometer a sharp twist, and set the automatic tapper going. Record the maximum temperature shown by the thermometer, determine the moisture content of the soil and correct for the amount of ice formed, as described under procedure A.

Since the freezing mixture surrounding the soil tube is at a lower temperature in this procedure, more ice is formed than in procedure A and the correction is less accurate. To compensate for this error a further correction may be applied, by deducting 0·05 from the pF value, calculated above. This latter correction is entirely empirical.

REFERENCES

(1) Bouyoucos, G. J.: *Soil Sci.*, 1935, *40*, 165–170.

(2) Briggs, L. J. and McLane, J. W.: *U.S.D.A. Bureau of Soils, Bull.* *45*, 1907.

(3) Briggs, L. J. and Shantz, H. L.: *U.S.D.A. Bur. Plant Indus., Bull.* *230*, 1912.

(4) Coutts, J. R. H.: *J. Agric. Sci.*, 1930, *20*, 407–413.

(5) Coutts, J. R. H.: *J. Agric. Sci.*, 1932, *22*, 203–211.

(6) da Costa, J. V. B.: *J. Agric. Sci.*, 1938, *28*, 654–662.

(7) Hardy, F.: *J. Agric. Sci.*, 1923, *13*, 340–351.

(8) Janert, H.: *Imp. Bur. Soil Sci. Techn. Comm.* *27*, 1933.

(9) Keen, B. A. and Coutts, J. R. H.: *J. Agric. Sci.*, 1928, *18*, 740–765.

(10) Keen, B. A. and Raczkowski, H.: *J. Agric. Sci.*, 1921, *11*, 441–449.

(11) Marshall, T. J. and Williams, R. F.: *J. Austr. Inst. Agric. Sci.*, 1942, *8*, 109–111.

(12) Prescott, J. A. and Poole, H. G.: *J. Agric. Sci.*, 1934, *24*, 1–14.

(13) Schofield, R. K.: *Trans. 1st Comm. Int. Soc. Soil Sci. Versailles*, 1934, 185–191.

(14) Schofield, R. K.: *Proc. 3rd Int. Congr. Soil Sci.*, 1935, *2*, 37–48.

(15) Schofield, R. K. and da Costa, J. V. B.: *J. Agric. Sci.*, 1938, *28*, 644–653.

(16) Taylor, J. K.: *J. Austr. Inst. Agric. Sci.*, 1936, *2*, 171–172.

(17) Tisdall, A. L.: *J. Counc. Sci. Ind. Res. Australia*, 1936, *9*, 295–300.

(18) Viehmeyer, F. J. and Hendrickson, A. H.: *Soil Sci.*, 1931, *32*, 181–193.

(19) Viehmeyer, F. J.. Israelson, O. W. and Conrad, J. P.: *Calif. Agric. Expt. Sta. Techn. Paper 16*, 1924.

(20) Viehmeyer, F. J., Oserkowsky, J. and Tester, K. B.: *Proc. 1st Int. Congr. Soil Sci.*, 1927, *1*, 512–534.

(21) West, E. S.: *J. Agric. Sci.*, 1931, *21*, 799–805.

CHAPTER V

SOIL COLOUR

Soil colour is one of the first properties noted in the field description of soils and, in order to secure uniformity in its description, it is necessary to establish standards for the several colour terms. This is now possible since the Maxwell rotating colour discs provide an effective and simple means of colour analysis. The Soil Colour Standards Committee of the American Soil Survey Association found that nearly all soil colours could be reproduced by spinning a disc composed of four variable segments (black, white, yellow and red) thus making possible the colour analysis of the soil in terms of these four standard colours. In Hutton's method (1) the colour segments are rotated at high speed in a horizontal plane and the soil is held over the rotating disc on a broad-bladed spatula or in a small dish. The circumference of the turntable is subdivided into 100 divisions so that the proportions of the four standard colours exposed, when adjusted to give a good colour match, can be read directly. The method has been improved by Shaw (4) who rotates the soil on the same spindle as the colour discs, to facilitate comparison and obviate the shadow effect due to its granular surface. He disperses the soil with water and paints the "mud" so obtained on to a disc of white blotting paper. When dry this is mounted with the colour discs on the turntable and clamped in place. It is difficult to coat blotting paper with coarse sandy soils, and the coating produced by heavy clay soils tends to crack on drying; the shadows in the cracks produce a darker shade. J. K. Taylor *(priv. comm.)* has improved the technique by using glue to attach the soil to the rotating disc, and his method, in its latest form, is described below.

The standard Maxwell colour discs used for this method are printed by the Munsell Color Company, Baltimore, U.S.A., and have been standardized by the U.S. Bureau of

Standards in terms of their spectral reflectance. These discs are $4\frac{1}{2}$ in. in diameter and have a radial slit so that all four of them can be mounted co-axially on a turntable, and any combination of the four colour segments can be produced. They are designated by the Munsell Color Co. as

Neutral	9/-	(white)
Neutral	1/-	(black)
Yellow	8/8	(yellow)
Red	4/9	(red)

With few exceptions all soils can be matched against these four standard colour discs. The exceptions include red and red-brown soils, which are always difficult and not always possible to match completely against these four standards. It is very probable that this difficulty in matching is due to the fact that the colour of these red soils has a greater degree of saturation than is provided by the standard red and yellow colour discs. Under these conditions a match would be obtained if white could be added to the colour of the soil under test.

The colour discs must be carefully handled to reduce wear and prevent scratches which will change their colour quality. The colour discs should also be protected from dust and light when not in use. Prolonged exposure to bright light causes fading. The discs must be replaced when they become worn, if the colour measurements themselves are to be kept standard. The speed of rotation of the turntable is immaterial so long as it is sufficiently fast to overcome flickering. A speed of about 2,500 r.p.m. is convenient.

For field description, Taylor proposes 23 colour classes. With one or two minor additions of intermediate colour classes, he considers these sufficient for the description of normal soils. Mean values for these classes, in terms of the standard colours, are given in Table 4. The mean values for white and red are not included since too few white soils have been examined and, as mentioned previously, it has not been possible to match red soils exactly, with the four standard colours. The United States Division of Soil Survey has adopted a series of 57 standard colour names for the field description of soil colour and colour charts showing these classes have recently been published (2). The colour charts are designed to facilitate the direct matching of the colour of the air-dry soil sample without recourse to the rotating colour discs.

TABLE 4.

Colour Classes used in the Description of Soil Colour (after J. K. Taylor).

Colour Class.	Black %	White %	Yellow %	Red %
Black	87·5	6	3	3·5
Very dark grey	79·5	8·5	6·5	5·5
Dark grey	69·5	14	9	7·5
Grey	57·5	19	13·5	10
Light grey	44·5	25·5	18	12
White	n.d.			
Brownish black	86	2	5	7
Very dark brown	79	0	8	13
Dark brown	72·5	1	9·5	17
Brown	56·5	4	16	23·5
Light brown	42	9	19·5	29·5
Dark greyish brown	75·5	5	9·5	10
Greyish brown	55	9	14·5	21·5
Light greyish brown	44	14	20·5	21·5
Dark yellowish brown	55·5	0·5	18·5	25·5
Yellowish brown	34·5	3	27·5	35
Yellow	45·5	4·5	26	24
Light yellow	23	12·5	36	28·5
Greyish yellow	46·5	8·5	25	20
Reddish brown	65·5	0·5	10	24
Light reddish brown	44	1·5	16	38·5
Chocolate	83	0	6	11
Red	n.d.			

Schofield (3) recommends the adoption of C.I.E. (Commission Internationale d'Eclairage) co-ordinates as a refinement in the definition of soil colour values. The C.I.E. system gives the geometrical representation of any colour. In general, this representation must be made in three dimensions and the co-ordinates may be spoken of as solid co-ordinates. C.I.E. values are given for the four standard Munsell colour discs and this paper should be consulted for the method of calculation, if it is desired to express any soil colour, already defined in terms of the standard colour discs, in the C.I.E. system. The most convenient instrument for the direct determination of the C.I.E. co-ordinates is the Trichromatic Colorimeter such as that of Guild or Donaldson. These instruments are too expensive for use in the ordinary soils laboratory, but are designed for use in commercial laboratories where the standardization of colour is an important daily routine.

The Determination of Soil Colour.

Apparatus:

Soil Colorimeter. A turntable is fitted to the spindle of an electric fan motor running at 2,500-3,000 r.p.m. The turntable is so constructed that the colour discs and soil disc can be clamped, by means of a metal washer and milled head thumb nut, on the spindle. The background, surrounding the colour discs, should be neutral grey in colour with a circle graduated into 100 divisions, so that the proportion of each colour segment exposed can be readily ascertained.

Munsell Colour Discs. These have been specified on p. 113.

Cardboard Discs. Cardboard discs, cut from 8 sheet white pasteboard, can be purchased. These should have a central hole for the spindle. Suitable dimensions are

External diameter, $1\frac{3}{4}$ in.
Diameter of hole for spindle, $\frac{1}{4}$ in.
Thickness, $\frac{1}{32}$ in.

Thinner discs are not suitable since they are more prone to curl when coated with soil.

Method:

Crush a small portion of air-dry soil in a mortar, so as to break up aggregates but not to grind the individual sand particles, and pass through a sieve with round holes 0·5 mm. in diameter. Add the portion remaining on the sieve to the fine fraction and mix thoroughly. This is done to produce a more uniform surface on the soil disc, and so minimize shadow effects. Shadow effects will, however, still be perceptible with coarse sandy soils.

Coat a white cardboard disc with hot glue of medium consistency and immediately dust on to it a thick layer of the finely ground soil. Press firmly with a spatula to ensure good adherence of the particles and leave for a few minutes to harden. Then gently tap off the surplus soil, place the disc on a piece of blotting paper, cover with a sheet of glass (e.g. an old photographic plate) so as to keep the disc in a flat position, and leave overnight to dry.

The soil must adhere evenly to, and fully cover the surface of the disc. If any glue shows through, or any soil has chipped off, the disc is unsuitable for colour matching. Chips in the surface produce a ring effect during spinning.

When dry, mount the soil disc on the turntable, taking care to avoid soiling the colour segments with any soil particles adhering to the edge of the disc. Adjust the colour segments to a probable value for the soil, clamp the soil disc and colour segments in position, and spin the colour wheel. Compare the colour produced by the four spinning colour segments with that of the soil. Stop the wheel, adjust the proportions of the four standard colour segments towards a closer match, and spin again. Repeat the operation until a good colour match is obtained, when the proportions of the four segments, as read from the circular scale, will define the colour of the soil in terms of black, white, yellow and red.

All colour comparisons should be carried out in bright but diffused daylight. The soil disc and colour standards must be observed from an eye position normal to the plane of rotation. Angular vision gives rise to false values.

REFERENCES

(1) HUTTON, J. G.: *Proc. 1st Int. Congr. Soil Sci.*, 1927, *4*, 164–172

(2) RICE, T. D., NICKERSON, D., O'NEAL, A. M. and THORP, J.: *U.S.D.A. Miscell. Pub. No. 425*, 1941.

(3) SCHOFIELD, R. K.: *Trans. 1st Comm. Int. Soc. Soil Sci.*, 1938, Bangor, *Vol. A*, 54–59.

(4) SHAW, C. F.: *Soil Sci.*, 1932, *33*, 183–185.

CHAPTER VI

STANDARD SOLUTIONS AND INDICATORS

STANDARD SOLUTIONS

For convenience, the strengths of the standard solutions most frequently used in soil analysis, together with suitable methods for their accurate standardization, are set out below. Since standard hydrochloric acid is more widely applicable than sulphuric acid it is used in preference to the latter. This avoids the necessity for keeping standard solutions of both acids. Tenth normal solutions of hydrochloric acid can be boiled for at least one hour without loss of acid, provided that the evaporated water is replaced. Half normal solutions show no loss when boiled for 10 minutes.

Standard Hydrochloric Acid.

Reagent hydrochloric acid (S.G. 1·18) contains approximately 418 g. of HCl per litre (11·5N). It is convenient to prepare, by dilution with water, a quantity (2–5 litres) of more dilute acid, of about 5N strength, to serve as a stock supply for the preparation of all standard solutions of hydrochloric acid. If this stock solution is kept in a well stoppered pyrex bottle its strength will remain constant indefinitely. Once prepared, and its strength determined, such a stock solution greatly facilitates the preparation of supplies of more dilute standard solutions.

To prepare 2N or N hydrochloric acid, transfer, by means of a measuring cylinder, sufficient of the stock solution to give a standard solution just slightly stronger than that required, when diluted in the proper sized volumetric flask. Standardize the solution against sodium borate or sodium carbonate, as described below, and adjust the strength to the exact normality required by the addition of the small amount of water, calculated from the titration. Check the standardization.

To prepare 0·1N, 0·05N or 0·02N hydrochloric acid

transfer, by means of a pipette, a suitable volume of the N acid (200 ml., 100 ml., or 40 ml. respectively) to a 2,000 ml. volumetric flask, dilute to volume and mix well. Check the standardization by titration against sodium borate or sodium carbonate as described below. Store the standardized solutions in well stoppered pyrex bottles.

STANDARDIZATION. Hydrochloric acid can be standardized by titration against weighed amounts of sodium borate or sodium carbonate. Sodium borate is preferred. By its use acids can be standardized more simply than with carbonate and yet at least as accurately. Sodium borate ($Na_2B_4O_7.10H_2O$) is easily obtained pure, has a high equivalent weight (190·7), and does not attract water during weighing. Reagent grade sodium borate is sufficiently pure for all standardizations except when the very highest accuracy is required. The salt should be kept in a desiccator, over deliquescent sodium bromide or a saturated aqueous solution of sodium chloride and sucrose. This maintains a relative humidity of 60 per cent. and ensures that the sodium borate has the right amount of water of crystallization. If a suitable reagent grade cannot be obtained, the purest borax should be recrystallized two or three times and dried over sodium bromide. Avoid undue exposure of the moist salt to the air as it absorbs a little carbon dioxide when moist.

Standardization by means of Sodium Borate:

On a tared watch glass weigh out a quantity of sodium borate, sufficient to neutralize 30–35 ml. of the acid to be standardized. Transfer it to a 250 ml. beaker flask, washing the watch glass with about 30 ml. of water. Suitable amounts of sodium borate are 5·7–6·7 g. for normal solutions or 0·57–0·67 g. for tenth normal solutions. For solutions more dilute than 0·05N it is preferable to make a solution containing a known amount of sodium borate and titrate a suitable aliquot.

Add 2–3 drops of methyl red indicator solution to the sodium borate in the beaker flask and run in, from a burette, the acid to be standardized until the first definite shade of pink persists. If the highest accuracy is required prepare a reference solution containing approximately the same concentration of boric acid and sodium chloride that exists in the solution at the end of the titration. Add 2–3 drops of methyl red to 60–70 ml. of this solution, contained in a second beaker flask, and place it near the solution being titrated. Then carry out the titration until the colour of the indicator corresponds to that in this reference solution.

For the standardization of solutions of hydrochloric acid stronger than 0·2N use dimethyl yellow in place of methyl red. In titrating these stronger solutions it is essential to use a reference solution approximating in composition to the boric acid and sodium chloride present at the equivalence point. Thus, in titrating N hydrochloric acid, use a reference solution which is molar in boric acid and 0·5 molar in sodium chloride.

The normality of the acid is calculated as follows:

$$\text{Normality} = \frac{\text{Weight of sodium borate}}{\text{Volume of acid used}} \times \frac{1}{0{\cdot}1907}.$$

Standardization by means of Sodium Carbonate:

Transfer a suitable quantity of purified anhydrous sodium carbonate or sodium bicarbonate to a weighed platinum crucible and ignite in a muffle furnace at 270–300° C. for 30 minutes. At this temperature bicarbonate is completely converted to carbonate and the preparation rendered anhydrous, without any decomposition to sodium oxide. If more than 1 g. of sodium bicarbonate is ignited, stir it frequently with a platinum wire, to ensure complete decomposition to carbonate. When the ignition is complete, cover the crucible, cool in a desiccator and weigh, to obtain the amount of anhydrous sodium carbonate.

After weighing, wash the contents of the crucible, quantitatively, into a 250 ml. beaker flask, add one or two drops of dimethyl yellow indicator solution and titrate with the acid to be standardized, until the first signs of the acid colour of the indicator persist.

According to Kolthoff, greater accuracy is obtained by the following procedure:

Run in the acid from a burette until the colour of the indicator begins to deviate from the water tint. Boil the solution for two minutes, cool and continue the titration until the acid colour of the indicator just persists.

In each case the working titre of the acid, referred to dimethyl yellow, is calculated as follows:

$$\text{Normality} = \frac{\text{Weight of sodium carbonate}}{\text{Volume of acid used}} \times \frac{1}{0{\cdot}0530}.$$

Standard Sodium Hvdroxide.

Sodium hydroxide solutions prepared directly from the best grade of reagent still contain 1–2 per cent. of the total alkali as carbonate. To prepare solutions practically free from carbonate one of the two methods described below should be used. Solutions of sodium hydroxide rapidly absorb carbon dioxide from the atmosphere and they should be kept in bottles permanently attached to the burette and protected by soda-lime guard tubes.

For dilution, water, free from carbon dioxide, should be used. This can be obtained by boiling ordinary distilled water for 5–10 minutes. However, the amount of carbon dioxide in "equilibrium water" is so small that for most purposes it is negligible and water, aerated for 8–16 hours by bubbling a rapid stream of air through it, can therefore be used for diluting sodium hydroxide solutions. The air should preferably be drawn from outside the laboratory.

PREPARATION OF SODIUM HYDROXIDE SOLUTIONS FREE FROM CARBONATE. Dissolve 500 g. of sodium hydroxide in 600 ml. of water in a hard glass flask, close with a rubber stopper or a piece of tin foil, and leave to stand for 24 hours. Sodium carbonate is practically insoluble in a solution of sodium hydroxide of this strength. Filter through a Buchner funnel fitted with a 9 cm. Whatman No. 50 filter paper, protecting the solution as much as possible from carbon dioxide of the atmosphere. The filtration is slow and takes 4–5 hours. Transfer the filtrate to a ceresine bottle, or to a hard glass flask coated on the inside with a thick layer of paraffin wax, and close with a rubber stopper. Such a solution contains about 0·65 g. of sodium hydroxide per millilitre. For the preparation of standard solutions pipette out sufficient of this concentrated solution and dilute to the required volume with boiled or aerated distilled water. In pipetting the concentrated solution avoid disturbing any carbonate which may have precipitated and settled to the bottom during storage. Kolthoff prepares the concentrated solution of sodium hydroxide as described above and filters it, out of contact with the air, through a sintered glass funnel.

If a trace of calcium in the standard solution does not interfere, sodium hydroxide, free from carbonate, can be prepared by the following method, due to Kolthoff. Make a solution of sodium hydroxide about 1·1N and add about 50 ml. of milk of lime to each litre. Shake for one hour and then leave to settle for several days. When quite clear remove the supernatant liquid, by siphoning, and dilute to the strength required.

STANDARDIZATION. Sodium hydroxide solutions can be standardized by titrating against standard hydrochloric acid, using any of the common indicators. However, a direct standardization is preferable, and can be easily carried out, since several substances are available as primary standards. For solutions more dilute than about 0·2N, potassium bi-iodate ($KIO_3.HIO_3$) is probably the most convenient substance. It has a high equivalent weight (389·9) and can readily be obtained in a pure anhydrous form. If necessary it can be purified by recrystallization from water and dried at a temperature not exceeding 120° C. Potassium bi-iodate is quite stable and non-hygroscopic. Dilute standard solutions, prepared from it, also maintain their strength indefinitely and this greatly facilitates the standardization of very dilute alkali solutions. Potassium bi-iodate is the acid salt of a strong acid, iodic acid, and a wide choice of indicators, from dimethyl yellow to phenolphthalein, is therefore available for its titration.

Potassium acid phthalate ($C_6H_4.COOH.COOK$) and hydrazine sulphate ($N_2H_4.H_2SO_4$) can also be recommended. These give values identical with those obtained by the use of potassium bi-iodate. These two substances are preferred to potassium bi-iodate for standardizing solutions stronger than 0·2N, for reasons of economy.

Potassium acid phthalate (equivalent weight 204·1) is

readily obtained pure. It is non-hygroscopic and may be weighed directly. If the highest degree of accuracy is required in the standardization it can be dried for one hour at 110–115° C. before weighing. If it is necessary to recrystallize the salt, crystallization should not be allowed to take place below 20° C., for at lower temperatures a more acid salt separates. The second dissociation constant of phthalic acid is $3 \cdot 9 \times 10^{-6}$. It is therefore necessary to use an acid sensitive indicator (phenolphthalein or one in which the colour change occurs at a higher pH value) for its titration with sodium hydroxide.

Hydrazine sulphate (Equivalent weight 130·1) is also readily obtained in a pure form, or can be easily purified by recrystallization and drying at a temperature not exceeding 150° C. As the second dissociation constant of hydrazine is very small, its salts are strongly hydrolyzed in water and they can be accurately titrated to the basic salts, according to the following equation:

$$2N_2H_4.H_2SO_4 + 2NaOH = (N_2H_5)_2SO_4 + Na_2SO_4 + 2H_2O$$

In this titration the sharpest end point is given by methyl red and this indicator should be used when standardizing sodium hydroxide solutions against hydrazine sulphate. It should be noted that the equivalent weight of hydrazine sulphate is 130·1 and not 65 as given by Kolthoff and Furman (*"Volumetric Analysis" Vol. II*).

Direct titration of any of these substances gives the working titre of the sodium hydroxide, referred to the particular indicator used. If the solution is reasonably free from carbonate the difference between the dimethyl yellow and methyl red values on one hand, and phenolphthalein on the other hand, is quite small.

Standardization by means of Potassium Bi-iodate:

Weigh out a quantity of pure dry potassium bi-iodate, sufficient to neutralize 30–35 ml. of the sodium hydroxide solution to be standardized, and dissolve it in about 35 ml. of carbon dioxide-free water in a 250 ml. beaker flask. For the standardization of tenth normal sodium hydroxide 1·2–1·4 g.

is a suitable amount. Add 2–3 drops of methyl red indicator and run in the sodium hydroxide, from a burette, until the indicator just changes from red to yellow. Note the burette reading. Add 2 or 3 drops of phenolphthalein and continue the titration until the solution is just alkaline to this indicator. Less than one drop of alkali is usually required. By the use of the two indicators the working titre of the standard sodium hydroxide against methyl red and phenolphthalein is obtained in the one titration.

Calculate the normality of the sodium hydroxide against each indicator, as follows:

$$\text{Normality} = \frac{\text{Weight of potassium bi-iodate}}{\text{Volume of sodium hydroxide used}} \times \frac{1}{0\cdot3899}.$$

Standardization by means of Potassium Acid Phthalate:

Weigh out a suitable quantity of pure dry potassium acid phthalate and dissolve it in about 35 ml. of carbon dioxide-free water. Titrate this solution with the sodium hydroxide to be standardized, using phenolphthalein as indicator. For the standardization of normal sodium hydroxide 6·1-7·1 g. of potassium acid phthalate is a suitable amount.

The normality of the sodium hydroxide, against phenolphthalein,

$$= \frac{\text{Weight of potassium acid phthalate}}{\text{Volume of sodium hydroxide used}} \times \frac{1}{0\cdot2041}.$$

Standardization by means of Hydrazine Sulphate:

Weigh out a suitable quantity of pure dry hydrazine sulphate and dissolve it in about 35 ml. of carbon dioxide-free water. Titrate this solution with the sodium hydroxide to be standardized, using methyl red as indicator. For normal sodium hydroxide, 3·8–4·6 g. of hydrazine sulphate is a suitable amount. For 0·1N solutions take about one-tenth of this amount.

The normality of the sodium hydroxide, against methyl red,

$$= \frac{\text{Weight of hydrazine sulphate}}{\text{Volume of sodium hydroxide used}} \times \frac{1}{0\cdot1301}.$$

Standard Oxalic Acid.

Standard solutions of oxalic acid are prepared by dissolving accurately weighed amounts of pure dry oxalic acid in water and diluting to the proper volume. If carefully prepared no other standardization is necessary. If the oxalic acid is for use in permanganate titrations, and not in acidimetry, the addition of 20 ml. of sulphuric acid per litre increases its stability. Oxalic acid solutions should always be kept in dark bottles in a cupboard.

0·1N Oxalic Acid.

Dissolve 12·604 g. of pure dry oxalic acid in water, add 40 ml. of concentrated sulphuric acid and when cold dilute to 2 litres.

0·05N Oxalic Acid.

Dissolve 6·302 g. of pure dry oxalic acid in water, add 40 ml. of concentrated sulphuric acid and when cold dilute to 2 litres.

Standard Potassium Permanganate.

Standard solutions of potassium permanganate cannot be accurately prepared by dissolving the weighed amount of potassium permanganate directly in water, since ordinary distilled water usually contains traces of reducing substances, which affect the titre. Standard solutions are best prepared as follows:

Heat a suitable volume of water nearly to boiling, remove from the flame and add a weighed amount of potassium permanganate. Add 3·2 g. per litre for 0·1N solution or 1·6 g. per litre for 0·05N solution. When the permanganate has completely dissolved return the flask to the burner, introduce a couple of glass beads, and keep gently boiling for 15-20 minutes. Replace the evaporation loss towards the end of the period. Then allow to cool and filter through a sintered glass funnel. When quite cold standardize as described below and store the solution in a dark glass bottle. The filtration removes any precipitated manganese dioxide, which is a strong catalyst for the auto-decomposition of permanganate solutions. If the permanganate solution is not boiled to oxidize reducing substances present, the solution should be left for at least a week before filtration and standardization.

The standardization of permanganate solutions should be checked at least once a month.

Standardization by means of Sodium Oxalate:

Weigh out a quantity of pure dry sodium oxalate, sufficient to reduce about 35 ml. of the permanganate to be standardized, and dissolve it in 150 ml. of water in a beaker flask. Add 5 ml. of concentrated sulphuric acid and heat to 80-85° C. Remove from the flame and, while still hot, titrate it with the permanganate solution, adding the latter slowly and with constant stirring, until a faint pink blush just persists. For the standardization of 0·1N permanganate take about 0·20 to 0·24 g. of sodium oxalate.

The normality of the permanganate

$$= \frac{\text{Weight of sodium oxalate}}{\text{Volume of permanganate used}} \times \frac{1}{0\cdot 0670}.$$

Standardization by means of Oxalic Acid:

Pipette out a suitable quantity of a standard solution of oxalic acid and titrate it with permanganate exactly as described above for sodium oxalate. In this case the normality of the permanganate

$$= \frac{\text{Volume of standard oxalic acid} \times \text{Normality factor}}{\text{Volume of permanganate used}}.$$

Standard Silver Nitrate.

Standard solutions of silver nitrate can be accurately prepared by dissolving a weighed amount of pure dry silver nitrate in water and diluting to the proper volume. Further standardization is unnecessary. The following strengths are useful:

0·1N Silver nitrate	16·989 g. per litre.
$\frac{N}{35\cdot 5}$ Silver nitrate	4·786 g. per litre.
$\frac{N}{71}$ Silver nitrate	2·393 g. per litre.

The two latter solutions are equivalent to 1 mg. and 0·5 mg. of chlorine respectively per millilitre.

Standard Potassium Chloride.

These solutions can also be accurately prepared by dissolving weighed amounts of pure dry potassium chloride in water and diluting to the required volume. Useful solutions are as follows:

0·01N Potassium chloride	0·7456 g. per litre
0·005N Potassium chloride	0·3728 g. per litre

INDICATORS

The indicator solutions used in the various titrations and for the colorimetric determination of pH are described together for convenience. This enables the strengths of all indicator solutions to be standardized. Most of the indicators used for pH work are made up as 0·04 per cent. solutions so that they can be used directly with standard colour discs. For carbonate and similar titrations dimethyl yellow (dimethyl-amino-azo-benzene) is recommended in place of methyl orange since the colour change is somewhat more perceptible. Sofnol red can be used wherever methyl red is specified.

Dimethyl Yellow (Red to yellow, pH 2·9–4·0). Dissolve 0·1 g. of dimethyl yellow in 100 ml. of 90 per cent. alcohol.

Methyl Red (Red to yellow, pH 4·2–6·3). Dissolve 0·2 g. of methyl red in 50 ml. of alcohol, add 50 ml. of water and filter the solution.

Phenolphthalein (Colourless to red, pH 8·3–10·0). Dissolve 0·5 g. of phenolphthalein in 50 ml. of alcohol and add 50 ml. of water.

Thymolphthalein (Colourless to blue, pH 9·3–10·5). Dissolve 0·5 g. of thymolphthalein in 50 ml. of alcohol and add 50 ml. of water.

Indicator Solutions for pH Determinations and Special Titrations.

Make up aqueous solutions of these indicators, containing 0·04 g. of indicator per 100 ml. in each case. To dissolve these indicators in water it is necessary to use an equivalent amount of sodium hydroxide to convert them to the sodium salts. In an agate mortar grind 0·1 g. of the indicator with the amount of 0·1N sodium hydroxide shown in the last column of the following table and dilute to 250 ml.

Indicator	Colour Change	pH Range	Amount of 0·1N NaOH necessary
Brom Cresol Green	yellow-blue	3·8-5·4	1·43 ml.
Brom Cresol Purple	yellow-purple	5·2-6·8	1·85 ml.
Brom Phenol Blue	yellow-blue	3·0-4·6	1·49 ml.
Brom Thymol Blue	yellow-blue	6·0-7·6	1·60 ml.
Cresol Red	yellow-red	7·2-8·8	2·62 ml.
Phenol Red	yellow-red	6·8-8·4	2·82 ml.
Thymol Blue	red-yellow yellow-blue	1·2-2·8 8·0-9·6	2·15 ml.

CHAPTER VII

CALCIUM CARBONATE

Carbonate present in soils is generally determined and reported as calcium carbonate even though it may be partly dolomitic, some of the calcium being replaced by magnesium. Several methods are available, depending upon the purpose of the determination and the accuracy required. If the values are required for the correction of exchangeable calcium or organic carbon determinations in calcareous soils, the highest accuracy is desirable and Hutchinson and MacLennan's method is recommended (p. 130). Frequently, however, more rapid methods are desirable, even though the results are less accurate. A knowledge of the approximate amount of calcium carbonate present in calcareous soils is required before starting mechanical analyses. Also it is useful to have a rough idea of the amount of calcium carbonate present as a guide to the amount of sample to be taken for its more accurate determination. For soil survey purposes, too, an approximate knowledge of the amount of calcium carbonate is frequently sufficient. Two rapid methods are therefore described. In the first of these, a modification of Passon's method developed by H. R. Skewes, carbon dioxide is liberated by the action of hydrochloric acid on the soil, in a closed system, and the increase in pressure is measured (p. 132). Provided that no great temperature fluctuations occur the results are accurate to within 0·05 unit for small amounts of carbonate (up to 1 per cent.) and to within 0·1 unit for larger amounts. In the second method (p. 135) the soil is treated with an excess of hydrochloric acid and the excess of acid titrated with sodium hydroxide using brom thymol blue as indicator. To reduce sampling errors normal solutions are used. Some of the acid added is used in replacing the exchangeable cations. Iron and aluminium are also dissolved but as they are reprecipitated at the end point of the titration they do not affect the determination. Values obtained by this method are approximately 1 unit high.

In the accurate determination of carbonate in soils, precautions must be taken to avoid the production of carbon dioxide by decarboxylation of the organic matter. This occurs readily if the hydrochloric acid used for decomposing the carbonates is too concentrated, or if elevated temperatures are used. Schollenberger (2) also noted that some organic matter was readily oxidized in those soils containing manganese dioxide. He avoided errors from this source by adding ferrous chloride to the dilute hydrochloric acid used and decomposing the carbonates at a low temperature (boiling under reduced pressure). Shaw (3) used stannous chloride for a similar purpose. Ferrous chloride is not quite as effective as stannous chloride but, for practical reasons, it is preferred. Its solution is more convenient to prepare and it does not foul the apparatus like stannous chloride.

In Hutchinson and MacLennan's method (1) the soil is decomposed by cold dilute hydrochloric acid, *in vacuo*, and the carbon dioxide evolved is absorbed in standard sodium hydroxide. The absorbed carbon dioxide is precipitated as barium carbonate, by the addition of an excess of barium chloride, and the alkali remaining is titrated with hydrochloric acid. In the presence of a large excess of barium chloride, barium carbonate is almost insoluble, so that a sharp end point is obtained. Phenolphthalein can be used as the indicator. However, if much barium carbonate is precipitated, thymolphthalein is preferable, since its transition range is slightly more alkaline and it is therefore less affected by barium carbonate. Provided that ferrous chloride is used to prevent decarboxylation in the case of certain organic and manganiferous soils, this method gives results of a high order of accuracy. The apparatus required is simple and determinations can be carried out rapidly. For consistent results a high vacuum is desirable. To maintain this the rubber stoppers used must be in good condition and fit tightly. It is easier to maintain a high vacuum in an apparatus assembled from a large filter flask, as proposed by Williams (4), instead of from two round bottom flasks, as used in the original method. Soils containing dolomite must be very finely ground and left to react with the acid for a somewhat longer time than is usually necessary.

The Determination of Calcium Carbonate: Hutchinson and MacLennan's Method.

Apparatus:

The apparatus required is shown in Fig. 11. A is a 50 ml. separating funnel the stem of which reaches to the bottom of the 150 ml. round bottom flask B. C is a bulb tube to prevent the mechanical carrying over of acid spray from B. D is a 1,000 ml. pyrex or hard glass filter flask and E is a stopcock for connexion to the vacuum pump.

Reagents:

Dilute Hydrochloric Acid. Dilute 100 ml. of concentrated hydrochloric acid to 2 litres.

Dilute Hydrochloric Acid containing Ferrous Chloride. Dissolve 3 g. of ferrous chloride ($FeCl_2.4H_2O$) in each 100 ml. of dilute hydrochloric acid immediately before use.

Barium Chloride. Dissolve 150 g. of barium chloride ($BaCl_2.2H_2O$) in water and dilute to 1 litre.

0·1N Hydrochloric Acid. See p. 117.

0·1N Sodium Hydroxide. See p. 120.

Indicator Solution. For phenolphthalein and thymolphthalein solutions see p. 126.

Method:

Weigh out 0·5-25 g. of soil, depending on the amount of carbonate present, and transfer it to the 150 ml. round bottom flask. The amount of soil taken should contain from 0·15 to 0·2 g. of calcium carbonate but, if only a small amount is present, not more than 25 g. of soil should be taken. If less than 15 g. is taken, it is necessary to grind the soil to pass a 0·5 mm. screen to reduce sampling errors. If the pH of the soil is less than about 7·6 it is generally safe to take 25 g. for the carbonate determination.

Pipette 50 ml. of 0·1N sodium hydroxide (carbonate-free) into the 1,000 ml. flask, add four or five drops of indicator solution (thymolphthalein or phenolphthalein) and close the flasks tightly with the rubber stoppers. Evacuate the system as completely as possible by means of an electric or good filter pump. When completely evacuated close the stopcock E and add about 50 ml. of dilute hydrochloric acid to the separating funnel. For soils containing manganese dioxide or

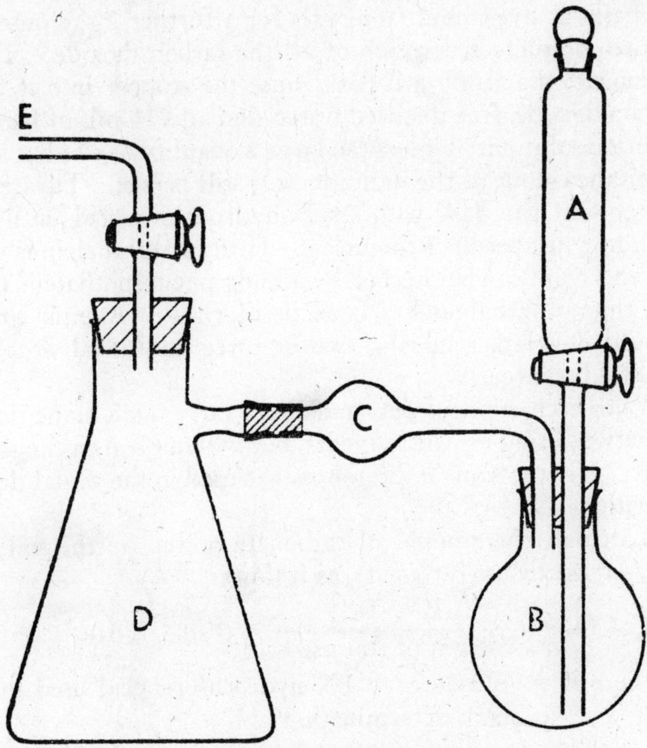

Fig. 11. Apparatus for the determination of carbonates by Hutchinson and MacLennan's method.

much organic matter, use dilute hydrochloric acid containing ferrous chloride. Cautiously open the stopcock of the separating funnel and slowly introduce the acid into the small round bottom flask containing the soil; avoid too vigorous a reaction at the start. When only a few drops of acid remain in the funnel close this stopcock. After a few minutes gently shake the flask to ensure complete decomposition of all carbonate present. Repeat this shaking four times in all during 20 minutes. Then connect the top of the separating funnel to a gas washing tower, containing 40 per cent. potassium hydroxide, and slowly draw in carbon dioxide-free air until the vacuum is destroyed. This operation should take about 10 minutes. The air is drawn in through the acid-soil mixture and helps to sweep out the last traces of carbon dioxide. Shake the flasks at five minute intervals for a further 20 minutes to ensure complete absorption of all the carbon dioxide. Then disconnect the 1,000 ml. flask, rinse the stopper into it with carbon dioxide-free distilled water, and add 10 ml. of barium chloride solution. Unless too large a quantity of soil has been taken the colour of the indicator will still persist. Titrate the contents of the flask with 0·1N hydrochloric acid until the indicator just becomes colourless. If thymolphthalein is used the end point can be checked by adding phenolphthalein after the thymolphthalein has been decolorized. A pink colour should develop, requiring two or three additional drops of acid to discharge it.

With each series of determinations carry out a blank determination, using all the reagents, but without soil in the small flask. Use the same indicator as that used in the actual determinations.

Calculate the amount of carbonate present in the soil, expressed as calcium carbonate, as follows:

$$\% \ CaCO_3 = \frac{B - T}{\text{Weight of soil used}} \times 0{\cdot}005 \times 100$$

where B = volume of 0·1N hydrochloric acid used in the blank determination

and T = volume of 0·1N hydrochloric acid used in the actual determination.

The Determination of Calcium Carbonate: Modified Passon's Method.

For soils containing 0·2–1 per cent. of calcium carbonate, the average error of this method varies from 10 per cent. to 4 per cent. of the total carbonate present, decreasing as the amount of carbonate increases. Thus for soils with 0·2 per cent. and 1 per cent. of carbonate the absolute error approximates to 0·02 and 0·04 of a unit, respectively. For soils with 1-5 or more per cent. of carbonate the average error declines from 4 to 2 per cent. of the total carbonate present. The method is recommended where an accuracy of this order is sufficient.

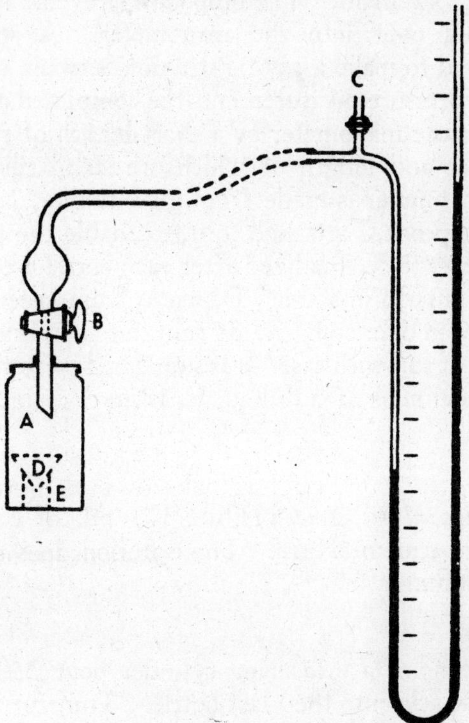

Fig. 12. Modified Passon's apparatus for the determination of carbonates.

Apparatus:

The apparatus required is easily assembled from commercially available components. It is illustrated in Fig. 12. When carrying out large numbers of determinations several units, consisting of bottle, lid, and stopcock, are desirable. They can all be connected in turn to the one manometer.

The bottle A consists of an ordinary 6 oz. wide mouthed bottle, taking a screw-on lid. The screw-on lid is made of tin plate, strengthened by a flat plate of metal on top, and protected from corrosion by a coating of colourless synthetic enamel. A stainless steel tube, one inch long, is soldered into the centre of the lid and a stopcock, B, and bulb tube, 2 mm. bore, is rigidly fixed into the stainless steel tube by means of de Khotinsky's cement. The bulb tube prevents spray from being carried over into the manometer. A soft rubber washer is used to make a gas-tight union between the lid and bottle. For pressure measurements the completed assembly is connected to the manometer by a short length of rubber tubing, the bore and length of which are standardized. The mercury manometer is made from glass tubing, 2 mm. bore, and has a stopcock C attached to it, to enable the pressure in the manometer to be equalized after each bottle is connected.

The soil is held in a cone, D, made from sheet celluloid. It is 39 mm. in diameter, and 34 mm. high, sufficiently large to hold 5 g. of soil with ease. It is supported by resting loosely on an open cylinder of celluloid, E, 18 mm. diameter and 25 mm. high.

Reagents:

2N Hydrochloric Acid. Dilute 175 ml. of concentrated hydrochloric acid to 1 litre. This solution does not require standardization.

Method:

By means of a measuring cylinder pour 25 ml. of 2N hydrochloric acid into the glass bottle. Transfer 5 g. of soil to the celluloid cone and support it in the wide mouthed bottle by means of the celluloid cylinder, so that the soil does not come in contact with the acid. Screw the lid tightly into position, using a soft rubber washer to ensure a gas-tight joint. Close the stopcock and shake the bottle to dislodge the soil into the acid. Shake mechanically for 10 minutes to ensure complete decomposition of the soil carbonates by the acid. Then connect the bottle to the manometer and equalize the pressure in the manometer by momentarily opening stopcock C. Open stopcock B and read the pressure in the manometer due to the carbon dioxide liberated.

Calibrate the apparatus by making determinations on several known amounts of pure calcium carbonate added to sufficient of a carbonate-free soil to give a total weight of 5 g. The effective volume of the bottle and connecting tube is an important factor in the standardization and must be kept the same in all units used. The bottles have a volume of 195 ml. and this is reduced to 168 ml. when the soil, celluloid containers and acid are added. The change in volume due to the use of soils of different specific gravity is negligible. The volume of the acid must be kept the same in all determinations. For soils very high in calcium carbonate 25 ml. of 2N acid is not sufficient to decompose all the carbonate. For these soils 25 ml. of a stronger acid should be used, or it is permissible to reduce the amount of soil to 2·5 g. In this case multiply the manometer readings by two.

All determinations should be carried out at a temperature as close as possible to that at which the apparatus was calibrated. A correction can be applied if the temperature differs appreciably. For each increase of 1° C. above the calibration temperature subtract one eightieth of the amount of calcium carbonate found.

The Determination of Calcium Carbonate: Rapid Titration Method.

This method yields approximate values only and is not recommended where results closer than about one unit in the percentage of calcium carbonate are required.

Reagents:

N Hydrochloric Acid. Dilute 175 ml. of concentrated hydrochloric acid to 2 litres. This solution does not require standardization.

N Sodium Hydroxide. This need not be specially carbonate-free. Dissolve slightly more than 80 g. of sodium hydroxide in 2 litres of water. Standardize against hydrazine sulphate as described on p. 123 and dilute until exactly N.

Brom Thymol Blue Indicator Solution. See p. 127.

Method:

Weigh out 5 g. of soil and transfer to a tall 150 ml. beaker. For soils with more than 30 per cent. of calcium carbonate take 2·5 g. only. By means of a pipette with an enlarged jet, add 100 ml. of N hydrochloric acid, cover with a clock glass and stir vigorously several times, during a period of one hour. Then allow to settle and pipette off 20 ml. of the supernatant liquid. Transfer to a small Erlenmeyer flask, add 6-8 drops of brom thymol blue indicator solution and titrate with N sodium hydroxide. With some soils the colour of the indicator may fade as the end point is approached. If this occurs add more indicator and complete the titration. Carry out a blank determination to obtain the titre of the hydrochloric acid.

The percentage of calcium carbonate is given by the expression:

$$\% \, CaCO_3 = (\text{Blank titration} - \text{Actual titration}) \times 5$$

If the sodium hydroxide solution is not exactly normal this simple relationship between percentage calcium carbonate and the difference in titration values can be maintained by taking a weight of soil equal to 5 g. times the normality factor of the sodium hydroxide, instead of 5 g. Thus if the sodium hydroxide is 0·90N carry out the determination on 4·5 g. of soil.

REFERENCES

(1) Hutchinson, H. B. and MacLennan, K.: *J. Agric. Sci.*, 1914, *6*, 323–327.

(2) Schollenberger, C. J.: *Soil Sci.*, 1930, *30*, 307–324.

(3) Shaw, W. M.: *J. Assoc. Off. Agric. Chem.*, 1931, *14*, 283–292.

(4) Williams, R.: *J. Agric. Sci.*, 1932, *22*, 838–844.

CHAPTER VIII

THE ANALYSIS OF THE HYDROCHLORIC ACID EXTRACT

At one time considerable value was attached to the hydrochloric acid digestion of the soil since it was thought that the method could discriminate between the weathered and the unweathered minerals present. The former fraction was supposed to represent the total source of plant nutrients. However, hydrochloric acid extraction is entirely empirical; it does not differentiate between the different categories of minerals present and the amounts brought into solution depend on such factors as the strength of acid and the time of digestion. Except for the determination of potash and phosphoric acid, digestion with hydrochloric acid is now only occasionally used.

In most methods hydrochloric acid of constant boiling point (S.G. 1·10) is used. Hilgard digested the soil for five days while Hall, whose method was adopted by the Agricultural Education Association in 1905, recommended digestion in a boiling water bath for 48 hours. In 1929 the International Society of Soil Science adopted the van Bemmelen-Hissink method in which the soil is boiled with concentrated hydrochloric acid until the constant boiling point is reached, when a reflux condenser is fitted and the digestion continued for a further two hours. On account of the unnecessary manipulative difficulties in this method the Agricultural Education Association (2) provisionally adopted, in 1931, a simplified digestion and dispensed with the reflux condenser. In the A.E.A. provisional method the soil is gently boiled for one hour with hydrochloric acid. A tall beaker, covered with a clock glass to minimize evaporation, is used. Values obtained by this digestion correspond approximately with those given by the official method of the I.S.S.S.

The amount of potash extracted varies considerably with the time of digestion and increased amounts are obtained up to at least five days' digestion. However, even this prolonged treatment fails to obtain the whole of the potash in solution. If such a value is required, fusion or treatment with hydrofluoric acid is necessary. The amount of potash brought into solution during one to two hours' digestion with hydrochloric acid is very much less than that obtained by digestion for 48 hours. J. G. Baldwin *(priv. comm.)* found that only 22-68 per cent. of the potash extractable in 48 hours was in solution at the end of one hour's digestion. It is apparent that the potash extracted does not belong to any particular category in the soil but is derived from the decomposition of silicate minerals slowly attacked by the boiling acid. It is considered that Hall's method of digestion is still the most useful, since so many comparative values have been determined by it. Furthermore, after 48 hours' digestion, the amounts of potash coming into solution are small and a state of equilibrium has been more nearly approached than after one hour's digestion.

Phosphoric acid, unlike potash, is readily extracted by hydrochloric acid and it is generally supposed that the bulk, if not all, of the soil phosphorus is extracted by any of the above-mentioned methods. McLean (4) concluded that 48 hours' digestion was amply sufficient to extract all of the acid soluble phosphorus and that this amount represented a definite category of soil phosphorus. It is the same as the amount brought out by a direct digestion of the soil for 45-75 minutes with sulphuric and nitric acids, according to the method of the Halle Station.

Apart from the presence of organic matter and the large amounts of iron and aluminium relative to the other constituents, the analysis of the hydrochloric acid extract presents no great difficulties. The scheme described below is based largely on current practice in silicate analysis. The basic acetate precipitation of the sesquioxides, followed by an ammonia reprecipitation, is necessary to prevent absorption of magnesia. Potash is separated by ignition of an aliquot of the hydrochloric acid solution and extraction with water, as first proposed by Neubauer. An excess of calcium salts is necessary to prevent some of the potassium remaining in the residue in an insoluble form. However, Beatter (1) and Martin and Griffith (3) have noted that in some cases hot water does not extract potassium completely from the ignited residue. To recover the last of the potash they dissolve the residue from the hot water extraction in dilute hydrochloric acid and precipitate iron and aluminium with ammonia.

Potash is conveniently determined by the perchlorate method. Sulphates must be removed prior to separation of the potassium perchlorate. If it is not desired to remove sulphates the cobaltinitrite method, described on p. 178, can be used as a gravimetric method since the amounts of potash are usually sufficient to give weighable precipitates.

When calcium salts are not unduly high, as in soils with less than 4 per cent. of calcium carbonate, phosphoric acid can be determined directly in an aliquot of the soil extract (p. 152) instead of igniting and extracting with hot water to remove potash and the excess of soluble calcium salts (p. 150). The introduction of "Celite" as a filter aid (E. C. Orton, *priv. comm.*) assists considerably in the final phosphate filtration.

Hydrochloric Acid Extraction: **Hall's Method.**

Place 50 g. of the air-dry soil in a 500 ml. pyrex Erlenmeyer flask and add 175 ml. of concentrated hydrochloric acid. Place a small glass funnel in the neck to act as a condenser. Boil for a few minutes over a flame so as to reduce the strength of the hydrochloric acid to the constant boiling strength. Then digest in a boiling water bath for 48 hours.

Dilute with 100-150 ml. of hot water, then filter through a 9 cm. Buchner funnel, fitted with a Whatman No. 50 filter paper, and wash with hot water containing 50 ml. of concentrated hydrochloric acid per litre. (This acid is necessary to prevent the hydrolysis of ferric and aluminium salts in hot dilute solutions.) The washing should be continued until the filtrate amounts to nearly 800 ml. Transfer the filtrate to a litre measuring flask, and when cold dilute to the mark and mix well.

Use suitable aliquots of this solution for the determination of

Fe_2O_3 and TiO_2
Mn_3O_4
$Fe_2O_3 + Al_2O_3$, CaO and MgO
K_2O and P_2O_5

according to the methods described below.

If potash and phosphoric acid only are required, digest 20 g. of soil with 75 ml. of concentrated hydrochloric acid and, after filtering, dilute to 500 ml.

Hydrochloric Acid Extraction: A.E.A. (1931) Provisional Method.

Place 20 g. of air-dry soil in a 500 ml. tall beaker, add 200 ml. of hydrochloric acid (constant boiling point 110° C.), cover with a clock glass and boil gently for one hour. Cool the acid liquid, filter and make the filtrate and washings up to a convenient volume. Determine potash and phosphoric acid in suitable aliquots.

The Determination of Iron, Aluminium, Manganese, Calcium, Magnesium, Potassium, and Phosphoric Acid in the Hydrochloric Acid Extract.

(1) FERRIC OXIDE AND TITANIUM DIOXIDE.

(a) *Ferric Oxide*, Fe_2O_3: Evaporate 50–75 ml. of the hydrochloric acid extract in a silica basin on the water bath until the volume is reduced to about 25 ml. Cover with a clock glass, and then add 20 ml. of fuming nitric acid, diluted if necessary with a little ordinary nitric acid to prevent too vigorous a reaction. Digest for twenty minutes to half an hour, then remove the clock glass and rinse it into the basin. Add 15 ml. of dilute sulphuric acid (1 + 1) and continue the evaporation on the bath. An excessive amount of sulphuric acid must be avoided as it greatly retards the rate of reduction by hydrogen sulphide later. After about one hour transfer to a sand bath, and cautiously evaporate until dense fumes of sulphuric acid have been produced for five minutes. If any insoluble matter, such as calcium sulphate, separates, the basin should be supported just above the sand bath by means of a triangle. This prevents loss of liquid by bumping, which would otherwise occur.

When the contents of the basin have cooled, add about 60-80 ml. of water, and warm on the water bath, until as much as possible is in solution. If a large quantity of calcium sulphate is present it will not be completely soluble, but this does not matter. Transfer to a 250 ml. Erlenmeyer flask.

Wash the basin thoroughly with hot water, then dilute the liquid in the flask to 150–200 ml.

When cold, pass hydrogen sulphide gas into the solution for 7–10 minutes. Disconnect from the gas generator and heat until nearly boiling. Test for the presence of any ferric salt by removing two or three drops of the solution and adding them to a solution of potassium thiocyanate contained in a watch glass or white porcelain dish. If reduction is complete, as indicated by no red colour being produced, connect the flask again to the hydrogen sulphide generator and bubble the gas in slowly, surrounding the flask with cold water. Continue passing the gas until nearly cold. Should the reduction have been incomplete, as may occasionally be the case, the hydrogen sulphide must be passed into the hot liquid until a second test shows the absence of any ferric salt; then cool as above, continuing to pass the gas until cold.

Filter through an 11 cm. Whatman No. 44 filter paper into a 500 ml. Erlenmeyer flask, keeping the filter paper full, to avoid any oxidation. Wash the flask and filter paper six times with water containing hydrogen sulphide. Frequently the filtrate becomes opalescent due to finely divided sulphur, but this will not matter as it will be completely oxidized in the subsequent boiling. Again test the filtrate for any ferric salt. If any should have become oxidized during the filtration, the solution must be warmed, treated with hydrogen sulphide until reduced, and then cooled as before. It is unnecessary to filter the liquid again.

Now pass carbon dioxide (freed from any possible traces of hydrogen sulphide by bubbling through a solution of copper sulphate and then water) into the flask, and boil the solution. Continue to boil for about fifteen minutes, without interrupting the stream of carbon dioxide. Boiling must be continued for some time after the elimination of all hydrogen sulphide, but the liquid must not be concentrated to more than half its original volume. Cool, by placing the flask in a dish of cold water, continuing to pass carbon dioxide until quite cold. Then rinse the tube into the flask with cold distilled water (previously boiled), and titrate with 0·1N potassium permanganate, recently standardized against pure sodium oxalate, until a pink blush just persists. The end point is quite sharp and the colour remains for at least a minute if the soil organic matter has been properly oxidized in the preliminary treatment with fuming nitric acid.

$$1 \text{ ml. of } 0{\cdot}1N\ KMnO_4 = 0{\cdot}0080 \text{g. } Fe_2O_3$$

Reserve the liquid after the titration, for the determination of titanium.

(b) *Titanium Dioxide, TiO_2:* To the liquid in the flask, after the titration with potassium permanganate, add 10 ml. of concentrated sulphuric acid, and concentrate, by boiling, until its volume is about 50 ml. The addition of two or three glass beads will promote even boiling. Add 5 ml. of 20 vol. hydrogen peroxide and transfer the solution to a volumetric flask so that the colour will be of suitable intensity for comparison in a colorimeter. A 100 ml. flask is generally convenient. When cold add a further 1–2 ml. of hydrogen peroxide and dilute to the mark. Mix well by inverting the closely stoppered flask several times. Then compare the colour produced, in a suitable colorimeter, with that developed by a known amount of a standard solution of titanium sulphate in another 100 ml. flask. If much calcium sulphate, or other insoluble matter, is present in the test solution it is necessary to clear some of the solution for the colour comparison, either by centrifuging or filtering through a dry Whatman No. 44 filter paper, rejecting the first runnings.

Take the average of eight consecutive colorimeter readings. The amount of TiO_2 in the test solution is given by the following expression:

$$S \times \frac{D_s}{D_u} \times \frac{V_u}{V_s}$$

where S = amount of TiO_2 in the standard colour solution
D_s = depth of the standard colour solution
D_u = depth of the unknown colour solution
V_u = volume of the unknown colour solution
and V_s = volume of the standard colour solution.

Make a correction for the colour due to ferric sulphate by subtracting 0·01 per cent. from the percentage of TiO_2 found, for each 5 per cent. of Fe_2O_3 present.

Standard Titanium Sulphate Solution: Dissolve about **2 g.**

of the purest titanium dioxide in 10 ml. of concentrated sulphuric acid, together with sufficient hydrofluoric acid, in a platinum basin. Evaporate to fuming five times successively, adding about 10 ml. of dilute sulphuric acid (1 + 1) each time. When all the hydrofluoric acid has been expelled, take up in 15 ml. of sulphuric acid and 60–80 ml. of water and filter into a 2-litre measuring flask. Add 150 ml. of concentrated sulphuric acid, cool, and dilute to the graduation mark. Transfer to a stoppered reagent bottle. Determine the actual strength of the standard solution by precipitating duplicate 50 ml. portions with ammonia, filtering, washing, and igniting as TiO_2.

To prepare standard colour solutions, pipette suitable volumes into 100 ml. measuring flasks, add 10 ml. of sulphuric acid, dilute, cool, add 5 ml. of hydrogen peroxide and adjust the volume to the mark.

(2) MANGANESE OXIDE (Mn_3O_4).

Evaporate 100 ml. of the hydrochloric acid extract nearly to dryness in a silica basin on the water bath. Cover with a clock glass and add 15 ml. of fuming nitric acid. Digest on the bath for about twenty minutes, and then remove the clock glass and rinse it into the basin. Then add 40 ml. of dilute sulphuric acid (1 + 1) and leave on the briskly-boiling bath for about one hour. Transfer to a sand bath and continue the evaporation cautiously until fumes of sulphuric acid are just produced, and all the chlorine is eliminated.

When the contents of the silica basin are cold, add 2 ml. of phosphoric acid and 30–50 ml. of water. Add 0·3 to 0·5 g. of potassium periodate, and bring the contents of the dish to the boil, stirring to prevent bumping. Keep just boiling for one minute after the development of the permanganate colour. When sufficiently cool transfer the contents of the dish to a volumetric flask of suitable size (50 ml. to 250 ml., according to the amount of manganese present), and wash the dish thoroughly with small portions of hot water. Place the flask in a boiling water bath for 10–15 minutes. Then remove and allow to cool. When quite cold dilute to the mark and mix the contents well.

Prepare a standard manganese colour solution by pipetting an appropriate amount of the standard manganous sulphate into a volumetric flask, adding 15 ml. of concentrated sulphuric acid, 2 ml. of phosphoric acid and diluting to about 60–70 ml. Add 0·3–0·5 g. of potassium periodate, heat in a boiling water bath for fifteen minutes after the development of the permanganate colour and then remove. When cold, dilute to the mark and mix well.

Compare the intensity of colour of the test solution with that of the standard solution by means of a colorimeter. For good colour comparisons the test solution should not be more than 40 per cent. stronger or 25 per cent. weaker than the standard colour solution.

When much insoluble matter, such as silica or calcium sulphate, is present in the test solution it must be removed before attempting the colour comparison. If this cannot be done by decantation, centrifuging for 3 or 4 minutes, or filtration through a sintered glass funnel, will be found satisfactory. Filtration through filter paper is not permissible on account of the reduction of the permanganate.

The amount of Mn_3O_4 present in the test solution is given by the following expression:

$$S \times \frac{D_s}{D_u} \times \frac{V_u}{V_s}$$

where S = the amount of Mn_3O_4 in the standard colour solution

D_s = depth of the standard colour solution

D_u = depth of the unknown colour solution

V_u = volume of the unknown colour solution

and V_s = volume of the standard colour solution.

Standard Manganous Sulphate Solution: This should contain the equivalent of 1 mg. of Mn per 10 ml.

Dissolve 0·5756 g. of the purest, dry, potassium permanganate in 500 ml. of water in a 2,000 ml. measuring flask. Add 40 ml. of concentrated sulphuric acid and reduce the permanganate by the cautious addition of sodium metabisulphite solution, until the manganese solution just becomes colourless. Oxidize the excess of sulphurous acid by the addition of a little nitric acid. When cool, dilute to the 2,000 ml. graduation mark, mix well, and store in a stoppered reagent bottle.

(3) IRON, ALUMINIUM, CALCIUM, AND MAGNESIUM.

(a) *Elimination of Silica:* Pipette 25–50 ml. (depending on the relative amounts of iron and calcium present) into a silica basin, and evaporate to dryness on the water bath. Cool for a few moments and then add 15 ml. of fuming nitric acid, cover with a clock glass and replace on the water bath. After fifteen minutes' digestion, remove the clock glass, rinse it into the basin, evaporate the contents to dryness, and leave on the bath for a further half to one hour to render the silica insoluble.

Take up in 30 ml. of dilute hydrochloric acid (1 + 9). Warm for a few minutes until all the soluble matter is in solution. Filter through a 9 cm. Whatman No. 44 filter paper into a 400 ml. beaker. Wash twice with cold water and then four times with hot water containing 50 ml. of hydrochloric acid per litre. Complete the washing with hot water alone and reject the filter paper containing the silica.

(b) *Iron and Alumina,* $Fe_2O_3 + Al_2O_3$ $(+ P_2O_5 + TiO_2)$. *Basic Acetate Separation:* Concentrate the filtrate and washings from the silica separation on the water bath until the volume is reduced to about 50 ml. When quite cold, add a freshly-prepared cold 20 per cent. solution of sodium carbonate, the beaker being covered to prevent loss by spray. Add the sodium carbonate gradually at first, and finally drop by drop, until the liquid in the beaker has just darkened in colour, but no precipitate has formed. If, after the addition of the last drop of sodium carbonate and after rinsing the cover glass and the sides of the beaker, there is a precipitate, then add one, or if need be, two drops of dilute hydrochloric acid (1 + 3). If this fails to clear the solution, the precipitate must be redissolved by the smallest possible amount of dilute acid, and diluted sodium carbonate again added more carefully drop by drop from a tube, until the liquid has just darkened in colour.

The volume at this stage should not exceed 75–100 ml. Add 6 to 8 ml. of 20 per cent. sodium acetate solution, and then fill the beaker to about 350–375 ml. with hot water. For heavy clay soils, containing moderate to large amounts of iron and aluminium, it is necessary to use 10–12 ml. of sodium acetate. Heat to boiling while still covered with the clock glass, and boil gently for three minutes, but no longer. Allow to stand a few minutes until most of the precipitate has settled, and then filter through a suitable size (12·5 cm. or 11 cm.) Whatman No. 41 filter paper. Collect the filtrate in a 600 or 800 ml. beaker. Wash the original beaker and precipitate three or four times with hot 0·2 per cent. sodium acetate.

Sodium carbonate and sodium acetate effect a much better separation from magnesium than do ammonium carbonate and ammonium acetate. For a good separation it is desirable that ammonium salts be absent at this stage.

Transfer the filter and precipitate to the original beaker. Add 25 ml. of warm dilute hydrochloric acid (1 + 1), pouring it around the sides of the beaker and stirring rod to dissolve the precipitate. Add 75 ml. of water, a few drops of methyl red or sofnol red indicator solution, and a little macerated filter paper. Warm until all the iron and aluminium are dissolved. Then add dilute ammonia (1 + 1) until precipitation again occurs and the colour of the indicator just changes to the alkaline shade. Add a further 2–3 drops of dilute ammonia and allow to stand until the colour of the indicator in the supernatant liquid can be seen. Heat to boiling point and boil for 30–40 seconds. If the red colour of the indicator returns add a further drop of dilute ammonia, but only sufficient just to restore the alkaline colour. Allow the precipitate to stand for some minutes until the bulk of it has settled; then filter through a 12·5 cm. Whatman No. 541 filter paper. Wash well with hot water, collecting the filtrate in the same beaker as that containing the filtrate from the first precipitation. When the washing is complete, drain the filter and precipitate by applying suction to the stem of the funnel. Transfer the precipitate to a weighed silica crucible.

Concentrate the combined filtrates, by leaving the beaker on the water bath overnight, until the volume is reduced to about 50–100 ml. Then make it just ammoniacal, boil, and filter through a 9 cm. Whatman No. 41 filter paper to remove the traces of aluminium which escaped precipitation earlier. Wash the beaker and filter with hot water and collect the filtrate in a 400 ml. beaker. Add the filter and precipitate to the crucible containing the main portion of the iron and aluminium precipitate, ignite carefully in a muffle furnace and weigh as $Fe_2O_3 + Al_2O_3 + TiO_2 + P_2O_5$. To obtain the amount of Al_2O_3 deduct the amounts of Fe_2O_3, TiO_2, and P_2O_5 found separately.

(c) *Elimination of Manganese:* After the removal of the last traces of iron and aluminium, concentrate the filtrate and washings on the water bath until the volume is again reduced to about 50 ml. Cool in a dish of water. When quite cold add sufficient bromine water (generally 30–50 ml.) to colour the liquid fairly strongly, and then add a very little dilute ammonia (1 + 9) until just alkaline to litmus. Cover the beaker with a clock glass, and boil for a short time (three minutes). Again cool in water and add bromine water and ammonia as before, and boil again to complete the precipitation of the manganese. Make just acid with dilute acetic acid (20 per cent.), and filter while hot through an 11 cm. Whatman No. 44 filter paper. Wash well with hot water and collect the filtrate in a 600 ml. beaker. Reject the filter paper containing the precipitated manganese as this precipitate is not sufficiently pure to weigh directly for the Mn_3O_4 determination. It usually only amounts to 1–3 mg.

(d) *Calcium Oxide, CaO:* Boil the filtrate from the manganese separation, which should not exceed 300–350 ml., and add 10–15 ml. of hot 10 per cent. ammonium oxalate solution. Then add sufficient ammonia (40 ml. of 1 + 1) to make the liquid quite alkaline. Allow the precipitated calcium oxalate to stand overnight.

Filter through an 11 cm. Whatman No. 44 filter paper and collect the filtrate in a second 600 ml. beaker. Wash three times, by decantation, with hot water. Evaporate the filtrate to small bulk on the water bath.

Dissolve the precipitated calcium oxalate by pouring 25 ml. and then 10 ml. of warm dilute hydrochloric acid (1 + 4) on to the filter and collecting the filtrate in the beaker used for precipitation, washing down its sides. Then wash the filter with two lots of hot 5 per cent. hydrochloric acid and finally hot water alone. Add 2–3 ml. of saturated ammonium oxalate solution to the filtrate and raise to boiling. Precipitate by adding an excess of dilute ammonia (1 + 1), allow to stand overnight, and filter through the same filter as used previously. Collect the filtrate in the beaker containing the first filtrate, which by now has been concentrated nearly to dryness. Wash the beaker and filter well with hot water. Reserve the filtrate for the magnesium determination.

Pierce the filter paper with a pointed glass rod, and wash the calcium oxalate into the beaker used for the precipitation. Then wash the filter paper alternately with warm dilute sulphuric acid (1 + 4), made just pink with potassium permanganate, and warm water until all the oxalic acid is in solution. Use three lots of the acid in all. Warm the solution to about 70° C. and titrate with 0·1N potassium permanganate. Finally, add the filter paper, stir, and see that the pink colour is not discharged.

$$1 \text{ ml. of } 0{\cdot}1N\ KMnO_4 = 0{\cdot}0028 \text{ g. } CaO.$$

(e) *Magnesium Oxide, MgO:* When the filtrate from the calcium reprecipitation is quite cold, add 50 ml. of 95 per cent. alcohol and 30 ml. of 10 per cent. sodium phosphate solution. After a quarter of an hour add 30–50 ml. of concentrated ammonia and allow to stand overnight. Filter through an 11 cm. Whatman No. 44 filter paper, and wash two or three times with dilute ammonia (1 + 9), discarding the filtrate.

Then dissolve the precipitated magnesium ammonium phosphate by pouring 15 ml. and 10 ml. of warm dilute nitric acid (1 + 4) through the filter, collecting the filtrate in the beaker in which the precipitation was made. Wash the filter thoroughly with warm water. When cold, add a few drops of sodium phosphate solution, 25 ml. of alcohol, and 40–50 ml. of ammonia to reprecipitate the magnesium ammonium phosphate. Leave overnight and then filter through a 9 cm. Whatman No. 44 filter paper. Wash with dilute ammonia water (1 + 9). Transfer the filter and precipitate to a weighed silica crucible, ignite in the muffle at a bright red heat, and weigh as $Mg_2P_2O_7$.

$$\text{Weight of precipitate} \times 0{\cdot}3621 = \text{wt. of } MgO.$$

(4) POTASH AND PHOSPHORIC ACID.

(a) *Potash*, K_2O*:* To 100 ml. of the hydrochloric acid extract, add sufficient of a 2 per cent. solution of barium chloride to precipitate all the sulphate (generally 3–5 ml. for ordinary soils), and also, if the soil did not effervesce when treated with hydrochloric acid, add 5 ml. of a 4 per cent. solution of calcium carbonate dissolved in a slight excess of hydrochloric acid. Evaporate to dryness in a silica basin on the water bath.

Transfer to an air oven at about 100° C. and gradually raise the temperature to 120–140° C. till quite dry. Then gently heat over a large burner until all the ammonium salts have been removed and all the iron salts rendered insoluble. The ignition must be sufficient to render iron and aluminium salts insoluble in water but the temperature must not exceed a very dull red heat, otherwise potassium may be lost through volatilization or rendered insoluble by combination with silica and aluminium. If desired, the ignition can be completed in a muffle furnace at a low temperature.

When cool, add 10–15 ml. of hot water and break up the lumps in the basin with a glass stirring rod. Filter through a 9 cm. Whatman No. 44 filter paper into a pyrex basin of 150 ml. capacity. Wash the dish and filter with several small portions of hot water until all the potash has been extracted and the filtrate amounts to about 100 ml. Reserve the filter and residue for the phosphoric acid determination.

To the filtrate add sufficient 20 per cent. perchloric acid to convert all the chlorides present into perchlorates (say, 1 ml. for each per cent. of CaO, K_2O and Na_2O, and $1\frac{1}{4}$ ml. for each per cent. of MgO in the soil. Also allow 1 ml. for the barium chloride added, and if calcium chloride was also added, a further 2 ml.). Evaporate on the water bath. An excess of perchloric acid is denoted by the appearance of white fumes when the evaporation is nearly completed. Finish the evaporation on a sand bath or electric hotplate until dense white fumes of perchloric acid are evolved.

When the white fumes appear add 10–15 ml. of water to dissolve the perchlorates, and then 1 ml. of perchloric acid, and continue the evaporation nearly to dryness on the water bath. Finish the evaporation over a sand bath or hotplate until dense white fumes have been produced for some time and the liquid just sets to a pasty crystalline mass when cold.

When quite cold add 15 ml. of alcohol acidified with perchloric acid (500 ml. of 95–96 per cent. alcohol + 5 ml. of 20 per cent. perchloric acid). Break up all lumps with the stirring rod and stir well, then allow to settle. When most of the crystals have settled (say after 15–30 minutes, or it may be left overnight, if covered to prevent evaporation), decant the clear liquid through a weighed Gooch crucible charged with asbestos. Then dry the contents of the dish for a few minutes on the water bath, take up in 10–15 ml. of water, add $\frac{1}{4}$ ml. of perchloric acid, and evaporate nearly to dryness on the water bath, so that the mass is just pasty when cold. When quite cold, add 10 ml. of the acidified alcohol, stir well to break up the crystals, and leave for a few minutes. Filter through the same Gooch crucible as used previously. Wash by decantation with 5 ml. of the acidified alcohol, and drain the dish and crucible well to remove most of this before adding the next wash liquid. Then transfer the potassium perchlorate crystals from the dish to the crucible, using two lots of about 15 ml. each of 95 per cent. alcohol, which has been saturated with potassium perchlorate.

Wash with a further two lots of 15–20 ml. of this wash liquid and drain well. Dry the crucible at 140° C. for one hour in an air oven, cool in a desiccator, and weigh as $KClO_4$.

$$K_2O = \text{weight of precipitate} \times 0 \cdot 3401.$$

For correct results by the perchlorate method sulphates must be removed completely before starting the perchlorate separation. Precautions must also be taken to protect the acid solutions from ammonia fumes during evaporation. Ammonia is readily absorbed and leads to erroneously high results as it is co-precipitated with the potassium perchlorate.

(b) *Phosphoric Acid, P_2O_5 (Lorenz Method):* Replace the filter paper, containing the residue after the extraction of the potash with hot water, in the original silica basin, and ignite to remove the filter paper. Add 30 ml. of dilute hydrochloric acid (1 + 1), and one drop of concentrated sulphuric acid. The latter is necessary to precipitate traces of barium. Cover with a clock glass, and digest on the sand bath for 15–20 minutes. Remove and rinse the cover glass into the basin and evaporate the contents to dryness on the water bath. Continue the heating on the bath for a further half to one hour to render the silica insoluble. Take up in 2·5 ml. of concentrated nitric acid and 20–25 ml. of water, warm to dissolve all the soluble salts, and filter through a 9 cm. Whatman No. 44 filter paper, collecting the filtrate in a tall 150 ml. beaker. Wash thoroughly with hot water, containing 25 ml. of nitric acid per litre, until the volume of the filtrate amounts to about 120 ml. Place the beaker on the water bath and evaporate the filtrate to dryness. Take up in 30 ml. of the Acid Reagent II (see below). Heat the beaker to incipient boiling, remove from the flame and stir for a moment to avoid overheating of the sides of the beaker. With constant stirring add 30 ml. of Lorenz Reagent I. After standing for 2–5 minutes stir well for half a minute, and then place the beaker aside, out of direct sunlight, and allow to stand overnight.

Before filtration add 0·02–0·04 g. of Celite filter aid 503 to the contents of the beaker and stir well. Connect an unweighed Gooch crucible (25 ml. capacity), fitted with a 1·8 cm. circle of Whatman No. 42 or 44 filter paper, to a filter pump and moisten the filter paper. The filter paper must just cover the holes but not touch the sides of the crucible. Pour the solution and precipitate from the beaker into the crucible and rinse the beaker twice with 20–25 ml. portions of the ammonium nitrate reagent, using a wash bottle to rinse the precipitate into the crucible. Then clean the sides of the beaker with a rubber-tipped stirring rod and rinse it into the crucible, using another two portions, each of 20–25 ml. of the ammonium nitrate. Drain the crucible completely between each addition of the washing liquid. Finally wash three times with acetone, filling the crucible once and half-filling it twice, sucking dry between each addition.

Wipe the outside of the crucible and place it in a vacuum desiccator, which must not contain any desiccating agent, evacuate to 100–200 mm. and after half an hour weigh. After weighing, dissolve the phosphomolybdate precipitate from the crucible by washing with dilute ammonia (1 + 20) and warm water. This leaves the Celite filter aid. Wash three times with small amounts of acetone, again dry in a vacuum desiccator and weigh. The loss in weight corresponds to the yellow precipitate.

$$P_2O_5 = \text{weight of this precipitate} \times 0{\cdot}03295.$$

When a soil contains less than 4 per cent. of calcium carbonate, phosphoric acid can be determined directly in the hydrochloric acid extract. Evaporate 100 ml. of the extract on a water bath until the silica is rendered insoluble. Take up in 2·5 ml. of nitric acid and 20–25 ml. of water, filter and proceed from this stage exactly as in the previous method. If more than 4 per cent. of calcium carbonate is present the calcium sulphate produced by interaction with the sulphuric acid interferes with the determination, unless the amount of phosphoric acid is so high that a much smaller aliquot can be taken.

The following reagents are required for the Lorenz method:

Lorenz Reagent I (Sulphate-Molybdic Acid). Dissolve 100 g. of ammonium sulphate in 1 litre of nitric acid of S.G. 1·36 at 15° C. in a 2-litre flask. Dissolve 300 g. of ammonium molybdate in hot water and transfer to a litre measuring flask, cool the solution to about 20° C., and dilute to the mark. Mix well and pour this solution, in a thin stream, and with constant agitation, into the solution contained in the 2-litre flask. Allow to stand for 48 hours at room temperature, then filter and store in a stoppered reagent bottle in a cool, dark place. If kept in a refrigerator the solution is stable indefinitely and does not deposit molybdic acid. If kept for long periods at ordinary summer temperatures precipitation slowly occurs and the reagent may become too weak to give satisfactory results.

Acid Reagent II. Add 56 ml. of concentrated sulphuric acid and 240 ml. of water to 1 litre of nitric acid of S.G. 1·20.

Ammonium Nitrate Solution. Make up a 2 per cent. aqueous solution for washing. If the solution is not acid to litmus, add a few drops of nitric acid per litre.

Acetone. This should be non-alkaline and free from

residue. The acetone washings are kept, dehydrated with potassium carbonate, decanted, acidified with a few drops of sulphuric acid and redistilled.

REFERENCES

(1) Beatter, E.: *Chem. and Ind.*, 1934, *53*, 712.
(2) Hendrick, J.: *Agric. Progress*, 1931, *8*, 134–135.
(3) Martin, W. S. and Griffith, G. A.: *Chem. and Ind.*, 1934, *53*, 830.
(4) McLean, W.: *J. Agric. Sci.*, 1936, *26*, 331–336.

CHAPTER IX

EXCHANGEABLE IONS AND EXCHANGE CAPACITY

The phenomenon of ionic exchange in soils is not restricted to cations; both cation exchange and anion exchange are now recognized. Cation exchange has been commonly called "Base exchange" by most workers in this field, although they have had a clear conception of the reaction as one involving cations and not bases. Gedroiz, in some of his later papers translated by Waksman, frequently used the terms "cation" and "base" synonomously and referred to "the bases or cations entering into the exchange process" as "humic bases or cations" and "zeolitic bases or cations." He also distinguished between hydrogen and the other exchangeable ions as "exchangeable hydrogen" and "exchangeable metallic cations." Now that anion exchange is coming into prominence, more attention has been focussed on this discrepancy in the terminology and, in the present work, the expression "Cation Exchange" is adopted in preference to "Base Exchange." This involves a corresponding change in three other recognized expressions. The old phrases together with the expressions in the terminology adopted are as follows:

Base exchange capacity	Cation exchange capacity
Total exchangeable bases	Total exchangeable metal ions
Percentage base saturation	Percentage metal ion saturation

CATION EXCHANGE

As cationic exchange is a reversible reaction, complete replacement can only be effected if the replaced ions are removed from the sphere of action. Equilibrium between the soil and the displacing solution is very rapidly established, even in the cold. Both the clay and humic fractions of a soil contribute to its exchange capacity. Weight for weight, the organic exchange complex possesses an exchange capacity several times that of the inorganic fraction.

The exchangeable cations in the soil can be replaced by

(a) leaching with concentrated salt solutions (e.g. ammonium chloride, ammonium acetate, sodium chloride, barium chloride).

(b) leaching with dilute acids (e.g. 0·04N hydrochloric acid, 0·5N acetic acid).

(c) electrodialysis.

Dilute hydrochloric acid was used in one of Gedroiz's original methods, but its use is not always permissible since, in some soils significant amounts of the exchange complex itself are destroyed. Neither can it be conveniently used for soils containing large quantities of calcium carbonate. To reduce the attack on the exchange complex, Williams (15) proposed the use of a weakly dissociated acid instead of hydrochloric acid and he leached soils with 0·5N acetic acid to replace the exchangeable ions. Numerous soils have been encountered in which very incomplete replacement of the exchangeable ions has been obtained with this reagent, even when considerably larger volumes of leachate have been collected. Its use can only be recommended for sandy soils, low in adsorbed metal ions.

Electrodialysis has been used to remove the exchangeable cations from soils and this method would appear to offer many advantages, since the ions are obtained free from the large amounts of neutral salt or acid used in their replacement by other methods. However, owing to the difficulty in securing complete removal of the magnesium ion in many soils, this method is not suitable for quantitative work.

For the determination of the individual cations, the best methods available at the present time involve leaching with concentrated salt solutions and are based on the original methods devised by Hissink (5) and Gedroiz (4). The determination of the exchangeable cations in soils of the humid regions, devoid of salts and calcium carbonate is straightforward. In Hissink's method the soil is leached with a normal solution of ammonium chloride and the individual ions, calcium, magnesium, sodium and potassium, are determined in the leachate. The ammonium chloride used should be a high grade reagent so that the blank corrections, due to impurities present in it, will be small. Particular attention should be paid to its freedom from traces of sodium. Schollenberger (13) proposed the use of a normal solution of ammonium acetate, neutralized to pH 7, in place of ammonium chloride (pH about 4·6) as the replacing solution partly because, being more alkaline in reaction, it will displace hydrogen ions more completely in the determination of exchangeable hydrogen and exchange capacity, and also because it is easily removed from the leachate, prior to the determination of the individual ions, simply by evaporation and gentle ignition. This second claim, however, amounts to little in practice since ammonium chloride is very conveniently removed and with even greater rapidity than ammonium acetate, by evaporation with nitric acid. Extraction with normal ammonium acetate gives values in general very similar to, although slightly lower than, those obtained by the use of ammonium chloride.

The determination of exchangeable ions in soils containing calcium carbonate is not so simple and straightforward. Ammonium chloride dissolves appreciable amounts of calcium carbonate so that the calcium in the leachate will represent the exchangeable calcium as well as some calcium derived from the carbonate dissolved. In Gedroiz's method for carbonate soils, the amount of carbonate in the soil is determined before and after extraction and the amount of calcium found is corrected for an amount equivalent to the carbonate dissolved. It is assumed that the latter consisted entirely of calcium carbonate. Such a method, based on difference determinations, is not capable of great accuracy in the presence of much calcium carbonate, since the errors of all the determinations fall upon the value for exchangeable calcium. In Hissink's method for the determination of the exchangeable cations in calcareous soils, magnesium, sodium and potassium are determined in the ammonium chloride extract as usual. Another portion of the soil is leached with normal sodium chloride, in which calcium carbonate is much less soluble, and two separate litres of leachate are collected. The whole of the exchangeable calcium is replaced and contained in the first litre, together with calcium derived from the solution of calcium carbonate. The calcium in the second litre is derived from the solution of calcium carbonate only and it is assumed to be equal in amount to that dissolved during the preparation of the first litre of leachate. The difference between the amounts of calcium contained in the first and second litres thus represents the amount of exchangeable calcium.

For the determination of calcium and magnesium in soils containing carbonates, alcoholic potassium chloride and aqueous and alcoholic barium chloride have been suggested at different times, since calcium carbonate is less soluble in these reagents than in sodium chloride. However, the values obtained are no better than those given by the sodium chloride method and, in the case of barium chloride, analytical difficulties, in the complete separation of barium from calcium, are introduced.

In Williams's method, carbonates and total calcium are determined in the same portion of soil, the former by liberation of carbon dioxide by acetic acid, *in vacuo*, and the latter by continued leaching of the soil with more acetic acid. The difference between the total amount of calcium found and the amount equivalent to the carbonates present corresponds to the exchangeable calcium. It is assumed that no magnesium carbonate is present. The method is suitable for soils with small amounts of carbonates but, in strongly calcareous soils, the proportion of carbonate calcium to exchangeable calcium is such that the experimental errors of the determination would reduce the accuracy of the results. Moreover, as previously mentioned, extraction with acetic acid does not always remove the whole of the exchangeable cations in heavy clay soils.

Kelley (6) considers that the accurate determination of exchangeable calcium and magnesium in alkali soils is impossible by any known method, on account of the presence of carbonates and silicates of calcium and magnesium in such soils. He determines the sum of exchangeable potassium and sodium by usual methods, and the total exchangeable metal ions, by an ammonia absorption method and regards the difference in these two values as exchangeable calcium and magnesium. No attempt is made to separate calcium from magnesium.

The difficulties in the determination of exchangeable calcium and magnesium in carbonate soils are indeed great, but it is considered that for many soils the most probable values can still be obtained by slight modifications of Hissink's sodium chloride method, as described on p. 170. By omitting the treatment with warm sodium chloride, the conditions of extraction of the first and second litres are made more strictly comparable, so that the amounts of carbonate dissolved in the two successive litres of leachate should be more nearly equal. The exchange reaction proceeds with rapidity, even in the cold, and practically the whole of the exchangeable ions are replaced in the first litre of leachate. Both calcium and magnesium are determined in each litre so that a correction for the magnesium, as well as the calcium, dissolved from the carbonates present is obtained. This correction only holds, of course, if there is sufficient carbonate originally present in the soil, in a suitable physical condition, to saturate the two litres of extract equally. When only small amounts are present it is possible that more may be dissolved by the first litre than by the second, especially if the carbonate occurs in a nodular form. It is difficult to specify the minimum amount of carbonate that should be present, since it depends partly on its state of subdivision, but all the conditions governing its rate of solubility should be as nearly alike as possible throughout the whole leaching. In general, if there is less than 0·3 per cent. of calcium carbonate (6 m. e. %) in the soil, it is preferable to make the extraction with normal ammonium chloride, since this reagent will dissolve the whole of the carbonate when present in such small amounts. The value for exchangeable calcium is then obtained by subtracting the calcium present as carbonate from the total calcium found. On account of the small amount of carbonate, the error due to the assumption that it is wholly calcium carbonate is not serious. For soils with more than this amount of carbonate, the extraction should be made with sodium chloride, although some soils with 0·3 to about 0·8 per cent. of calcium carbonate may give lower amounts than usual in solution in the second litre. If more accurate values are required the sum of the exchangeable calcium and magnesium should be determined indirectly, by an ammonia absorption method, such as that of Chapman and Kelley (3).

Further difficulties in the determination of the exchangeable cations occur when soluble salts are present in the soil. Many of the earlier investigators either removed the soluble salts by washing with water before extracting the exchangeable cations or made a separate determination of the water soluble cations, deducting the amounts found from the values obtained on leaching the soil with the replacing solution. Corrections made in either of these ways are open to objection, since hydrolysis and dispersion of the exchange complex occurs as the salts are removed. Finely divided material may pass through the filter paper and the permeability decreases so much, particularly in heavy clay soils, that it becomes very difficult, if not impossible, to complete the extraction. Gedroiz pointed out that, if gypsum is present, this preliminary leaching with water is likely to lead to changes in the exchangeable cations, owing to the replacement of other ions by calcium derived from the calcium sulphate. Prescott suggested leaching with alcohol for the removal of soluble salts, since hydrolysis and dispersion of the exchange complex does not occur in this medium. Chlorides are readily soluble in aqueous alcohol but sulphates are much less soluble. A preliminary leaching with 40 per cent. alcohol is very effective in removing salts when they consist largely of chlorides. If sulphates are present, Walkley removes soluble salts by extracting the soil with an equal weight of water before leaching with 40 per cent. alcohol. The preliminary leaching with water assists in the removal of greater amounts of sulphates than can be removed by leaching with alcohol alone. However, when sulphates occur to any greater extent than can be removed by Walkley's procedure, no satisfactory method can, at present, be recommended for the determination of the individual ions.

If it is suspected that sulphates have not been completely removed before commencing the extraction of the exchangeable cations, the amount of sulphate remaining should be determined, by precipitation as barium sulphate, in an aliquot of the ammonium chloride leachate. This determination will assess the maximum error, due to its incomplete removal.

The analytical determination of calcium, magnesium, sodium and potassium in ammonium chloride extracts, and calcium and magnesium in sodium chloride extracts is straightforward and details of the methods recommended are given on pages 172 to 184.

Ammonium chloride is easily removed by evaporation of an aliquot of the leachate to dryness, taking up in predetermined amounts of water and nitric acid and digesting, for a short time, on a water bath. In this way the vigorous reaction can be controlled and the ammonium chloride is rapidly and completely destroyed, without loss of the other constitutents. A single evaporation removes ammonium salts almost completely, certainly quite sufficient for the determination of calcium, magnesium and sodium. For the determination of potassium, however, it is essential to remove completely the last traces of ammonium. This evaporation with nitric acid also destroys most of the organic matter although ignition is necessary to ensure the subsequent precipitation of iron and aluminium hydroxides.

Before the determination of calcium and magnesium in the ammonium chloride extract of acid soils the small amounts of iron and aluminium present must be precipitated as hydroxides. Manganese is also eliminated at the same time by carrying out the precipitation in the presence of ammonium persulphate. The filtrate is then made more strongly ammoniacal and calcium precipitated as oxalate. It is reprecipitated to secure complete separation from magnesia, and titrated with standard permanganate in the usual way. Magnesium is determined by precipitation and reprecipitation, either as magnesium ammonium phosphate or magnesium hydroxyquinolate. The latter method is recommended. If this separation is used the excess of persulphate must first be destroyed, by the addition of sodium bisulphite, to prevent decomposition of the hydroxyquinoline reagent, by oxidation. Magnesium hydroxyquinolate, when precipitated in the presence of an excess of the reagent, adsorbs hydroxyquinoline, so leading to high results. This source of error is eliminated if the precipitate is dissolved and reprecipitated after the addition of a few drops only of the reagent.

Iron, aluminium (and manganese except in traces) arc not extracted from alkaline soils by ammonium chloride so that the step involving their separation can be omitted in extracts from such soils. Iron, aluminium and manganese are also absent in the sodium chloride extracts from carbonate soils. Calcium and magnesium are determined in the sodium chloride extract exactly as in the ammonium chloride extract, except that the first precipitations of each are made without removal of the sodium chloride. If soluble organic matter is present it is destroyed by acidifying and boiling with hydrogen peroxide before the precipitation of calcium. Failure to remove organic matter leads to the discoloration of the calcium oxalate precipitate and high values for calcium. The trace of oxidizing substances, which remains after the use of hydrogen peroxide, is destroyed by the addition of a few drops of sodium bisulphite solution, before the precipitation of magnesium as hydroxyquinolate.

Sodium should always be determined by one of the uranyl acetate methods since the amounts present are usually so small that the older indirect methods are subject to too great an error. Kahane's method, as modified (8), is recommended. It is particularly suitable for the accurate determination of such small amounts of sodium as are found in soil extracts. Sodium uranyl magnesium acetate contains only 1·5 per cent. of sodium, so that quite small amounts of this element yield weighable amounts of precipitate. Calcium, magnesium, and the small amounts of iron, aluminium, manganese, ammonium and potassium present in the ammonium chloride extracts after evaporation with nitric acid do not interfere with the determination, which can therefore be carried out directly on this solution without any other preliminary separation. The sodium uranyl magnesium acetate is precipitated completely in 30 minutes in the alcoholic solution used. If much exchangeable potassium is present the precipitate must not be left for more than 1–2 hours before filtration, otherwise a small amount of a potassium uranyl magnesium acetate salt slowly separates and leads to slightly high values for sodium. Sulphates only interfere with the method if both calcium and sulphate together are present in amounts greater than the amount of calcium sulphate which will remain in solution in the reagent. Phosphates interfere since uranyl phosphate is insoluble under the conditions of precipitation and would contaminate the precipitate. However, the amount of phosphate present in an ammonium chloride extract is insufficient to affect the determination. The small amounts of silica present do not affect the determination. It is collected along with the sodium uranyl magnesium acetate precipitate and, after drying and weighing, the sodium salt is dissolved from the Gooch crucible by hot water, leaving the insoluble silica. The decrease in the weight of the crucible corresponds to the sodium uranyl magnesium acetate. A blank determination must always be carried out to correct for the trace of sodium present, even in the best reagents.

For the determination of potassium equally good results can be obtained either by a modified cobaltinitrite method (9) or the perchlorate method. On account of the small quantities of potassium usually present the volumetric cobaltinitrite method is recommended. Potassium cobaltinitrite, as precipitated, does not correspond to the formula $K_2NaCo(NO_2)_6$ usually quoted. The proportion of potassium to sodium is usually less than indicated by this formula and varies with the amount of potassium present in the solution at the time of precipitation. As this variation is consistent, under standardized conditions of precipitation, the amount of potassium can be determined with accuracy and Table 5 (p. 180) shows the permanganate equivalent of the precipitate for all amounts of potassium between 0 and 18 mg. The potassium cobaltinitrite precipitate should be washed either with a freshly prepared saturated aqueous solution of the salt (this solution is only stable for 2–3 hours) or with 35 per cent. alcohol. The precipitate is appreciably soluble in 10 per cent. acetic acid and in 2·5 per cent. sodium sulphate, two solutions frequently recommended for washing.

Whichever method is used for the determination of potassium, particular care must be taken to exclude traces of ammonium salts throughout the determination, because both ammonium cobaltinitrite and ammonium perchlorate are only slightly soluble and are easily co-precipitated with the corresponding potassium salt. During the evaporation of acid solutions appreciable amounts of ammonia may be absorbed, if this reagent is being used in the same laboratory.

After the addition of a small amount of a calcium salt to assist in the subsequent extraction of the potash, the bulk of the ammonium salts in the ammonium chloride extract is destroyed, by evaporation with nitric acid as usual, and the last traces removed by very careful ignition in a muffle furnace at a temperature just below a dull red heat. This procedure eliminates the difficulties experienced by Milne and no loss of potassium occurs.

From this stage the potassium determination must be carried out in a laboratory free from ammonia fumes. Ammonia must also be absent from the reagents used, since a blank determination does not correct fully for its presence. This applies to both the perchlorate and cobaltinitrite methods because ammonium is readily co-precipitated with potassium. It is not, however, precipitated quantitatively; a greater proportion of the amount present is precipitated the greater the amount of potassium present in the determination. Thus a bigger proportion of the ammonium impurity is co-precipitated in an actual determination (in the presence of potassium) than in a blank determination. This was found particularly to be the case in one series of analyses by the cobaltinitrite method in which an unsuitable batch of cobalt nitrate reagent was used. Ammonia is a serious impurity in those grades of cobalt nitrate specially purified from traces of iron and nickel. Such reagent is not suitable for use in the cobaltinitrite method. Ordinary C.P. quality cobalt nitrate is usually free from ammonia.

In the determination of exchangeable manganese, precautions must be taken since this element occurs in soils as the bivalent exchangeable form and as higher oxides. The amounts of exchangeable manganese, except in strongly acid soils, are usually quite small. However, the higher oxides are easily reduced to bivalent manganese under some conditions and this must be avoided in the determination of exchangeable manganese. Ammonium chloride is not permissible as the replacing agent, since soil organic matter can reduce considerable quantities of the active oxides of manganese at the reaction of this solution. Normal ammonium acetate, neutralized to pH 7, appears to be the most suitable extracting agent for the determination of exchangeable manganese. Manganese is determined colorimetrically, by oxidation with potassium periodate, after removal of the ammonium acetate.

The amounts of each individual exchangeable cation are expressed in milliequivalents per 100 g. of soil. The sum of these amounts for calcium, magnesium, sodium and potasssium represents the total exchangeable metal ions (Hissink's S value). Besides expressing this value in milliequivalents per 100 g. of soil, the relative proportions which each of the four ions contribute to it are expressed on a percentage basis, so that Ca + Mg + Na + K = 100.

The determinations of exchangeable hydrogen and cation exchange capacity involve the definition of a fully saturated or neutralized soil. The considerable amount of work which has been done on the subject of cationic exchange in soils now gives us a clearer understanding of the mechanism involved in the reaction and helps in deducing a definition of exchangeable hydrogen and total cation exchange capacity more fundamental than many of the conventional definitions previously accepted.

The exchange complex consists of weak insoluble polybasic acids, or "acidoids," in which, in the unsaturated or fully acid condition, hydrogen ions are dissociated over the surface of large insoluble anions. In soils in the natural state these hydrogen ions are partly or wholly replaced by other cations. Both organic and inorganic acidoids occur in the exchange complex. Like all weak acids they are not fully neutralized (that is, their acidic hydrogen ions are not fully replaced by metallic cations) at the neutral point (pH 7), but only at some more alkaline reaction corresponding to their equivalence points. The practical determination of the equivalence point is beset with numerous difficulties on account of the indefinite and heterogeneous nature of the big molecules involved. Total cation exchange capacity is an expression of the total amount of exchangeable metallic cations present at the equivalence point, or point of complete neutralization. It is therefore equivalent to the total amount of acidic hydrogen present in these weak acids when fully unsaturated. The exchangeable hydrogen present in any soil is thus defined as the amount of acidic hydrogen that must be replaced by metallic cations to bring the soil to the fully saturated condition.

The concept of saturation was first introduced by Hissink but the exact reference point at which a soil is to be regarded as fully saturated with metallic cations has been very differently defined by different workers since that time. Hissink considered that a soil was not fully saturated until it was brought to equilibrium with a salt solution containing barium hydroxide, so neutralizing the absorption complex at a high pH value. Hissink's method for determining total exchange capacity (Hissink's T value) was based on this. Most American workers have adopted the neutral point (pH 7) as the reference point, bringing the soil into equilibrium with such solutions as ammonium acetate or barium acetate adjusted to this value.

Bradfield and Allison (1) proposed that a fully saturated soil should be defined as one which has reached equilibrium with a surplus of calcium carbonate at the partial pressure of carbon dioxide existing in the atmosphere and at a temperature of 25° C. since, in soils of the humid regions, calcium is the most important exchangeable metallic cation. In such soils the pH value never rises naturally above pH 8·4, any metal ions present in excess of the amount necessary to neutralize the absorption complex to this value appearing as calcium carbonate. However, in soils of the arid and semi-arid regions, magnesium and sodium are important exchangeable ions and pH values above 8·4 are common. Such soils would be more than 100 per cent. saturated according to Bradfield and Allison's definition.

From its definition, the exchange capacity of a soil will be greater the more alkaline the reference point selected. It would appear that there is a maximum value at about pH 9 and that beyond this reaction exchange capacity decreases, owing to a breakdown of the exchange complex. Although the chemical equivalence point is the only fundamental reference point for the definition of a saturated soil, in the absence of any specific determination or definition of it, the natural point of equilibrium with calcium carbonate, proposed by

Bradfield and Allison, could well be considered the most logical reference point. In all determinations of total cation exchange capacity, or of exchangeable hydrogen, the reference point should be clearly indicated.

The expression, exchangeable hydrogen, represents the acidic hydrogen present in the soil and is, by definition, related to the total exchangeable metal ions and total cation exchange capacity by the following expression:

Exchangeable metal ions + Exchangeable hydrogen = Total cation exchange capacity.

A determination of any two of these values thus gives the remaining value. Degree of saturation (the value $\frac{S \times 100}{T}$, originally proposed by Hissink) or percentage metal ion saturation then corresponds to:

$$\frac{\text{The actual quantity of adsorbed metal ions (S)}}{\text{The total exchange capacity of the soil (T)}} \times 100$$

The most common methods used for the determination of total exchange capacity involve the leaching of the soil with a salt solution, adjusted to a suitable pH value, until all the ions (including the hydrogen ions) are replaced by the cation of the salt solution. The total amount of this cation adsorbed by the soil is then determined. Soluble salts originally present in the soil do not interfere with the determination. In Chapman and Kelley's method (3) the soil is leached with neutral ammonium acetate, the excess of ammonium acetate removed by washing with methyl alcohol, and the amount of adsorbed ammonia determined by aerating the soil for 12 hours with sodium carbonate. In many soils the small difference between this value and that determined at pH 8–9 indicates that the equilibrium between the exchangeable hydrogen and the replacing ion in the salt solution differs very considerably from that in an aqueous suspension in the absence of excess salt. Owing to the large concentration of the displacing ion a greater proportion of the hydrogen ions in the soil is replaced by the ammonium ion and the value obtained for the exchange capacity of the soil would appear to correspond to a reference pH value higher than that of the salt solution used.

In Puri's method (11) the soil is brought to equilibrium with a solution containing sodium hydroxide and sodium chloride and the sodium adsorbed is displaced by barium hydroxide and determined volumetrically as carbonate.

For the determination of exchangeable hydrogen several methods have been proposed. In one of Parker's methods (7) the soil is titrated with barium hydroxide until a pH value of 7 is reached; in another method the exchangeable hydrogen is replaced by leaching with a solution of barium acetate, adjusted to pH 7, and titrated electrometrically, in the leachate to the same pH. Methods, in which equilibrium is reached in a concentrated salt solution, are subject to the same criticism as that of the similar methods used for the determination of total exchange capacity, since the reference point apparently corresponds to a reaction value more alkaline than that of the salt solution used. In Bradfield and Allison's residual carbonate method (1) the soil is treated with a known amount of lime water and the excess is converted to carbonate and brought into equilibrium with the carbon dioxide of the atmosphere. The excess of calcium carbonate is determined; the difference between it and the amount of calcium added as hydroxide corresponds to the amount of exchangeable hydrogen neutralized. Buffer solution methods, in which the amount of base absorbed by a soil in coming into equilibrium with a buffer solution of known pH, have been used by Schofield (12), Bradfield and Allison (1) and Piper (10). In these methods the amount of base absorbed is determined by titrating the buffer solution before and after shaking with the soil. The amount of base absorbed corresponds to the exchangeable hydrogen neutralized in coming to equilibrium at the pH of the buffer solution used. In the meta-nitrophenol method, described on p. 185, the determination is carried out with buffer solutions of two different values so that the results can be interpolated to a final pH value of 8·4. Exchangeable hydrogen determined by this method agrees with that defined by Bradfield and Allison.

Instead of determining the total exchangeable metal ions individually, their total may be determined, as in Bray and Willhite's method (2) in which the exchangeable ions are displaced as acetates by leaching with ammonium acetate. On evaporation and ignition the excess of ammonium acetate is destroyed and the metal acetates are converted to carbonates or oxides, which are determined by dissolving them in standard hydrochloric acid and titrating the excess of acid. The presence of soluble salts in the soil does not interfere with this determination. In Parker's method (7) total exchangeable cations are determined by replacing them by leaching the soil with barium chloride. The amount of adsorbed barium is then determined, by leaching with ammonium chloride, presumably after the removal of the excess of barium chloride with alcohol. The amount of adsorbed barium is corrected, by titration of the barium chloride leachate, for any exchangeable hydrogen displaced, and this corrected value is then equivalent to the total exchangeable cations. Parker's method is satisfactory for acid soils but a practically similar method proposed by de'Sigmond and Iyengar (14) for soils containing calcium carbonate and gypsum gives erroneous values in the presence of calcium carbonate. Barium chloride reacts with calcium carbonate giving some barium carbonate and this is partly dissolved in the subsequent leaching with the ammonium salt. Barium from this source is thus included with the exchangeable barium.

Schofield has suggested a rapid method for the determination of total exchangeable cations. A small amount of soil is shaken with 0·05N hydrochloric acid and the residual acidity determined by titration. This method, described on p. 189, gives approximate results for the amount of metal ions present in acid soils or in soils with less than about 1 per cent. of calcium carbonate. The result, however, depends upon the proportion of soil to acid and the main value of this method is exploratory, in selecting soils for the determination of the individual exchangeable cations.

The Determination of Exchangeable Cations.

A. Preparation of the Soil Extract.

(1) REMOVAL OF SOLUBLE SALTS.

Reagents:

40 % Alcohol. Add 1,200 ml. of water to 1 litre of absolute or 96 per cent. alcohol.

Method:

When soluble salts are present weigh the appropriate amount of soil for the subsequent ammonium chloride or sodium chloride extraction and transfer it to a 400 ml. beaker. Add 100 ml. of 40 per cent. alcohol and allow to stand for half an hour with occasional stirring. Filter through a 9 cm. Buchner funnel fitted with a Whatman No. 50 filter paper and wash with several further 50 ml. portions of the same alcohol until the filtrate is chlorine free; 3–5 such washings are usually sufficient. Finally wash with 50 ml. of absolute alcohol to dry the soil and assist in its easy removal from the funnel. Return the soil to the beaker, washing the funnel and filter paper into the beaker. This washing is done with water if the soil is to be subsequently extracted with ammonium chloride, the volume being restricted to about 125 ml. The ammonium chloride extraction is then continued as described below. If the soil is to be used for a sodium chloride extraction, then this washing is performed with N sodium chloride, 200 ml. being used in all and the determination continued as usual.

When appreciable amounts of soluble sulphates are present, weigh out the appropriate amount of soil and transfer it to a 400 ml. beaker as before. Add an equal weight of water, stir well and leave for half an hour. Then filter through a 9 cm. Buchner funnel, drain thoroughly and continue the washing with several portions of 40 per cent. alcohol exactly as described in the previous paragraph.

(2) LEACHING WITH NORMAL AMMONIUM CHLORIDE SOLUTION. *(Procedure to be used for all soils in which calcium carbonate is absent or less than 0·3 per cent., and for the determination of exchangeable potassium and sodium in calcareous soils.)*

Reagents:

2N Ammonium Chloride. Make a solution of ammonium chloride containing 107 g. in each litre. The ammonium chloride used should be a high grade reagent and should contain less than 0·002 per cent. residue on ignition at a low temperature.

N Ammonium Chloride. Dilute the above solution with an equal volume of water.

Method:

Fifty g. of soil are to be used when more than 20–25 m.e. % of exchangeable bases are present and 100 g. when the amount is less than 10 m.e. %. Schofield's rapid 0·05N hydrochloric acid method (p. 189) is very useful in obtaining approximations of the amount of exchangeable cations present.

Transfer 50–100 g. of soil (depending on the texture of the soil and the probable amount of the exchangeable cations present) to a 400 ml. beaker, add 125 ml. of water and allow to stand for a short time to disperse. Stir well and add an equal volume of 2N ammonium chloride. Place in a water bath at 70° C. and leave there for one hour, stirring at intervals. Remove from the bath and allow to stand overnight. Then decant through an 18·5 cm. Whatman No. 44 filter paper and transfer the soil quantitatively to the filter using a jet of N ammonium chloride solution. Collect the filtrate in a litre measuring flask. Continue to leach the soil with small quantities (about 50–60 ml.) of N ammonium chloride solution, allowing the filter to drain completely between each addition, until nearly one litre of filtrate has been collected. Add 2 ml. of formalin (A.R.) to the filtrate (to prevent mould growth) and adjust to the litre mark. Use this filtrate for the determination of the individual cations. If it is not quite clear refilter it, without dilution, through a 9 cm. Whatman No. 50 filter paper.

(3) LEACHING WITH NORMAL SODIUM CHLORIDE SOLUTION. *(Procedure for the determination of exchangeable calcium and magnesium in calcareous soils.)*

Reagents:

N Sodium Chloride. Dissolve 58·5 g. of sodium chloride for each litre of solution required.

Method:

Transfer 30 g. of the soil to a 400 ml. beaker, add 200 ml. of N sodium chloride, stir well at intervals and allow to stand overnight at room temperature. Then filter through a 15 cm. Whatman No. 44 filter paper, collecting the filtrate in a litre volumetric flask. Continue the leaching, by the repeated addition of small amounts (about 50 ml.) of N sodium chloride, allowing the filter to drain completely each time. When one litre of filtrate has been collected, change the flask and continue the leaching until a second litre has been collected. If these filtrates are not quite clear, refilter through 9 cm. Whatman No. 50 filter papers.

Determine calcium and magnesium in aliquots of each solution, as detailed below. The differences in the amounts found in the first and second litres of extract correspond to exchangeable calcium and magnesium.

(4) LEACHING WITH NORMAL AMMONIUM ACETATE AT pH 7. *(Procedure for the determination of exchangeable manganese, total exchangeable metal cations and total exchange capacity.)*

Reagents:

N Ammonium Acetate. Prepare a 2N solution of acetic acid and a 2N solution of ammonia, standardizing each by tenfold dilution of a small portion and titration against 0·1N alkali or acid using methyl red as the indicator. Mix equal volumes of the two solutions; if accurately prepared the ammonium acetate will have a pH of 7·0. Check this either colorimetrically, in a comparator using brom thymol blue as indicator, or preferably electrometrically, using the glass or quinhydrone electrode. If the solution is not close to pH 7·0 add a little more 2N acid or ammonia until the desired pH is obtained. The solution does not have to be exactly normal but its pH must be 7·0. [Note, 2N acetic acid (approx.) = 576 ml. of glacial acetic acid (S.G. 1·052) diluted to 5 litres. 2N ammonia (approx.) = 540 ml. of concentrated ammonia (S.G. 0·88), or 750 ml. of S.G. 0·91, diluted to 5 litres.]

Method:

Transfer 25–100 g. of soil to a 400 ml. beaker, add 250 ml. of N ammonium acetate solution, stir well and allow to stand overnight, at room temperature. Then filter through a 15 or 18·5 cm. Whatman No. 44 filter paper and leach with further portions of ammonium acetate solution, exactly as described in the procedure for the preparation of the ammonium chloride extract.

B. The Analysis of the Ammonium Chloride Extract for Calcium, Magnesium, Sodium and Potassium.

(1) GENERAL METHOD FOR THE REMOVAL OF AMMONIUM CHLORIDE.

Pipette an aliquot of the ammonium chloride extract into a silica basin (or pyrex basin when it is not necessary to ignite the residue) and evaporate to dryness on the water bath. When completely dry turn off the bath and allow to cool for 5–10 minutes. Then add 3½ ml. of water and 5 ml. of concentrated nitric acid for each 50 ml. of extract originally taken. Since the proportion of water to acid is an important factor in controlling the rate of decomposition it may be more convenient to mix these beforehand and add 8–9 ml. of the dilute acid per 50 ml. of extract. Cover with a clock glass and allow the decomposition to proceed. Under these conditions the reaction proceeds quite smoothly and no loss occurs. If it shows signs of becoming too vigorous add a small amount of water to the basin; if too much water is added the reaction becomes unduly prolonged. After a time (5–10 minutes) slowly raise the bath to boiling. When the decomposition is complete, rinse the clock glass and remove it from the basin. Continue the evaporation to dryness.

(2) DETERMINATION OF CALCIUM AND MAGNESIUM.

Pipette duplicate portions, each of 150 ml., of the ammonium chloride extract into 200 ml. silica basins and destroy the ammonium salts as above. When this operation is completed remove each basin from the water bath and ignite for 5–10 minutes in a muffle at a very dull red heat (red just perceptible in the darkened muffle). When cool add 4 ml. of concentrated hydrochloric acid and 15 ml. of water, warm on a water bath until solution is complete and transfer to a 150 ml. tall shaped beaker, washing the basin with hot water.

Removal of Iron, Aluminium and Manganese.

Add 5 ml. of an 8 per cent. ammonium persulphate solution, heat to boiling and add 8–10 ml. of dilute ammonia (1 + 1) to give a slight excess. Filter while hot through a 7 cm. Whatman No. 41 filter paper. Use a larger filter paper if much precipitate is obtained. Collect the filtrate in a 250 ml. squat shaped beaker, and wash the filter and precipitate several times with hot water. The precipitate consists of ferric and aluminium hydroxides and hydrated manganese dioxide.

Calcium.

Add 10 ml. of concentrated ammonia to the filtrate after separating the iron, aluminium, and manganese as above; heat to boiling and add 10 ml. of hot 10 per cent. ammonium oxalate solution to precipitate the calcium. Continue the heating for about a minute after the formation of the precipitate; then allow to stand overnight. Filter through a 9 cm. Whatman No. 44 filter paper and wash twice with hot water, collecting the filtrate in a 400 ml. beaker.

Dissolve the precipitated calcium oxalate by adding four lots, each of 5 ml., of hot dilute hydrochloric acid (1 + 4) to the filter, by means of a dip pipette, and collect the filtrate in the beaker in which the precipitation was made. Wash well with hot water. Add 2 ml. of 2 per cent. ammonium oxalate, heat to boiling and reprecipitate by the addition of 20 ml. of dilute ammonia (1 + 1). Keep just boiling for about a minute, as before, until the precipitate becomes granular. Cover and allow to stand overnight. Filter through the filter paper previously used, and collect the filtrate and the first two washings in the 400 ml. beaker containing the filtrate from the first precipitation. Reserve this for the magnesium determination.

Complete the washing of the precipitate with further portions of hot water until chlorine free. When this washing has been completed discard the filtrate, replace the 250 ml. beaker in which the precipitation was made, pierce the filter paper with a pointed glass rod and wash as much as possible of the precipitate from the filter, using a jet of hot water. Then wash the filter paper alternately with warm dilute sulphuric acid (1 + 4), made just pink with two or three drops of standard potassium permanganate per 200 ml., and warm water until all the oxalic acid is removed. Three portions of the acid, each about 15 ml., should be used. Then warm the contents of the beaker to about 75° C. and titrate with 0·05N potassium permanganate.

Since 1 ml. of 0·05N permanganate corresponds to 0·05 milligram equivalents of calcium, calculate the amount of exchangeable calcium in the original soil, expressed in m.e. % by multiplying the volume of permanganate used by one of the factors given below. Deduct any calcium, present as carbonate, from this value.

Soil extracted.	150 ml. extract ex 1 litre.	200 ml. extract ex 1 litre.	400 ml. extract ex 1 litre.
30 g.	—	—	0·417
50 g.	0·666	0·500	—
66·66 g.	0·500	0·375	—
100 g.	0·333	0·250	—

Magnesium: 8-Hydroxyquinoline Method.

To the filtrate from the calcium precipitation add 10 ml. of a 5 per cent. solution of sodium metabisulphite to reduce the ammonium persulphate and so prevent oxidation of the hydroxyquinoline reagent. Heat to about 80° C. and add a small excess of a solution of 8-hydroxyquinoline (3 per cent. hydroxyquinoline in 10 per cent. acetic acid) to the hot solution. In general 5 ml. will be sufficient to precipitate all the magnesium present in the extracts of ordinary soils and 10 ml. that in heavy subsoil extracts. A large excess of hydroxyquinoline must be avoided since it is only slightly soluble in alkaline solution. Stir well and add an additional 5 ml. of concentrated ammonia. An excess of hydroxyquinoline is denoted by the yellow colour of the solution after the precipitate has formed. Allow to stand until cool (or overnight). Filter through a 9 cm. Whatman No. 41 filter paper and wash 3 times with dilute ammonia (1 + 20) rejecting the filtrate. Dissolve the precipitate in 3 lots, each of 5 ml. of warm dilute hydrochloric acid (1 + 5) and wash the filter paper completely with hot water, collecting the filtrate in the original beaker in which the precipitation was made. Add 5 drops of the hydroxyquinoline reagent, heat to about 80° C. and reprecipitate by the addition of 25 ml. of dilute ammonia (1 + 1). Stir well and allow to stand for 2 hours. Filter through a Gooch crucible fitted with a circle of Whatman No. 42 filter paper, or a sintered glass crucible, wash well with dilute ammonia (1 + 20) and dry overnight at a temperature not exceeding 105° C. Cool in a desiccator and weigh as $Mg(C_9H_6ON)_2.2H_2O$. To obtain the amount of exchangeable magnesium, expressed in m.e. %, multiply the weight of precipitate by the appropriate factor in the following table:

Soil extracted.	150 ml. extract ex 1 litre.	200 ml. extract ex 1 litre.	400 ml. extract ex 1 litre.
30 g.	—	—	47·8
50 g.	76·5	57·4	—
66·66 g.	57·4	43·1	—
100 g.	38·3	28·7	—

Magnesium: Phosphate Method.

To the filtrate from the calcium precipitation add 5 ml. of 10 per cent. solution of sodium phosphate ($Na_2HPO_4.12H_2O$) or, if the soil is likely to be rich in magnesium, add 10 ml. Stir well and add 25 ml. of concentrated ammonia to complete the precipitation. Cover the beaker and allow to stand overnight. Filter through a 9 cm. Whatman No. 44 filter paper and wash twice with dilute ammonia (1 + 9). Place the beaker in which the precipitation was made under the funnel and dissolve the precipitate with 2–3 lots, each of 5 ml., of warm dilute nitric acid (1 + 4) and wash repeatedly with hot water. Add a few drops of the sodium phosphate reagent to the filtrate and, when quite cold, reprecipitate the magnesium ammonium phosphate by the careful addition of ammonia, making the solution just alkaline at first, and after 30 minutes adding an excess, 25 ml. being added in all. Again allow to stand overnight and filter through the filter paper previously used. Transfer the precipitate from the beaker and wash several times with dilute ammonia (1 + 9). Ignite the filter and precipitate until a white residue is obtained, cool in a desiccator and weigh as $Mg_2P_2O_7$.

To obtain the amount of exchangeable magnesium, expressed in m.e. %, multiply the weight of precipitate by the appropriate factor in the following table:

Soil extracted.	150 ml. extract ex 1 litre.	200 ml. extract ex 1 litre.	400 ml. extract ex 1 litre.
30 g.	—	—	149·6
50 g.	239·4	179·5	—
66·66 g.	179·5	134·7	—
100 g.	119·7	89·8	—

(3) DETERMINATION OF SODIUM: KAHANE'S METHOD (MODIFIED).

Reagents:

Uranyl Magnesium Acetate Solution.

Uranyl acetate, crystals, sodium-free . . . . . . 32 g.
Magnesium acetate, crystals, sodium-free . . . . 100 g.
Acetic acid glacial . . . . 20 ml.
Absolute alcohol . . . . 453 ml.
(or 500 ml. of 90 % rectified spirit)
Water . . . . sufficient to dilute to 1,000 ml.

Dissolve the uranyl acetate and magnesium acetate in water by warming, add the acetic acid and alcohol, cool and dilute to 1 litre. Leave in a cool dark place for 1–2 days and then filter off the precipitate of sodium from the reagents. Store in a pyrex bottle to prevent absorption of sodium from ordinary glass. The reagent is sensitive to direct sunlight but, if kept in the dark, it is stable for several years. Only reagents practically sodium-free should be used in making the above solution as if much sodium precipitate is formed the reagent is significantly weakened and is liable to yield low results.

Alcohol saturated with Sodium Uranyl Magnesium Acetate. Shake up some 96 per cent. alcohol with a few grams of precipitated sodium uranyl magnesium acetate in a stoppered bottle and leave for 24 hours. Filter the nearly clear supernatant liquid through a Whatman No. 44 filter paper and store in a second bottle with a further supply of the crystalline triple salt. This gives a stock supply of alcohol saturated with the sodium salt and after filtration it is always ready for use as the washing solution. Always add fresh additions of alcohol to the first bottle and leave for 24 hours or longer before transferring to the second bottle. Some impurity, present as a trace in the alcohol, reacts with the sodium uranyl magnesium acetate forming a brownish amorphous compound and the first treatment removes this impurity. The sodium uranyl magnesium acetate in the second bottle then retains its bright colour and crystalline form.

Method:

Pipette duplicate portions (25 ml.–100 ml. depending on the probable amount of sodium present) of the ammonium chloride extract into 150 ml. pyrex basins and destroy the ammonium chloride by evaporation with nitric acid as previously described. The amount of extract taken must not contain more than 5 mg. of sodium otherwise precipitation is incomplete unless more than 15 ml. of uranyl magnesium acetate reagent are used.

When the residue from the nitric acid evaporation has been evaporated to dryness, remove the basin from the water bath and allow to cool. Take up in 6 ml. of cold water. It is not necessary to filter this extract unless it is very turbid, since a small amount of turbidity does not interfere and is corrected for later. If filtration is necessary, wash well and re-evaporate the filtrate to dryness in a 100 ml. pyrex glass basin and take up again in 6 ml. of cold water. Add 15 ml. of uranyl magnesium acetate reagent to the contents of the basin and stir for a quarter of a minute or until a precipitate forms. Cover, and allow to stand for 30 minutes but not longer than two hours. Prolonged standing leads to the precipitation of small amounts of a potassium salt when much exchangeable potassium is present. Filter through a small Gooch crucible charged with asbestos. Wash the precipitate of sodium uranyl magnesium acetate twice with 2 ml. portions of the reagent and then 5 times with 96 per cent. alcohol saturated with the triple salt, transferring the precipitate quantitatively to the Gooch crucible. Then dry the crucible in an oven at 105° C. for not more than one hour, cool in a desiccator, and weigh as $Na(UO_2)_3Mg(CH_3COO)_9.8H_2O$. After weighing wash the crucible with several portions of hot water and dry again. This second weight gives the weight of the crucible plus any insoluble residue that was not removed before the sodium precipitation, and the difference corresponds to the sodium uranyl magnesium acetate.

Carry out a blank determination on all the reagents used.

As an alternative to oven-drying, ether may be used. In this case, wash the crucible, or crucible and precipitate, three times with ether, wipe dry on the outside and allow to stand near the balance for 15 minutes before weighing.

To obtain the amount of exchangeable sodium, expressed in m.e. %, multiply the weight of precipitate by the appropriate factor below, after making the necessary correction for the blank.

Soil extracted.	25 ml. extract ex 1 litre.	35 ml. extract ex 1 litre.	50 ml. extract ex 1 litre.	75 ml. extract ex 1 litre.	100 ml. extract ex 1 litre.
50 g.	52·2	37·3	26·1	17·4	13·04
66·66 g.	39·1	27·9	19·6	13·0	9·78
100 g.	26·1	18·6	13·0	8·7	6·52

(4) DETERMINATION OF POTASSIUM.

Cobaltinitrite Method.

Reagents:

Sodium Nitrite. Dissolve 350 g. of sodium nitrite (potassium-free) in water and dilute to 1 litre.

Cobalt Nitrate. Dissolve 200 g. of cobalt nitrate (potassium- and ammonium-free) in water and dilute to 1 litre. Cobalt nitrate specially purified from nickel and iron is not suitable since this grade usually contains ammonium.

Saturated Sodium Chloride Solution. Dissolve 350 g. of sodium chloride in 1 litre of warm water, filter and cool.

35 % Alcohol. Dilute 420 ml. of absolute alcohol with 600 ml. of water.

0·05N Potassium Permanganate. See p. 124.

0·05N Oxalic Acid. See p. 124.

Method:

Pipette 100–150 ml. of the ammonium chloride extract into a 200 ml. silica basin and, if the soil is non-calcareous, add 3–5 ml. of a 4 per cent. solution of calcium chloride. Destroy ammonium chloride as usual by evaporation with nitric acid and then ignite for 7–10 minutes in a muffle at a temperature just below a very dull red heat, to remove the last traces of ammonium salts. From this stage until after the precipitate of cobaltinitrite has been filtered, carry out all operations in a room free from ammonia fumes, since traces of ammonia are readily absorbed, especially during the evaporation of the acidified solution.

After ignition take up in 15 ml. of hot water and filter through a 7 cm. Whatman No. 44 filter paper, collecting the filtrate in a 100 ml. pyrex basin. Break up any lumps remaining in the silica basin and continue the washing, warming the basin, until all soluble salts have been extracted. Add 6 drops of concentrated hydrochloric acid to the filtrate and evaporate to dryness on the water bath.

When cold take up the evaporated residue in 1·5 ml. of glacial acetic acid and 10 ml. of saturated sodium chloride solution, added in that order. Stir and after 3–4 minutes add 5 ml. of 35 per cent. sodium nitrite, stirring again. When solution is complete, say after 5–10 minutes, but not longer, add 5 ml. of 20 per cent. cobalt nitrate from a dipping pipette with an enlarged aperture. This addition must be made rapidly (taking not more than 2–3 seconds) and with constant stirring, since the desired composition of the precipitate formed depends on the rapid addition of precipitant. Stir for 40–60 seconds, cover, and allow to stand overnight in a cool place.

Filter through a small Gooch crucible charged with asbestos. The asbestos used should be previously digested with acidified permanganate followed by an excess of oxalic acid and then washed thoroughly with water. This asbestos can be used repeatedly. Transfer the precipitate to the crucible and wash, four times in all, with 8–10 ml. portions of 35 per cent. alcohol. Finally wash the sides of the crucible with three very small lots (about 2 ml. each) of cold water to remove the alcohol.

Pipette a suitable amount of 0·05N potassium permanganate into a 400 ml. beaker, dilute to about 150 ml. and add 5 ml. of concentrated sulphuric acid. Then add the crucible and precipitate to this acidified permanganate, stir to keep the precipitate beneath the surface of the liquid and warm gently. If the colour appears likely to be discharged remove the beaker from the flame and add a further measured amount of standard permanganate to ensure an excess. Heat nearly to boiling, remove and rinse the crucible and continue heating to boiling. Remove from the flame and after a few minutes add a small excess of 0·05N oxalic acid. Warm until all oxides of manganese have been dissolved and titrate the excess of oxalic acid with the standard permanganate. The difference between the amounts of permanganate and oxalic acid, expressed in ml. of 0·05N solution corresponds to the amount of permanganate reduced by the cobaltinitrite precipitate. From this titration value the amount of K_2O present can be calculated since

$$K_2O \text{ (in mg.)} = 0{\cdot}354 \times \text{volume of } 0{\cdot}05N\ KMnO_4 + 0{\cdot}00034 \times (\text{volume of } 0{\cdot}05N\ KMnO_4)^2$$

or it may be calculated more conveniently by using the simplified formula

$$K_2O \text{ (in mg.)} = \text{volume of } 0{\cdot}05N\ KMnO_4 \times (0{\cdot}354 + 0{\cdot}00034 \times \text{volume of } 0{\cdot}05N\ KMnO_4).$$

Values for the latter factor ($0{\cdot}354 + 0{\cdot}00034 \times$ volume of $0{\cdot}05N\ KMnO_4$), for all titrations from 1 to 48 ml. are given in Table 5.

TABLE 5.

Values for the factor ($0{\cdot}354 + 0{\cdot}00034 \times$ volume of $0{\cdot}05N\ KMnO_4$) for the determination of potassium by the cobaltinitrite method.

Titration.	Factor.	Titration.	Factor.	Titration.	Factor.
1 ml.	0·35434	17 ml.	0·35978	33 ml.	0·36522
2 „	0·35468	18 „	0·36012	34 „	0·36556
3 „	0·35502	19 „	0·36046	35 „	0·36590
4 „	0·35536	20 „	0·36080	36 „	0·36624
5 „	0·35570	21 „	0·36114	37 „	0·36658
6 „	0·35604	22 „	0·36148	38 „	0·36692
7 „	0·35638	23 „	0·36182	39 „	0·36726
8 „	0·35672	24 „	0·36216	40 „	0·36760
9 „	0·35706	25 „	0·36250	41 „	0·36794
10 „	0·35740	26 „	0·36284	42 „	0·36828
11 „	0·35774	27 „	0·36318	43 „	0·36862
12 „	0·35808	28 „	0·36352	44 „	0·36896
13 „	0·35842	29 „	0·36386	45 „	0·36930
14 „	0·35876	30 „	0·36420	46 „	0·36964
15 „	0·35910	31 „	0·36454	47 „	0·36998
16 „	0·35944	32 „	0·36488	48 „	0·37032

To calculate the amount of exchangeable potassium, in m.e. %, deduct the amount of potash found in the blank determination and multiply the weight of K_2O, in milligrams, by the appropriate factor from the following table:

Soil extracted.	100 ml. extract ex 1 litre.	150 ml. extract ex 1 litre.
50 g.	0·425	0·283
66·66 g.	0·318	0·212
100 g.	0·212	0·141

Perchlorate Method.

Pipette 150–250 ml. of the ammonium chloride extract into a silica basin and proceed exactly as for the cobaltinitrite method except that it is necessary to precipitate completely any sulphates that might be present by the addition of 3 ml. of 2 per cent. barium chloride ($BaCl_2.2H_2O$) before evaporating with nitric acid. A large excess of barium chloride must be avoided but sufficient must be used to precipitate sulphates completely. After ignition, extraction with water, and filtration, as before described, collect the filtrate in a 100 ml. pyrex basin and add 5 ml. of 20 per cent. perchloric acid. Evaporate on the water bath and finally on a hot plate until dense white fumes of perchloric acid are evolved. If no fumes are produced add more perchloric acid (1–2 ml.) and water and re-evaporate, since it is necessary to have an excess of perchloric acid at this stage. When these fumes have been produced, cool, dissolve the perchlorates in 10–15 ml. of water and add a further 10 drops of perchloric acid, and continue the evaporation nearly to dryness on the water bath. Finish the evaporation over a hot plate until dense white fumes have been produced for some time and the liquid just sets to a pasty crystalline mass when cold.

When quite cold add 15 ml. of 96 per cent. alcohol containing 0·2 per cent. perchloric acid. Break up all the lumps with a stirring rod, stir well, then cover the basin and allow to stand for half an hour. When most of the crystals have settled, decant as much as possible of the clear liquid through a Gooch crucible, charged with asbestos. Then place the basin on the top of a warm water bath to eliminate the last of the alcohol, take up in about 10 ml. of warm water and add 5 drops of perchloric acid. Evaporate just to dryness as before and when cold take up in 10 ml. of the acidified alcohol, stirring well to break up lumps. After a few minutes decant through the Gooch crucible previously used. Wash with 5 ml. of acidified alcohol, draining the dish and crucible completely. Then transfer the crystals of potassium perchlorate to the crucible using 96 per cent. alcohol previously saturated with potassium perchlorate as the wash liquid. Wash three times in all, using about 15 ml. of this wash liquid each time. Drain the crucible well and dry for one hour in an oven at 140° C. Cool in a desiccator and weigh. Remove the potassium perchlorate from the crucible by washing with hot water, dry and weigh again. The difference in weight corresponds to $KClO_4$.

To calculate the amount of exchangeable potassium, in m.e. %, deduct the weight of potassium perchlorate found in the blank determination and multiply by the appropriate factor from the following table:

Soil extracted.	150 ml. extract ex 1 litre.	200 ml. extract ex 1 litre.	250 ml. extract ex 1 litre.
50 g.	96·2	72·1	57·7
66·66 g.	72·1	54·1	43·3
100 g.	48·1	36·1	28·9

(5) DETERMINATION OF SULPHATE.

It is desirable to examine the ammonium chloride extract of certain soils for the presence of the sulphate ion so as to ascertain whether significant amounts of soluble sulphates have been included in the determination of the exchangeable metal ions.

Transfer as much of the ammonium chloride extract as is available to a silica basin and remove ammonium salts by evaporation with nitric acid as usual.

Take up the residue in 5 drops of concentrated hydrochloric acid and 10 ml. of warm water. Filter through a 7 cm. Whatman No. 44 filter paper and wash with hot water, collecting the filtrate in a 100 ml. beaker. The filtrate should not exceed 50 ml. Heat to boiling and precipitate any sulphate by the addition of 1 ml. of N barium chloride. Keep gently boiling for a few minutes, or stand in a warm place, and leave overnight. Filter any precipitate of barium sulphate through a 7 cm. Whatman No. 44 filter paper, wash well, ignite and weigh as $BaSO_4$.

C. The Analysis of the Sodium Chloride Extract for Calcium and Magnesium.

CALCIUM. Transfer 400 ml. of the sodium chloride extract to a 600 ml. beaker and add $2\frac{1}{2}$ ml. of concentrated hydrochloric acid. If the extract shows any colour due to dispersed soil organic matter add 5 ml., or in the case of strongly coloured extracts, 10 ml. of 30 per cent. hydrogen peroxide (perhydrol). Cover with a clock glass and boil cautiously for 10 minutes. Remove from the flame and, after a few minutes, slowly add 35 ml. of dilute ammonia (1 + 1). Any hydrogen peroxide remaining decomposes vigorously hence the necessity to add the ammonia carefully and with stirring. If no hydrogen peroxide has been used, add 35 ml. of dilute ammonia (1 + 1) and proceed with the precipitation.

Raise the contents of the beaker to the boil and add 10 ml. of hot 10 per cent. ammonium oxalate solution. Boil gently for about one minute and allow to stand overnight. Filter the precipitated calcium oxalate through an 11 cm. Whatman No. 44 filter paper and collect the filtrate in another 600 ml. beaker. Wash twice with hot water. Dissolve, reprecipitate and titrate the calcium oxalate exactly as described under calcium in the ammonium chloride extract (p. 173).

MAGNESIUM. Concentrate the filtrate from the first calcium precipitation, by evaporation in a water bath, until its volume is reduced to about 200 ml., and then collect the filtrate from the second calcium precipitation in the same beaker. If sodium chloride crystals have formed in the filtrate, boil until solution is complete. Precipitate magnesium either by the hydroxyquinoline or the phosphate method.

8-Hydroxyquinoline Method. To the filtrate from the calcium precipitation add 10 ml. of concentrated ammonia and, if perhydrol was used previously, a few drops of a 5 per cent. solution of sodium metabisulphite. Heat to about 80° C. and add 10 ml. of a solution of 8-hydroxyquinoline solution (3 per cent. hydroxyquinoline in 10 per cent. acetic acid) or sufficient to give a small excess, as denoted by the yellow colour of the solution after the precipitate has formed. A large excess is to be avoided. In second litre extracts 5 ml. will be sufficient. Allow to stand until cool (or overnight). When only small amounts of magnesium are present stir at intervals to ensure complete precipitation. Filter through a 9 cm. Whatman No. 41 filter paper and proceed as described under magnesium in the ammonium chloride extract (p. 174).

Phosphate Method. When the filtrate from the calcium precipitation is cool add 10–15 ml. of a 10 per cent. solution of sodium phosphate ($Na_2HPO_4.12H_2O$) and 30 ml. of concentrated ammonia. Stir well at intervals until a precipitate is formed, especially if only small amounts of magnesium are present. Cover and allow to stand overnight in a cool place. Filter through an 11 cm. Whatman No. 44 filter paper and wash twice with dilute ammonia (1 + 9), discarding the filtrate. Dissolve and reprecipitate as described under magnesium in the ammonium chloride extract (p. 175).

D. The Analysis of the Ammonium Acetate Extract.

MANGANESE. Transfer an aliquot of the ammonium acetate extract, corresponding to 10 g. of soil, to a silica basin, evaporate to dryness on the water bath and ignite for 5–10 minutes in a muffle furnace at a dull red heat to destroy organic matter. Take up in 2 ml. of concentrated hydrochloric acid and 10 ml. of water, dissolve and evaporate to dryness on the water bath. Take up in 10 ml. of water and 2 ml. of nitric acid and re-evaporate to expel hydrochloric acid. Take up in 2 ml. of concentrated phosphoric acid and 25 ml. of water and add about 0·3 g. of potassium periodate. Boil gently and dilute with 25–50 ml. of water after the appearance of the permanganate colour. Boil again for about half a minute and transfer to a 50–150 ml. volumetric flask, depending on the intensity of the colour. Dilute nearly to volume and place the flask in a boiling water bath for 15 minutes. When cold dilute to the graduation mark, stopper and mix well.

Prepare a standard colour solution by pipetting 5–20 ml. of a standard manganese sulphate solution (p. 144; 10 ml.=1 mg. of manganese) into a 100 ml. measuring flask, add 25 ml. of water, 2 ml. of phosphoric acid and about 0·3 g. of potassium periodate. Heat the flask in the boiling water bath until the permanganate colour appears, dilute nearly to volume and replace in the water bath for 15 minutes. When cold, dilute to volume and mix well.

By means of a colorimeter, compare the colour of the unknown solution with that of the standard colour solution. If the colour is very weak Nessler tubes are more convenient. For good comparisons in a colorimeter the unknown solution should not be more than 40 per cent. stronger or 25 per cent. weaker than the standard solution.

The amount of manganese present, in milligrams, is given by the expression:

$$S \times \frac{D_s}{D_u} \times \frac{V_1}{V_2}$$

where S = amount of manganese (in mg.) in the standard colour solution

D_s = depth of the standard colour solution

D_u = depth of the unknown colour solution

V_1 = volume of the unknown colour solution

and V_2 = volume of the standard colour solution.

Having taken an amount of extract corresponding to 10 g. of soil, the amount of manganese, in m.e. %, present in the original soil then corresponds to:

Mn (in mg.) present in the unknown colour solution x 0·182.

CALCIUM, MAGNESIUM, SODIUM AND POTASSIUM. These can be determined in the ammonium acetate extract exactly as described for their determination in the ammonium chloride extract. Ammonium acetate is destroyed by evaporation and gentle ignition instead of evaporation with nitric acid.

The Determination of Exchangeable Hydrogen: Meta-nitrophenol Method.

Reagents:

m-Nitrophenol Buffer Solutions (0·06N in m-nitrophenol). To prepare a suitable range of these solutions dissolve pure m-nitrophenol at the rate of 8·34 g. per litre, in lime water of different strengths. Convenient strengths for ordinary soils are 0·025, 0·030, and 0·035N in lime. For strongly acid soils more basic solutions can be used with advantage. Shake the lime water and m-nitrophenol at intervals, over 1–2 days, to ensure complete solution. m-Nitrophenol dissolves only slowly in the less basic solutions. Store the buffer solutions in resistant glass bottles.

0·05N Hydrochloric Acid. See p. 117.

Brom Cresol Green Indicator Solution. See p. 127.

Method:

Pipette 100 ml. of a m-nitrophenol buffer solution of suitable basicity into a dry pyrex test tube (200 mm. x 32 mm.) and add a weighed amount of soil (3–10 g.). Make two separate determinations, using either the same buffer solution with two different weights of soil or alternatively, two different buffer solutions with the same weight of soil. Choose the basicity of the buffer solution or the weight of soil used, so that, after shaking, 25 ml. aliquots of one solution will contain lime, equivalent to somewhat more than 11·5 ml. of a 0·05N solution and similar aliquots of the other solution will contain somewhat less than 11·5 ml. of 0·05N lime water. The differences from this value may be 2 to 2·5 ml.

Close the tube with a rubber stopper and shake overnight in a mechanical shaker. Then filter the contents, without dilution, through a dry 11 cm. Whatman No. 30 filter paper collecting the filtrate in a dry 125 ml. Erlenmeyer flask. Reject the first portion. Titrate 25 ml. aliquots of the filtrate with 0·05N hydrochloric acid, using 5 drops of brom cresol green as indicator. The colour change at the end point is distinct, although a residual colour remains. Titrate 25 ml. of the original buffer solution, to serve as a blank determination. Use the colour change at the end point of this titration as the reference standard for all the other titrations.

When 25 ml. aliquots are used the amount of lime absorbed is calculated as follows:

Calcium absorbed (in m.e.%) =

$$\frac{B - T}{20} \times \frac{100}{25} \times \frac{100}{W}$$

$$\text{or } \frac{B - T}{W} \times 20$$

where B = blank titration, in ml. of 0·05N acid
T = actual titration, in ml. of 0·05N acid
and W = weight of soil taken.

From the two separate determinations, two values are obtained giving the amount of base absorbed by the soil corresponding to the final basicities of the buffer at equilibrium. Soils more alkaline than the buffer solution used will actually give up calcium to the buffer solution and the amount of calcium absorbed will then have a negative value. The basicity of the buffer at equilibrium is expressed by the volume of 0·05N hydrochloric acid used to neutralize the 25 ml. aliquot in each determination. From the two values for calcium absorbed at the two equilibrium basicities, obtain, by direct interpolation, the amount of calcium that would be absorbed at a final basicity value of 11·5 ml. This corresponds to the amount of exchangeable hydrogen originally present in the soil.

The value so calculated is subject to a small error since the amount of moisture present in the air-dry soil dilutes the buffer solution slightly, thus reducing its titration value against the standard acid by an amount slightly greater than corresponds to the amount of calcium absorbed. A correction should therefore be made, in calculating the amount of calcium absorbed, by deducting an amount, $y \times \frac{\textit{Titration value}}{500}$ from each of the values previously calculated. In the above expression y represents the percentage of moisture in the air-dry soil and *Titration value* corresponds to the amount of 0·05N hydrochloric acid (in ml.) used in the titration. The corrected values are then used for the interpolation for exchangeable hydrogen.

The Determination of Total Cation Exchange Capacity.

Accepting Bradfield and Allison's definition (p. 165) of a fully saturated soil, total cation exchange capacity is the sum of the exchangeable metal ions and exchangeable hydrogen, as determined above. If, however, one of the other proposed reference points for a saturated soil is preferred an ammonia absorption method may be used for the determination of cation exchange capacity.

Reagents:

N Potassium Sulphate. Dissolve 87 g. of potassium sulphate in water and dilute to 1 litre.

60 % Alcohol. Add 520 ml. of water to 1 litre of absolute alcohol.

0·05N Hydrochloric Acid. See p. 117.

0·05N Sodium Hydroxide. See p. 120.

Methyl Red Indicator Solution. See p. 126.

Method:

Carry out this determination on the soil remaining on the filter paper after leaching with N ammonium acetate as described on p. 171. Wash the soil repeatedly with 60 per cent. alcohol until the excess of ammonium acetate is removed. This may be determined by adding a small quantity of ammonium chloride to the first lot of alcohol used for washing and then leaching with alcohol until the filtrate gives no test for chloride. When free from chloride discard the alcohol washings. Then remove the adsorbed ammonium by washing once with 0·1N potassium sulphate and continuing with N potassium sulphate until 1 litre of filtrate has been collected. This leaching is somewhat slow, especially for heavy soils, and it is necessary to start with tenth normal potassium sulphate to avoid precipitation of the sulphate by the alcohol retained in the soil. One litre of leachate is generally sufficient to displace all of the exchangeable ammonium from light to medium soils but, with other soils, continue the leaching beyond one litre and distil it separately until no more ammonia is recovered.

Transfer an aliquot of the filtrate to a litre Erlenmeyer flask, add 2–3 g. of magnesia and distil the ammonia into a measured amount of 0·05N hydrochloric acid. Titrate the excess of hydrochloric acid with 0·05N sodium hydroxide, using methyl red as indicator. In the case of heavy soils distil the leachings in excess of 1 litre with magnesia and add the amount of ammonia recovered to that present in the first litre of leachate.

The total amount of ammonia adsorbed by the soil and displaced by the potassium sulphate corresponds to the total cation exchange capacity of the soil.

The Determination of Total Exchangeable Metal Cations: Bray and Willhite's Method.

Transfer a suitable aliquot of the neutral ammonium acetate extract (p. 171), corresponding to 10 g. of soil, to a large silica basin and evaporate to small volume on the water bath. Transfer the solution at this stage to a 100 ml. silica or platinum basin and continue the evaporation to dryness, finishing the operation over a hot plate. Ignite in a muffle furnace, gently at first and then for 20 minutes at a medium to full red heat, so converting the acetates of the exchangeable metal ions to carbonates or oxides. If a silica basin is used support it on a triangle to prevent the bottom being overheated.

When cold, take up in 20–50 ml. of 0·05N hydrochloric acid, or sufficient to ensure an excess of acid at this stage, warm gently and leave to stand until solution of the alkalies is complete. Then titrate the excess of acid with 0·05N sodium hydroxide using methyl red as indicator. If the methyl red is adsorbed as the titration proceeds add a further drop or two just before the end point. Calculate the amount of total exchangeable metal ions originally present in the soil, in m.e. %, from the following expression:

$$(V_a - V_b) \times N \times \frac{1{,}000}{V_e} \times \frac{100}{W}$$

where V_a = volume of standard hydrochloric acid taken
V_b = volume of standard sodium hydroxide used in back titration
N = normality factor
V_e = volume of extract taken
and W = weight of soil extracted.

The Approximate Determination of Total Exchangeable Metal Cations: Schofield's Rapid Method.

Pipette 100 ml. of 0·05N hydrochloric acid into a pyrex test tube (200 mm. x 32 mm.), add a weighed amount of soil (1–5 g.), stopper and shake overnight. The amount of soil taken should not be sufficient to neutralize more than about 20 per cent. of the standard acid. In general, the larger amount may be taken for sandy loams or soils slightly podsolized. For heavy clays and base saturated soils only 1–2 g. should be used.

The values obtained by this method are equilibrium values and therefore depend on the proportion of soil to acid.

Filter through a dry 11 cm. Whatman No. 30 filter paper, collecting the filtrate in a dry 125 ml. Erlenmeyer flask, rejecting the first portion. Titrate a 25 ml. aliquot against standard lime water using brom thymol blue as indicator. As the end point is approached add 2 or 3 more drops of indicator to overcome absorption by any sesquioxide precipitate. Titrate a blank similarly.

Then the approximate value for total exchangeable metal ions, in m.e. %, is given by the expression:

$$(B - T) \times N \times \frac{100}{25} \times \frac{100}{W}$$

where B = blank titration, in ml. of standard alkali
T = actual titration, in ml. of standard alkali
N = normality of the standard alkali
and W = weight of soil taken.

If any calcium carbonate is present express its amount in milligram equivalents per cent., and deduct it from the value calculated for total exchangeable metal ions.

ANION EXCHANGE

So little work has been done on anion exchange that the methods at present in use must be regarded as tentative. As more experience is accumulated new and improved methods will no doubt be devised. The present methods are the counterparts of the corresponding methods used in cation exchange studies.

By analogy with cation exchange, the pertinent determinations are anion exchange capacity and the amounts of the individual exchangeable anions. Of the latter the hydroxyl and phosphate ions are by far the most important. In addition small amounts of the sulphate ion are probably involved but the chloride ion does not appear to enter into the reaction.

Like cation exchange capacity, the total anion exchange capacity of a soil is a function of the reaction at which it is determined. It, however, increases with increase of acidity. The reaction at which it is to be determined must therefore be defined. At the present time no particular reaction is widely advocated. Anion exchange capacity is very small at the neutral point, and there is, unfortunately, no natural reference point, such as that corresponding to the saturation with exchangeable metallic cations in equilibrium with an excess of calcium carbonate. It is, therefore, tentatively suggested that anion exchange capacity should be determined at pH 4. This empirical reference point is selected since it corresponds, approximately, to the reaction of a soil completely deprived of its exchangeable metallic cations. This reaction is reached by removing the exchangeable metal ions from the soil either by leaching with dilute acid or by electrodialysis. The value is an approximation only; soils with a wide silica : sesquioxide ratio become more acid than pH 4 when completely desaturated, while soils with a narrow ratio do not reach pH 4. Whatever value is taken for the reference point it should be clearly stated.

The method described for anion exchange capacity is analogous to the corresponding determination of cation exchange capacity. The soil is leached with a solution of ammonium phosphate, adjusted to pH 4 and normal in respect to the phosphate ion, until all the exchangeable anions are replaced by phosphate ions. The excess of ammonium phosphate is then removed and the adsorbed phosphate determined colorimetrically, after displacing it with sodium hydroxide. The method suffers from several imperfections. Ammonium phosphate is not buffered at pH 4 so that it is not an ideal leaching agent for the determination. Furthermore, the method is not applicable to soils containing calcium carbonate. The technique used by J. S. Hosking in these laboratories is described on p. 192.

It should be noted that if values for anion exchange capacity at any other reaction are required the determination can be carried out by changing the proportions of phosphoric acid and ammonia in the solution used for saturating the soil with phosphate.

Exchangeable phosphate is determined by treating the soil with sodium hydroxide solution, thus replacing the adsorbed phosphate by the hydroxyl ion. The phosphate is then determined colorimetrically (p. 195).

Before any method for the determination of exchangeable hydroxyl and sulphate can be recommended, further work is necessary. The definition of exchangeable hydroxyl depends, of course, on the reference point adopted. Leaching with ammonium phosphate, or with the salt of any other polybasic acid, may increase the cation exchange capacity of the soil, so this precludes the direct titration of replaced hydroxyl ions in the leachate.

Total Anion Exchange Capacity: Phosphate Adsorption Method.

Reagents:

N Ammonium Phosphate pH 4·0. Dissolve 384 g. of ammonium dihydrogen phosphate in water, filter and dilute to 10 litres. Check the pH of the solution by means of the glass or quinhydrone electrodes and if it does not correspond to pH 4·0 adjust to this value by the careful addition of a small quantity of N phosphoric acid.

60 % Alcohol. Add 520 ml. of water to 1 litre of absolute alcohol.

N Sodium Hydroxide. Dissolve 40 g. of sodium hydroxide in water and dilute to 1 litre.

Ammonium Molybdate. Add 100 ml. of a 10 per cent. aqueous solution of ammonium molybdate to 300 ml. of dilute sulphuric acid (1 + 1). Store in a dark bottle away from light.

Stannous Chloride. To 0·5 g. of powdered tin add 5 drops of a 4 per cent. solution of copper sulphate and 10 ml. of concentrated hydrochloric acid. Warm until solution is complete and dilute to 50 ml. This solution is stable for some time if kept in an atmosphere of carbon dioxide, provided that a little undissolved tin is left in the bottom of the bottle.

Standard Phosphate Solution. Dissolve 0·1433 g. of potassium dihydrogen phosphate in water and dilute to 1 litre. As a working standard dilute 20 ml. of this solution to 1 litre; 10 ml. then contains 0·02 mg. PO_4.

Sodium Sulphite Solution. Dissolve 5 g. of sodium sulphite in 100 ml. of water.

Method:

Transfer 1–5 g. of soil, depending on its anion exchange capacity, to a 30 ml. narrow mouthed bottle, preferably with a screw cap to resist pressure. Add 22·5 ml. of N ammonium phosphate solution. Stopper tightly and shake vigorously. Then place in a water bath, kept at a temperature of 50–60° C., shaking vigorously at intervals of 15 minutes. After 1–2 hours, remove the bottle from the bath and allow it to stand overnight. Then filter the contents through a Buchner funnel, fitted with a 4·25 cm. Whatman No. 50 filter paper, using slight suction. A hardened filter paper is necessary to prevent its subsequent disintegration. If the initial filtrate is turbid it must be returned to the funnel before continuing. Rinse the bottle and stopper with about 10 ml. of ammonium phosphate solution and then with a jet of the same solution, from a wash bottle. When the soil has been transferred quantitatively to the filter paper, continue the leaching with 5–10 ml. portions of ammonium phosphate, draining the soil completely between each addition, until the volume of the leachate amounts to about 100 ml.

Remove the excess of free ammonium salts, remaining in the soil in the Buchner funnel, by leaching with 5–10 ml. portions of 60 per cent. alcohol, giving particular attention to the soil adhering to the sides of the funnel. Fifty ml. is usually sufficient to remove all soluble phosphates. Then rinse the soil with two lots, each of 5 ml., of absolute alcohol, to dry the soil and assist in its removal from the funnel.

When sufficiently dry, transfer the soil and filter paper to a small beaker and wash the funnel with water to remove the last traces of soil. Dilute to about 200 ml., add 25–30 ml. of N sodium hydroxide and boil for 5 minutes. Then transfer the suspension to a 250–500 ml. volumetric flask, rinsing the beaker with hot water. When cold dilute to the graduation mark, stopper and mix well. Determine phosphate in an aliquot of this solution.

If the phosphate determination is to be made in Nessler tubes, a quantity of approximately 0·02 mg. PO_4 is convenient; if a colorimeter is to be used a larger amount, 0·10 mg., is desirable. The colour produced is influenced by the concentration and the amounts of reagents used. The amounts and dilution recommended should, therefore, be closely adhered to, both in the unknown and standard comparison solution. For accurate results the standard should not differ by more than 25 per cent. from the unknown solution. If necessary a preliminary determination should be carried out, in Nessler tubes, to determine a suitable volume of the unknown solution for the colour comparison.

If the aliquot still contains sufficient residual colour, after its final dilution, to influence the shade of the blue phosphomolybdate then the organic matter must be destroyed, by means of bromine water, before the blue colour is developed. To a 50 ml. aliquot add 5 ml. of a saturated solution of bromine water and, if not sufficiently alkaline, a few drops of N sodium hydroxide until the colour of bromine disappears. Stir and add N hydrochloric acid, drop by drop, until the solution is just acid. If free bromine is not apparent at this stage, repeat the treatment with a further 5 ml. of bromine water and sodium hydroxide. Remove the excess of bromine by the addition of 5 ml. of sodium sulphite solution and then boil to remove most of the excess of sulphurous acid. The solution should now be colourless and is suitable for the development of the blue phosphomolybdate colour.

To develop the blue colour, pipette a suitable quantity of the standard phosphate solution or the unknown solution, the latter freed from organic matter if necessary, into a 100 ml. volumetric flask, dilute to about 70 ml., add 2 ml. of ammonium molybdate solution and shake well. With constant shaking add 5 drops of stannous chloride solution. Then dilute to the mark and mix well. Make up both standard and unknown solutions at the same time and under the same conditions. After 10 minutes compare the colour of the unknown with that of a suitable standard using a colorimeter or, for weaker colours, Nessler tubes.

The anion exchange capacity at pH 4, in m.e. %, is then given by the expression:

$$S \times \frac{D_s}{D_u} \times \frac{V_1}{V_2} \times \frac{3 \cdot 157}{W}$$

where S = PO_4, in mg., in the standard colour solution
D_s = depth of the standard colour solution
D_u = depth of the unknown colour solution
V_1 = original volume of the soil extract
V_2 = volume taken for the colour development
and W = weight of soil taken.

The Determination of Exchangeable Phosphate.

Reagents:

All the reagents described on p. 192, except N ammonium phosphate, are required.

Method:

Transfer 1–5 g. of soil to a 30 ml. narrow mouthed bottle, add 22·5 ml. of 60 per cent. alcohol and allow to stand for one hour, with occasional shaking, to dissolve soluble salts. Filter through a small Buchner funnel, fitted with a 4·5 cm. Whatman No. 50 filter paper, rinsing the bottle with 10 ml. portions of alcohol. Wash twice more with 10 ml. portions of 60 per cent. alcohol and then twice with 5 ml. of absolute alcohol to dry the soil.

Transfer the soil and filter paper to a beaker, dilute to about 200 ml., add 25–30 ml. of N sodium hydroxide and continue the phosphate determination exactly as described for the determination of adsorbed phosphate (p. 193).

The amount of phosphate found corresponds to the exchangeable phosphate, although it may contain some organic phosphate from the decomposition of soil organic matter.

REFERENCES

(1) BRADFIELD, R. and ALLISON, W. B.: *Trans. 2nd Comm. Int. Soc. Soil Sci.*, 1933, Copenhagen, Vol. 4., 63–79.

(2) BRAY, R. H. and WILLHITE, F. M.: *Ind. Eng. Chem. (Anal. Ed.)*, 1929, *1*, 144.

(3) CHAPMAN, H. D. and KELLEY, W. P.: *Soil Sci.*, 1930, *30*, 391–406.

(4) Gedroiz, K. K.: *Zhur. Opit. Agron.*, 1918, *19*, 226–244. (Translation by S. A. Waksman).

(5) Hissink, D. J.: *Soil Sci.*, 1923, *15*, 269–276.

(6) Kelley, W. P.: *J. Agric. Sci.*, 1934, *24*, 72–92.

(7) Parker, F. W.: *J. Amer. Soc. Agron.*, 1929, *21*, 1030–1039.

(8) Piper, C. S.: *J. Agric. Sci.*, 1932, *22*, 676–687.

(9) Piper, C. S.: *J. Soc. Chem. Ind.*, 1934, *53*, 392T–396T; and 1935, *54*, 157T–158T.

(10) Piper, C. S.: *J. Coun. Sci. Ind. Res. Australia*, 1936, *9*, 113–124.

(11) Puri, A. N.: *Soil Sci.*, 1934, *37*, 105–108.

(12) Schofield, R. K.: *J. Agric. Sci.*, 1933, *23*, 252–254.

(13) Schollenberger, C. J.: *Science*, 1927, *65*, 552–553.

(14) Sigmond, A. A. J. de and Iyengar, M. A. S.: *Soil Research*, 1935, *4*, 217–222.

(15) Williams, R.: *J. Agric. Sci.*, 1928, *18*, 439–445.

CHAPTER X

NITROGEN

Nitrogen is determined in soils by one of the many modifications of the Kjeldahl method in which the organic matter is oxidized by sulphuric acid and the nitrogen converted to ammonia. The various modifications differ from the original in the addition of potassium or sodium sulphate to raise the temperature of the digest and in the use of other catalysts in place of mercury. The correct determination of nitrogen is not as simple and straightforward as is usually thought. The most frequent sources of error lead to low values.

The whole of the organic nitrogen is not converted to ammonia until some considerable time after the digest becomes "clear" and colourless, the actual time depending on the catalyst used. Selenium is more efficient than copper sulphate since it shortens the time of digestion required to obtain maximum values for nitrogen. Some analysts prefer a mixture of copper sulphate and selenium. While the digestion must be continued for a sufficient time after it has become colourless, to ensure conversion of the whole of the nitrogen to ammonia, it must not be unduly prolonged or an excessive amount of sulphuric acid may be lost. The overheating, resulting from this, leads to a loss of ammonia through dissociation of the ammonium sulphate.

Bal (2) has shown that some heavy clay soils give erroneously low values for nitrogen unless the soil is allowed to stand with water before digestion. If this preliminary soaking with water is omitted nitrogen within the clay aggregates is only partly attacked. Walkley (4) found that very fine grinding, in a ball mill, ensured complete digestion in similar types of soils. It is not necessary to use Bal's modification to obtain maximum values for nitrogen in all soils. Its general adoption is, however, recommended since it does not add materially to the time required for the determination and it ensures complete digestion of the organic nitrogen in all soils.

The usual methods do not include the whole of the nitrogen of the soil for most of the nitrate nitrogen is lost during the digestion. This loss may be disregarded for most soils since the amount of nitrate nitrogen is negligible in comparison with the organic nitrogen. If, however, it becomes necessary to include it, Ulsch's method should be used. In this method the nitrate is reduced, by finely powdered iron and dilute sulphuric acid, before commencing the digestion (1). For good results the iron must be in a very fine state of subdivision. The more usual salicylic acid method cannot be used with Bal's modification since nitration of the salicylic acid does not take place if more than a trace of water is present.

Bumping does not occur during distillation if the solution is transferred from the digestion flask by decantation, thus separating most of the sand. The addition of a small piece of zinc to the distillation flask promotes smooth boiling and the slow evolution of hydrogen minimizes the danger of the distillate sucking back. Some analysts use copper flasks for the distillation but borosilicate glass is to be preferred, since the contents can be seen.

The concentrated sulphuric acid and caustic soda are most conveniently stored in 5–10 litre bottles fitted with the arrangement, shown in Fig. 13, for delivering the required volumes automatically. A conical graduated separating funnel is supported from the neck of the bottle containing sulphuric acid at such a height that digestion flasks can be placed under its delivery stem. The height of the delivery end of tube B in relation to the separating funnel determines the amount of acid measured out each time. A is closed and pressure applied to the bottle until the acid has filled the separating funnel to a level slightly above the outlet of tube B. If now the pressure in the bottle is released, by opening A, the acid in the separating funnel above the level of B siphons out, leaving the required amount. The level of B must, of course, be above the level of the acid in the storage bottle for the siphon to function. The measured amount of acid can then be delivered directly into the digestion flask.

A similar arrangement is used for the caustic soda except that in this case a 100 to 150 ml. measuring cylinder is placed under the outlet tube. The height of the outlet tube in relation to the cylinder again determines the volume automatically delivered. A wooden block of suitable height enables the cylinder to be placed in the correct position each time. To reduce drainage the vertical part of the outlet end of the delivery tubes should be kept as short as possible.

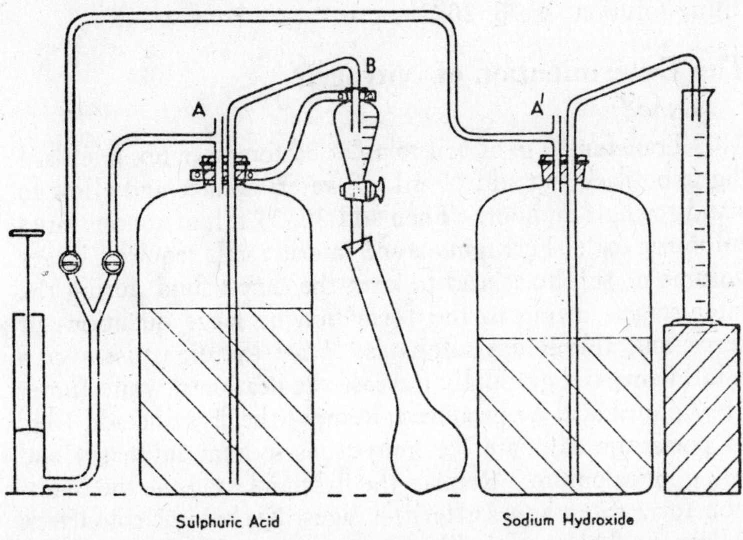

Fig. 13. Automatic measuring apparatus for sulphuric acid and sodium hydroxide.

The ammonia, which is distilled off from the digest, is usually absorbed in standard acid. In Winkler's modification (3) it is absorbed in an unstandardized solution of boric acid. Since boric acid is a very weak acid the absorbed ammonia can be determined by titration with standard hydrochloric acid, if a suitable indicator (brom phenol blue) is used. This gives a direct determination, in place of the usual indirect determination by difference, and only one standard solution is required. Standard sodium hydroxide is not required. Moreover, a sufficiently large excess of boric acid can be used so as to ensure complete absorption of the ammonia even when the amounts obtained are larger than anticipated. For successful absorption the temperature of the distillate must not exceed 50° C.

and it is preferably kept below 40° C. Air-cooled condensers cannot be used.

If the ammonia is absorbed in standard hydrochloric acid either methyl red or brom cresol green may be used. For the titration of tenth normal solutions the choice depends upon the particular colour change preferred by the analyst. For the relative merits of these two indicators when titrating more dilute solutions see p. 205.

The Determination of Nitrogen.

Method:

Transfer 10 g. of soil to a flat bottomed pyrex Kjeldahl digestion flask and add 10 ml. of water. Shake and allow to stand for half an hour. Then add 30–35 ml. of concentrated sulphuric acid. Ferruginous and lateritic soils require a larger volume of sulphuric acid to keep the digest fluid during the later stages, owing to the formation of large quantities of ferric and aluminium sulphates. Start the digestion over a small flame and gradually increase the heat until white fumes of sulphuric acid are produced. Remove the flask and add 10 g. of potassium sulphate (or anhydrous sodium sulphate) and 0·2 g. of selenium. Replace the flask and continue the digestion for 1 to 1½ hours after the digest has become colourless. Allow the flask to cool, dilute the contents with about 100 ml. of water and transfer the fluid part to a 1,000 ml. conical flask, leaving as much as possible of the sand behind. Wash the sandy residue with four or five lots of 50–60 ml. of water, decanting the washings into the conical flask after allowing the sandy residue to settle for a few seconds each time. Add a piece of granulated zinc and then 100–110 ml. of caustic soda solution (1 lb. of caustic soda + 1 litre of water) or sufficient to make the contents of the flask alkaline to phenolphthalein. Pour the caustic soda solution down the side of the flask so that it forms a heavy layer at the bottom. Place the stopper in the flask and connect it to the distillation apparatus. Mix the contents well by shaking and commence the distillation, collecting the ammonia in 25 ml. of 0·1N hydrochloric acid containing two or three drops of methyl red indicator solution. For soils high in nitrogen absorb the ammonia in 35 ml. of standard acid or take a smaller quantity of soil for the determination. Distil until about one-third of the liquid has passed over.

When the distillation is completed rinse the condenser tube which dips into the standard acid and titrate the excess of acid with tenth normal sodium hydroxide.

Carry out a blank determination in exactly the same manner but using about 0·2 g. of cane sugar in place of the soil, so as to correct for any nitrogen contained in the reagents.

The percentage of nitrogen in the soil, on the basis of a 10 g. sample =

$$(B - T) \times N \times 0{\cdot}14$$

where B = blank titration, in ml. of standard alkali
T = actual titration, in ml. of standard alkali
and N = normality of the standard alkali.

Winkler's Modification.

Proceed as before but collect the distillate in 50 ml. of 4 per cent. boric acid solution instead of in standard hydrochloric acid. When the distillation is completed add 3 drops of brom phenol blue indicator solution and titrate the ammonia absorbed with 0·1N hydrochloric acid.

The percentage of nitrogen in the soil, on the basis of a 10 g. sample =

$$(T - B) \times N \times 0{\cdot}14$$

where T = volume of standard acid used in the actual titration
B = volume of standard acid used in the blank titration
and N = normality of the standard acid.

REFERENCES

(1) Ashton, F. L.: *J. Agric. Sci.*, 1936, *26*, 238–248.
(2) Bal, D. V.: *J. Agric. Sci.*, 1925, *15*, 454–459.
(3) Markley, K. S. and Hann, R. M.: *J. Assoc. Off. Agric. Chem.*, 1925, *8*, 455–467.
(4) Walkley, A.: *J. Agric. Sci.*, 1935, *25*, 598–609.

CHAPTER XI

NITRATES, NITRITES AND AMMONIA

Nitrates and nitrites are readily soluble in water and the amount extracted is independent of the soil : water ratio. Ammonia, on the other hand, is present as an exchangeable ion. It is adsorptively bound to the exchange complex and it must, therefore, be replaced by an excess of some other ion if it is desired to obtain it in solution quantitatively.

On account of their solubility and mobility, nitrates and nitrites are subject to considerable redistribution within the profile, consequent on changes in the moisture conditions of the soil. Samples should therefore be taken to a sufficient depth, and from a sufficient number of sites, to reduce the experimental error of sampling to a value commensurate with the nature of the investigation. For most purposes a composite sample, representing 4–6 sites per plot of a tenth of an acre, is suitable but for some investigations it may be necessary to sample a greater number of sites per plot. The surface mulch, to the depth of ploughing (3½–4½ inches) should be sampled separately as it may be very much richer than the remainder of the profile, particularly under fallow conditions. The next sample can conveniently represent the rest of the surface soil to a depth of 9 inches and the remaining samples can be taken in successive 9-inch layers to a depth of 18–45 inches as required. When the soil is moist a semicylindrical sampling tool, similar to a cheese tester but about 2 cm. internal diameter, is very useful. It does not disturb the plot as much as a larger sampling tool and enables more sites to be sampled with ease. For dry soils or when deeper samples are required the 4-inch post hole auger is most suitable if care is taken to prevent contamination of the lower samples from the sides of the hole near the top.

Under favourable moisture conditions the amounts of nitrates, nitrites and ammonia fluctuate considerably in the soil, depending on the activity of the micro-flora and -fauna.

Ammonia is rapidly oxidized to nitrites, and nitrites to nitrates. For this reason valid determinations of these constituents can only be made in samples taken directly from the field. Richardson (5) found large increases in the ammonia-nitrogen when soils were allowed to air-dry before the determination; the amount of change varied from soil to soil. If it is impossible to carry out the analyses as soon as the soil reaches the laboratory the sample should be dried very rapidly in an oven at 55° C. However, Russell and Page (6) state that such drying leads, on the average, to an increase of 2 parts of nitrate-nitrogen per million parts of soil. Better results might be obtained if the oven is designed to enable a rapid stream of air to be drawn through it, so as to hasten the removal of moisture. According to Russell and Page, there is practically no increase in nitrate-nitrogen if the soil is dried in a vacuum oven for two hours at 55° C.

If the samples have to be transported to the laboratory the moist soil should be kept in tightly closed tins and microbiological activity inhibited by the addition of 1–2 ml. of toluene to each sample. There is, however, some doubt concerning the efficacy of toluene in preventing changes in the nitrate content and rapid drying should be adopted wherever possible, when the analyses cannot be undertaken without delay. For the determination of nitrates, reduction methods, in which the nitrate-nitrogen is reduced to ammonia, are preferred, although, in all such methods, nitrites are also reduced. However, nitrites are not often present in amount sufficient to disturb the determination. Nitrates can very conveniently be determined in the soil extract after the determination of ammonia by Olsen's method (p. 207). The phenoldisulphonic acid method gives a direct determination of nitrates but it is subject to numerous difficulties. To obtain accurate values by it, the soil extract must be quite clear and colourless, otherwise the differences in shade will make the colour matching very difficult. Chlorides also interfere and, if present in the soil in amounts greater than 0·002 per cent., the excess must be removed by precipitation with silver sulphate.

Nitrites are seldom present in soils in amounts sufficient to warrant their determination, except in special cases. In a moist soil they are so rapidly oxidized to nitrates that it is doubtful whether reliable values are obtained, even when the determination is carried out immediately the sample is collected. Bartholomew (1) found that considerable amounts of added nitrites were oxidized to nitrate during the preparation of the aqueous extract of the soil, while within one hour the oxidation was practically complete and only slight traces of nitrite-nitrogen could be detected. The oxidation is biological and does not occur in autoclaved soil.

When nitrites are present in sufficient quantity, as in soil composts, they may be determined by Shrikhande's method, in which use is made of the reaction between nitrites and hydroxylamine hydrochloride. When a soil extract containing nitrite is warmed with hydroxylamine hydrochloride the latter is decomposed according to the equation:

$$NH_2OH.HCl + NaNO_2 = NaCl + N_2O + 2H_2O.$$

As a result of the decomposition of the hydroxylamine hydrochloride the acidity of the solution diminishes and this decrease can be determined by titration, in the presence of a suitable indicator, such as phenolphthalein. However, since 1 ml. of 0·05N sodium hydroxide is equivalent to 0·715 mg. of nitrite-nitrogen, the method is not sufficiently sensitive for the determination of the nitrites present in normal soils. For details of the method reference should be made to the original paper (7).

Ammonia cannot be satisfactorily distilled directly from the soil in the presence of an alkali, because even a mild alkali like magnesia leads to some decomposition of soil organic matter, with liberation of ammonia from proteins. Little or no decomposition of protein occurs if the inorganic ammonia is removed, by aeration in the cold, after the addition of sodium chloride and sodium carbonate as in Mathews's method (2). A vigorous stream of air is drawn through the mixture for 5–6 hours and the ammonia liberated is absorbed in 0·02N acid. The method is tedious and requires special apparatus. Unless a steady stream of air can be maintained many fine-textured soils are liable to block up the tubes and aeration cannot be started again. McLean and Robinson (3) recognized the existence of ammonia in the soil in the exchangeable form and based their method for its determination on the well-known leaching method of Hissink for other exchangeable ions. They leach the soil with sodium chloride and recover the ammonia from the leachate by distillation in the presence of magnesia (p. 210). The most convenient method of extraction for ordinary soils is that due to Olsen (4) and slightly modified by Richardson (5). In this method (p. 208) the soil is shaken with a potassium chloride-hydrochloric acid buffer solution containing sufficient hydrochloric acid to give a reaction of pH 1·0 to 1·5 after interaction with the soil. The concentration of potassium and particularly the hydrogen ions in the buffer at this reaction is such that the ammonia is almost completely replaced. It is thus brought into solution quantitatively by simply shaking the soil with an excess of the buffer solution, and recovered by distillation of an aliquot of the filtered extract with an excess of magnesia. Nitrates can then be readily determined in the same extract, by reduction with Devarda's alloy after the distillation of the ammoniacal nitrogen. A potassium sulphate-sulphuric acid buffer is sometimes preferred to potassium chloride-hydrochloric acid for extracting the soil since it froths somewhat less during distillation.

Olsen's method is not applicable to soils containing more than a few per cent. of calcium carbonate on account of the excessive amounts of hydrochloric acid required to give a final reaction of pH 1·0–1·5. Ammonia must be determined in these soils by Mathews's or McLean and Robinson's methods and nitrates by leaching with water.

For the back titration of the excess of standard acid used to absorb the ammonia, brom cresol green is more suitable than methyl red because it is less affected by carbon dioxide. When titrations are made with solutions more dilute than tenth normal, and methyl red is used as the indicator, the solution should be boiled to remove carbon dioxide just before the end point of the titration is reached. When brom cresol green is used the necessity for boiling off carbon dioxide is avoided (unless more than ordinary amounts are present) by titrating until the colour of the indicator matches that of a buffer solution of pH 4·7–4·8, containing the same quantity of indicator.

According to Richardson, the magnesia used for the distillation of the ammonia should be freshly ignited otherwise the amount of carbon dioxide from it may introduce an error in the titration which is not eliminated in the blank determination.

The Determination of Nitrates.

When the soil does not contain appreciable quantities of organic matter the following procedure is recommended:

Break the soil sample, as brought in from the field, into pieces not more than three-eighths to half an inch in diameter and weigh 100–250 g. into a flat tray. At the same time weigh out two smaller samples for moisture determinations. Dry the portion for the nitrate determination rapidly in an oven at 55° C. for 12–16 hours. The drying enables the subsequent leaching to be carried out with a minimum puddling of the soil. Transfer the dried soil to a 9 cm. Buchner funnel, fitted with a Whatman No. 50 filter paper, and pour on sufficient distilled water, containing 10 drops of concentrated sulphuric acid, to cover the soil. After a few minutes soaking connect to the filter pump and continue to leach the soil with successive quantities of distilled water until the filtrate amounts to about 600 ml. Transfer the filtrate to a 1 litre Erlenmeyer flask, add 1 g. of magnesium oxide and evaporate until the volume is reduced to about 200 ml. Cool, and add in the following order 5 g. of zinc dust, 70 ml. of 30 per cent. sodium hydroxide, and 5 g. of powdered iron. Connect the litre flask to the condenser of the nitrogen distillation unit through an efficient splash trap. Allow the reduction to proceed in the cold for half an hour, continue for half an hour over a very small flame and boil off the ammonia in a third period of half an hour. Collect the ammonia in 10–35 ml. of 0·02N hydrochloric acid and titrate the excess of acid with 0·02N sodium hydroxide using methyl red or brom cresol green.

Two g. of Devarda's alloy may be used, instead of the zinc and iron powders, to bring about the reduction of the nitrates. In this case reduction is complete in a little more than half an hour. The heating must be very gentle until the initial reaction has moderated when it can be increased to drive off the whole of the ammonia. When Devarda's alloy is used for reduction a very efficient spray trap must be used to retain the fine spray. A glass bulb filled with glass beads or a glass column filled with small flint pebbles is satisfactory.

Where much organic matter is present in the soil extract as prepared above the following procedure is recommended:

Transfer the water extract of the soil to a 1 litre Erlenmeyer flask, add $2\frac{1}{2}$ ml. of 30 per cent. sodium hydroxide and 10 ml. of 3 per cent. potassium permanganate, cover with an inverted porcelain crucible lid, boil until the volume is reduced to about 150 ml. and keep just boiling for a further period of 2–4 hours. If the permanganate is completely decolorized add a little more until no appreciable change is noticeable in half an hour. Dilute to 300 ml., add 25 ml. of 30 per cent. sodium hydroxide, 2 ml. of rectified spirit and 3 g. of Devarda's alloy. Connect to the distillation apparatus through an efficient spray trap as before. After reduction in the cold for a few minutes heat cautiously and distil off the ammonia, collecting it in 0·02N hydrochloric acid as before.

In each case carry out a blank determination on all the reagents used, absorbing the ammonia in the same volume of standard acid as was used for the actual determinations.

Calculation of the Results:

The amount of nitrate-nitrogen in mg. per kg. (p.p.m.) is given by the following expression:

$$(B - T) \times N \times 14 \times \frac{1000}{W}$$

where B = blank titration, in ml. of standard alkali
T = actual titration, in ml. of standard alkali
N = normality of the standard alkali
and W = weight of soil taken.

Correct the weight of soil taken for the amount of moisture contained in it so as to express the results on an oven-dry basis.

The Determination of Nitrates After Olsen's Extraction for Ammonia.

Dilute the residue remaining in the distillation flask after the determination of ammoniacal nitrogen (p. 208) until the volume is again 300 ml. and add about 2·5 g. of finely powdered Devarda's alloy. Immediately reconnect the flask to the distillation unit through an efficient spray trap and heat gently so as to take 1¼ to 1½ hours to distil about 200 ml.

Collect the distillate in 10–15 ml. of 0·02N hydrochloric acid containing 20 drops of brom cresol green. When the distillation it completed titrate the excess of acid with 0·02N sodium hydroxide until the colour of the indicator matches that in a reference buffer solution of pH 4·7–4·8 (p. 205).

Carry out a blank determination on all the reagents used.

Calculation of the Results:

The expression for the calculation of the amount of nitrate-nitrogen is identical with that given for ammoniacal nitrogen on p. 210.

The Determination of Ammonia: Richardson's Modification of Olsen's Method (5).

Reagents:

2N Potassium Chloride. Dissolve 149 g. of potassium chloride in water and dilute to 1 litre.

2N Hydrochloric Acid. This is prepared with sufficient accuracy by diluting 175 ml. of concentrated hydrochloric acid to 1 litre.

0·02N Hydrochloric Acid. See p. 117.

0·02N Sodium Hydroxide. See p. 120.

Brom Cresol Green Indicator. See p. 127.

Buffer Solution pH 4·8. Mix 40 ml. of 0·2N acetic acid and 60 ml. of 0·2N sodium acetate. For use dilute 10 ml. to about 200 ml., add 20 drops of brom cresol green, and a few drops of a one per cent. solution of mercuric chloride to preserve the solution. The colour of the diluted buffer is stable for about one month.

Method:

Break up the freshly taken sample by hand or, if not too moist, pass it through a 4 mesh sieve; discard the larger roots and stones. Subsample by successive quartering and take 100 g. for the preparation of the extract and 50 g. for a moisture determination. Determine moisture by drying for 16–24 hours in an electric oven at 105° C.

Transfer the 100 g. sample to a 500 ml. bottle fitted with a good quality rubber stopper and add 100 ml. of 2N potassium chloride solution and sufficient 2N hydrochloric acid to bring the soil extract to the required acidity, after equilibrium has been established, of about pH 1·0–1·5. Also add about 7 drops of toluene and sufficient water to bring the total volume of solutions added to 200 ml. For most soils 10 ml. of 2N acid is sufficient. Organic and alkaline soils require more, while provision must also be made for any calcium carbonate present. If the necessary amount of acid is not known from previous determinations on the same soil it is most conveniently determined by shaking by hand a separate 10 g. sample of the soil with 10 ml. of 2N potassium chloride, 1 ml. of 2N hydrochloric acid and 9 ml. of water, filtering, and determining the pH of the filtrate by means of thymol blue. A bright red colour (pH 1·0–1·5) indicates that sufficient acid has been added but an orange or yellow colour denotes insufficient acidity and the test must be repeated with larger amounts of acid.

After the addition of the correct amount of hydrochloric acid to the main sample, stopper the bottle and shake for one hour in a mechanical shaking machine. If carbonates are present take the usual precautions to prevent pressure being developed from the carbon dioxide liberated, by shaking by hand and releasing the stopper once or twice.

After mechanical shaking pour the whole of the contents of the bottle, in one operation, on to a large (24 cm.) Whatman No. 12 fluted filter paper and collect the filtrate in an Erlenmeyer flask, rejecting the first 25–30 ml. The extraction should be carried out on the day of sampling, and the distillation as soon as possible after filtration. However, the presence of toluene makes it permissible to postpone distillation for a day or two if the flask containing the filtrate is kept tightly stoppered.

Transfer 100 ml. of the filtrate to a 750–1,000 ml. Erlenmeyer flask, permanently marked (with glass-writing ink) at the 300 ml. level. Dilute to this mark, add 4–5 g. of freshly ignited magnesia, connect to the distillation apparatus and distil until 150–200 ml. of distillate have passed over. The larger volume is necessary to ensure complete removal of the ammonia if a litre distilling flask is used. Keep the residue remaining in the distilling flask for the determination of nitrates (p. 207).

Collect the distillate in 10–15 ml. of 0·02N hydrochloric acid diluted sufficiently to cover the lower end of the condenser tube. When the distillation is completed, add about 20 drops of brom cresol green indicator solution and titrate the excess of acid with 0·02N sodium hydroxide until the colour of the indicator matches that in a comparison buffer solution of pH 4·7–4·8.

Carry out a blank determination on all the reagents used.

On account of the sensitivity of 0·02N acid to alkali derived from glass, only the best quality non-alkali glassware is permissible for the condenser tubes and titration flasks. Richardson prefers a block tin condenser, with a short length of non-alkali glass tubing dipping into the standard acid.

Calculation of the Results:

Determine the amount of oven-dry soil and the amount of water present in the 100 g. of moist soil taken for the determination. Then the amount of ammoniacal nitrogen present, in mg. per kg. (p.p.m.) is given by:

$$(B - T) \times N \times 14 \times \frac{200 + V}{100} \times \frac{1000}{W}$$

where B = blank titration, in ml. of standard alkali
T = actual titration, in ml. of standard alkali
N = normality of the standard alkali
V = volume of water in the moist soil taken
and W = weight of oven-dry soil taken.

The Determination of Ammonia: McLean and Robinson's Method (3).

Transfer 25–50 g. of the moist soil to a 400 ml. beaker, add 100 ml. of a cold normal solution of sodium chloride, stir well and leave to stand for half an hour. Decant the supernatant liquid through an 18·5 cm. Whatman No. 44 filter paper and collect the filtrate in a litre Erlenmeyer flask. Wash the soil once by decantation, with the normal sodium chloride solution and then transfer it completely to the filter. Continue the leaching until the volume of the filtrate approximates to 500 ml.

Add 3–4 g. of magnesia to the filtrate, connect the flask to the ammonia distillation apparatus and heat gently so that a volume of 150–200 ml. distils over in half to three-quarters of an hour. Collect the distillate in 10–15 ml. of 0·02N hydrochloric acid and titrate the excess of acid with 0·02N sodium hydroxide, using methyl red as the indicator. Carry out a blank determination using all the reagents.

For soils with high ammoniacal nitrogen, it is necessary to collect a second half litre of leachate or leach with a more concentrated (15 per cent.) solution of sodium chloride.

McLean and Robinson prefer B.D.H. Universal Indicator for the titration and titrate to a distinct blue colour (pH 8·5). This introduces error due to carbon dioxide and the most suitable indicator is probably brom cresol green as used in Richardson's modification of Olsen's method (p. 208).

Some alkaline soils give a strongly coloured extract on leaching with sodium chloride, due to the dispersion of humified organic matter. With these soils better results are obtained if the leaching is carried out with sodium chloride solution containing 10 ml. of N hydrochloric acid per litre.

Calculation of the Results:

The amount of ammoniacal nitrogen in mg. per kg. (p.p.m.) corresponds to:

$$(B-T) \times \text{Normality factor} \times 14 \times \frac{1{,}000}{\text{Weight of soil taken}}$$

where B = volume of standard sodium hydroxide used in the blank determination

and T = volume of standard sodium hydroxide used in the actual determination.

The value should be expressed on an oven-dry soil basis.

REFERENCES

(1) Bartholomew, R. P.: *Soil Sci.*, 1928, *25*, 393–398.
(2) Mathews, D. V.: *J. Agric. Sci.*, 1920, *10*, 72–85.
(3) McLean, W. and Robinson, G. W.: *J. Agric. Sci.*, 1924, *14*, 548–554.
(4) Olsen, C.: *Compt. Rend. Trav. Lab. Carlsberg*, 1929, *17*, 1–20.
(5) Richardson, H. L.: *J. Agric. Sci.*, 1938, *28*, 73–121.
(6) Russell, E. J. and Page, H. J.: *Chemists' Year Book for 1929*, 794–808.
(7) Shrikhande, J. G.: *Current Sci.*, 1939, 7, 369–370.

CHAPTER XII

ORGANIC MATTER

Because of its complex nature numerous difficulties beset the accurate determination of soil organic matter. The value for loss on ignition is not admissible since it includes water of constitution of the clay fraction. W. O. Robinson's method (8), in which soil organic matter is oxidized by a concentrated solution of hydrogen peroxide, gives a direct determination but can only be used in soils which do not contain manganese dioxide nor more than 1 per cent. of calcium carbonate. Organic matter is far more generally calculated from a determination of organic carbon, on the assumption that the organic matter of the average soil contains 58 per cent. of carbon. This conversion factor is sometimes referred to as the van Bemmelen factor but it is, in fact, of much older origin. The proportion of carbon in soil organic matter varies widely in different soils and, at best, this factor should only be regarded as an arbitrary approximation. Since organic carbon can be determined directly, and with considerable accuracy, by any of the usual methods of dry combustion, it is preferable to report it as such, rather than a value for organic matter derived from it on the above assumption.

Some soils contain fragments of elementary carbon, such as coal, charcoal or graphite. Carbon in this form should not be considered as part of the soil organic matter proper, but, unfortunately, the dry combustion method does not distinguish between the different forms of carbon. Carbonates, if present, can be removed by treatment with sulphurous acid before combustion, or a correction can be applied for the amount of carbonate-carbon. Sulphurous acid is the only common acid suitable for preliminary removal of carbonates because its reducing properties minimize the oxidation of organic matter during the process. The equipment for the dry combustion method is relatively expensive but with an electrically-heated furnace determinations can be carried out in rapid succession. It is not necessary to cool the tube between samples. Except for soils rich in organic matter, 25–35 minutes are sufficient for each determination. As the result of a co-operative study the Organic Carbon Committee of the International Society of Soil Science (6) concluded that the values obtained by a variety of dry combustion methods were so concordant that the choice of method depended upon laboratory convenience. Details of the dry combustion method in use at the Waite Institute are described below.

In W. O. Robinson's hydrogen peroxide method elementary carbon is only slightly attacked. The method therefore gives a good measure of organic matter in those soils for which it is applicable.

Wet methods of oxidation, such as digestion of the soil with chromic and sulphuric acids, give, in general, low recoveries owing to their failure to secure full oxidation of the carbon. In only one of the methods tested by the Organic Carbon Committee of the International Society of Soil Science was the treatment sufficiently drastic to give nearly complete oxidation of the whole of the carbon to carbon dioxide.

In recent years various simple and rapid titration methods have come into prominence. Schollenberger (9) originally proposed treating the soil with known amounts of chromic acid in hot sulphuric acid and determining the amount of chromic acid reduced by the soil organic matter. The method was modified by Walkley and Black (12) and further simplified by Walkley (11). It is described below in its latest form. This method is not affected by the presence of calcium carbonate in the soil and it is also able to discriminate between elementary carbon and soil organic matter proper. Chlorides interfere since they are oxidized by chromic acid with liberation of free chlorine. However, a correction may be applied if their amount is known since they react quantitatively (4Cl ≡ 2O ≡ C). Thus 11·83 g. of chlorine are equal in reducing power to 1 g. of carbon. This correction can be made with sufficient accuracy by subtracting one-twelfth of the amount of chlorine present in the soil from the apparent carbon value.

These rapid methods are of particular value in investigations in which large numbers of related soils are being compared. Caution should, however, be exercised in making comparisons between unrelated soils since the organic matter is oxidized to a different extent in different soil types. Conversion factors may be determined empirically, so that values obtained by these methods may be compared with those obtained by the dry combustion method. Such conversion factors vary with the type of soil and the exact details of manipulation. It would probably be better to regard the results obtained by these rapid titration methods as "single value" constants rather than attempt to convert them to organic carbon values.

For subdividing soil organic matter extraction with hot sodium hydroxide is used. This separates the material into a non-humified and a humified portion, the latter being soluble in the hot alkali. The humic matter consists of the dark coloured, high molecular colloidal fraction of the soil organic matter. Arnold and Page (1) found that hot sodium hydroxide dissolves somewhat more than twice as much humified organic matter as is dissolved in the cold. Humin is insoluble in cold sodium hydroxide but becomes hydrated and converted to humic acid when treated with a hot solution of alkali. In Eden's method (3) humic acid is determined by decomposing carbonates and humates in the soil with dilute hydrochloric acid, extracting the residue with hot sodium hydroxide and comparing the colour with that of a standard solution of Merck's *acidum huminicum.* During the preliminary treatment with acid some loss of a lightly coloured water soluble constituent, fulvic acid, occurs but the amount is seldom sufficient to affect the subsequent colour comparison. Details of the method are given on p. 227.

The colour of the humic acid solution extracted by Eden's method is not always the same shade as that of the solution prepared from *acidum huminicum.* Instead of using the latter substance as a standard, Joseph and Whitfeild (5) use a purified humus extract prepared from a soil similar to those in which humic acid is to be determined. Standard solutions of *acidum huminicum* rapidly deteriorate in colour and should not be used if more than 10 days old. Cornell and Leibbrant (2) recommend that humic matter should be determined accurately in a standard soil sample by Eden's method and subsamples should then be extracted and used as the working standard in each series of subsequent determinations. This procedure is easier than the frequent preparation and standardization of fresh solutions of *acidum huminicum.*

In Eden's method humic acid is determined from its tinctorial power. Colourless compounds brought into solution do not enter into the colour comparison. In Arnold and Page's method the total carbon in the sodium hydroxide extract is determined. By comparing the tinctorial power of their extracts with that of *acidum huminicum* they conclude that the sodium hydroxide extracts probably contain a considerable proportion of colourless organic compounds.

Robinson and Jones (7) suggested oxidation with hydrogen peroxide for discriminating between humified and non-humified organic matter since they found that 6 per cent. hydrogen peroxide did not appreciably attack structural organic matter. W. O. Robinson (8) found, however, that much undecomposed organic matter is attacked by hydrogen peroxide in the presence of soil. Hosking (4) also noted that the action of hydrogen peroxide on organic matter is influenced by soil reaction and only a fraction of the oxidizable organic matter is attacked in alkaline soils, unless the soil is first acidified. Complete oxidation of the oxidizable portion of the organic matter only occurs at reactions more acid than about pH 5·8.

Waksman and Stevens (10) do not agree with the use of sodium hydroxide extractions or hydrogen peroxide oxidations for the fractionation of soil organic matter and they suggest a proximate method of analysis. They separate soil organic matter into four fractions, namely ether and alcohol soluble; carbohydrates; proteins; and lignin-humus complex. The original paper should be consulted for details of their method.

The Determination of Organic Carbon: Dry Combustion Method.

In this method the soil is ignited in a current of purified air and the carbon dioxide, produced by the combustion of the organic matter, is absorbed in a suitable absorbent and weighed. If carbonates are present in the soil they are first removed by treatment with sulphurous acid. For the combustion a silica tube packed with copper oxide is preferred. Provided that it is not accidentally overheated it has a very long life before it is seriously attacked by the copper oxide, with formation of copper silicate. If the copper gauze plugs are wrapped in thin asbestos millboard before insertion in the tube, attack at these points is prevented.

Apparatus:

An electrically-heated combustion furnace with the heating units arranged in three sections of 4 in., 8 in., and 12 in. respectively is most convenient. The gas purification and absorption trains are shown diagrammatically in Fig. 14. Air is purified from carbon dioxide and water vapour by the absorption vessels *A–C*, and, after passing through the heated combustion tube, it is drawn through the absorption vessels *E–L*. These bulbs purify and dry the current of air and then absorb the carbon dioxide, produced in the combustion. Chlorine compounds are removed by the heated silver gauze in the combustion tube. Should any escape this, they are absorbed by the zinc tube *F*, which also absorbs other acidic vapours. The last traces of sulphur trioxide and oxides of nitrogen (from the combustion of sulphur and nitrogen in the soil organic matter) are retained in the concentrated sulphuric acid in the Vanier bulb *H*. The slow stream of air is conveniently drawn through the whole apparatus by means of a filter pump operating in conjunction with the device for maintaining a constant gentle suction (p. 62).

The details of the apparatus are as follows:

A. Gas washing bottle containing 40 per cent. potassium hydroxide solution.

B. Gas washing bottle containing concentrated sulphuric acid.

C. Stoppered U tube filled with Ascarite.

D. Silica combustion tube 104 cm. long and 18 mm. bore. This tube can be either transparent silica or translucent, with a 20 cm. transparent window section, so that the boat can be observed.

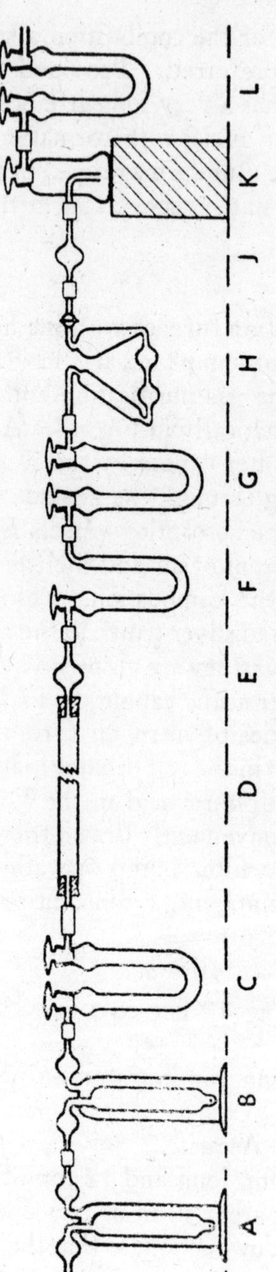

Fig. 14. Combustion train for the determination of organic carbon.

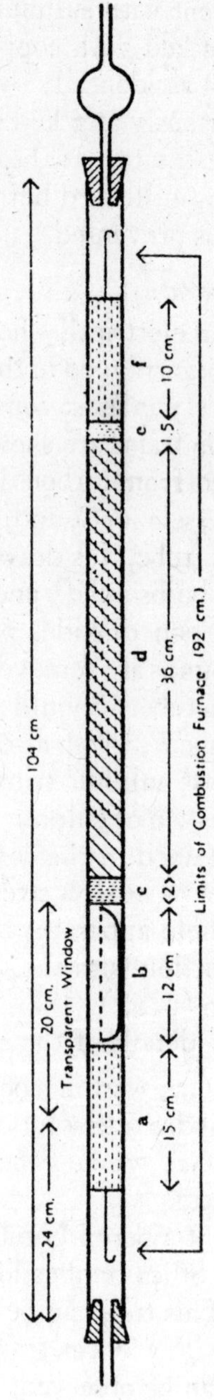

Fig. 15. Details of the filling used in the combustion tube.

E. Glass bulb to collect any condensed water vapour, thus increasing the life and efficiency of the calcium chloride tube *G.* Throughout the combustion keep this glass bulb cool by means of a piece of moistened filter paper. Once or twice each day, remove any water which collects in the bulb.

F. Stoppered U tube (Vanier Zinc Tube) filled with zinc wire or filings to remove acidic vapours.

G. Stoppered U tube filled with granular calcium chloride. Each time the calcium chloride is renewed in this tube, pass dry carbon dioxide through it for 20–30 minutes, to neutralize any basic calcium chloride. Then draw a stream of dry carbon dioxide-free air through the tube for a further 30 minutes, to remove all carbon dioxide.

H. Vanier Sulphuric Acid Bulb, filled with concentrated sulphuric acid. Renew the sulphuric acid in this bulb after every 10–15 determinations.

J. Glass bulb tube with a bulb of about 10–15 ml. capacity, to retain any fine sulphuric acid spray which might otherwise be carried over into the Ascarite bulb *K.*

K. Midvale Absorption Bulb, Stetser and Norton Modification, filled with Ascarite. To fill this bulb put a layer of about half an inch of glass wool in the bottom and then add Ascarite until the bulb is almost full. Tamp the Ascarite down lightly with a piece of wire, to secure an even passage of the carbon dioxide through the absorbent. Finally, cover the Ascarite with another small plug of glass wool. A drying tube is unnecessary with Ascarite as the absorbent acts as its own drying agent, having the same drying power as fresh concentrated sulphuric acid. After each refilling with Ascarite, put the new absorption bulb in position in the train between the Ascarite bulb in use *(K)* and *L*, for 1–2 hours, while other combustions are proceeding. Otherwise a freshly filled bulb is liable to give low results for the first time that it is used, on account of the loss of the water absorbed by the Ascarite during the filling of the bulb.

L. Stoppered U tube filled with granular calcium chloride, to act as a trap to prevent water vapour from the water pump reaching the Ascarite Bulb *K*.

Fill the silica combustion tube as follows:

Make the plug *c* (Fig. 15) by rolling a strip of copper gauze (40 mesh, 2 cm. wide) and wrapping asbestos paper around it until it just fits the bore of the tube. Place this plug in the tube immediately to the right of the transparent window section, in the position indicated in the diagram. The layer of asbestos paper prevents the copper fusing into the silica tube should this portion become accidentally overheated. Next pour a mixture of wire-form copper oxide and granular copper oxide, the latter screened to remove all material finer than 1 mm., into the tube and lightly pack it into position. This layer of copper oxide should occupy a length of 36 cm.; retain it in position by a second plug of copper gauze *e*, wrapped as before. After this plug is in position insert a loosely rolled piece of 60 mesh silver gauze, *f*, 10 cm. in length.

The combustion boat, *b*, should have a capacity of about 7 ml. if a preliminary sulphurous acid treatment is included to remove carbonates from the soil sample. Such a boat will be about 108 mm. long and can be either silica or porcelain. If no preliminary treatment with sulphurous acid is required, smaller boats can be used.

The distributor, *a*, is a copper gauze coil 15 cm. long, fitted with a projecting copper or nichrome wire for easy removal when changing the combustion boats.

After filling the combustion tube, heat it to a dull red heat for about two hours, drawing a slow current of air through it to oxidize the copper gauze coils to copper oxide and to decompose any organic matter (as in the asbestos paper).

Preparation of the Soil Sample:

For this determination use soil which has been ground to pass a sieve with round holes 0·5 mm. in diameter. If the calcium carbonate content of the soil is less than 0·1 per cent. it may be disregarded, since the amount of carbon dioxide derived from it will be equivalent to 0·01 per cent. organic carbon or less. Weigh 1–5 g. of soil, depending on the amount of organic matter present, into a combustion boat, which has been previously ignited and cooled. Transfer the boat to the combustion tube and proceed with the determination in the ordinary way, as described below.

If the calcium carbonate content of the soil is between 0·1 and 0·8 per cent., the combustion of the soil can be carried out directly, and a deduction made to correspond to the carbon dioxide derived from the calcium carbonate. Combustion should be carried out for $1\frac{1}{2}$ hours, instead of the usual period, so as to ensure complete decomposition of the calcium carbonate. In this case, transfer 1–3 g. of soil, depending on the amount of organic matter present, to the combustion boat, as before. Alternatively, treat the soil with 5 ml. of sulphurous acid, as described below for soils containing more than 0·8 per cent. of carbonate, to decompose the carbonate before the combustion. This is generally preferable.

For soils containing more than 0·8 per cent. of calcium carbonate, it is essential to decompose this carbonate by means of sulphurous acid before performing the combustion. To do this, weigh 1–3 g. of the finely ground soil, depending on the relative amounts of organic carbon and calcium carbonate present, into a combustion boat (preferably silica) and cautiously add 5 ml. of a strong solution of sulphur dioxide in water (8–12 per cent.). Allow the reaction to proceed for about two hours and then place the boat and its contents in an oven at 105° C. until dry. When cold add a further 5 ml. of sulphurous acid and leave for two hours before drying again. If any effervescence is noticed on the second treatment, make a third treatment with sulphurous acid, to secure complete decomposition of all carbonates. This is rarely necessary. After the final treatment carry out the combustion in the ordinary way.

Some soils cannot be treated in the boat with sulphurous acid, either on account of excessive calcium carbonate (25 per cent. or more) or because of excessive frothing. Treat 1–2 g. of such soils, in a small glazed basin, with an excess of sulphurous acid (10–20 ml.). Leave overnight, then dry at 105° C. When dry, carefully scrape the soil from the basin and transfer it completely to the combustion boat. Wash out the basin with three lots, each of about 2 ml., of sulphurous acid, adding the washings to the boat. Dry the boat and contents at 105° C., and carry out the combustion as usual. Alternatively, these highly calcareous soils may be treated very effectively as follows: Transfer 2–3 g. of soil to a small beaker, add 10 ml. of water and bubble sulphur dioxide through the suspension, by means of a capillary tube, until all the carbonate is destroyed. Then partly evaporate the contents of the beaker and transfer them to a combustion boat as before.

Combustion:

Place the boat, containing the soil sample, in the combustion tube in the position shown, and connect the Ascarite absorption bulb *K* (previously weighed against a tare). Then test the whole train to see that all the connexions are air-tight. If these are all correct, draw a slow stream of air through the apparatus and switch on the combustion furnace, but do not apply any heat to that portion of the tube containing the soil sample until the copper oxide portion, the silver gauze, and the copper gauze distributor, *a*, are heated just to a dull red heat. When this is so, heat the soil sample, gently at first and then more strongly, to burn off all the organic matter. When combustion is complete, as judged by the Ascarite absorption tube beginning to cool down to room temperature after the absorption of the carbon dioxide, allow the portion of the tube containing the soil to cool, keeping the remainder hot if other determinations are to be made. Continue the stream of air for a further five minutes and then remove the absorption bulb *K* for weighing. Connect a second weighed Ascarite bulb in place of the first and quickly remove the boat of burnt soil from the combustion tube. Replace it with another boat containing a fresh soil sample, and proceed immediately with the next determination. Stop the air stream while the absorption bulb and combustion boat are being changed.

Leave the absorption bulb in the balance case for 15–20 minutes before weighing, protecting the open side-tube by a piece of closed rubber tubing except during the actual weighing. The increase in weight corresponds to the carbon dioxide absorbed and the percentage of organic carbon in the soil is given by:

$$\frac{\text{Weight of carbon dioxide absorbed}}{\text{Weight of soil used}} \times \frac{12}{44} \times 100.$$

If calcium carbonate has not been removed, deduct any carbon dioxide derived from it from the amount weighed. When it is necessary to convert the values obtained for organic carbon to organic matter, use the conventional factor 1·72. However, as mentioned previously, it is preferable to report the value directly as "Organic Carbon."

Carry out blank determinations periodically, using pure sucrose as a standard. Theoretical results should be obtainable. For determinations on sucrose it is necessary to carry out the actual burning more slowly than is necessary for a soil.

Should the sulphuric acid in the Vanier bulb *G* become discoloured during a combustion determination, it indicates that volatile organic matter has passed over without being completely oxidized. This may be due to the copper oxide in the combustion tube not being sufficiently hot or the combustion being carried out at too great a rate. If this occurs repeat the determination, replacing the sulphuric acid with fresh acid.

The Determination of Organic Carbon: Walkley and Black's Rapid Titration Method.

In this method the soil is digested with chromic and sulphuric acids, making use of the heat of dilution of the sulphuric acid. The excess of chromic acid, not reduced by the organic matter of the soil is then determined by titration with standard ferrous sulphate. Nitrates interfere only if present in amounts in excess of one-twentieth of the carbon content. Carbonates, even when they constitute 50 per cent. of the soil, do not affect the results. Manganese dioxide may also exceed the carbon content by three or four times without introducing serious error. Interference due to significant amounts of chlorides can be overcome by the addition of an excess of silver sulphate to the sulphuric acid as described below or a suitable correction can be applied if the amount of chlorine is known (p. 214). Elementary carbon, such as charcoal or coal, is practically unattacked in this method so this source of error is eliminated.

Reagents:

N Potassium Dichromate. Dissolve 49·04 g. of reagent grade $K_2Cr_2O_7$ in water and dilute to 1 litre.

Sulphuric Acid. Not less than 96 per cent.

Phosphoric Acid. 85 per cent

Diphenylamine. Dissolve 0.5 g. diphenylamine in a mixture of 100 ml. conc. sulphuric acid and 20 ml. water.

N Ferrous Sulphate. Dissolve 278·0 g. of reagent grade $FeSO_4.7H_2O$ in water, add 15 ml. of concentrated sulphuric acid and dilute to 1 litre. Standardize by titrating against 10·5 ml. of N potassium dichromate, as described in the method given below. The ferrous sulphate solution is quite stable if kept under an atmosphere of hydrogen in the reservoir bottle of an automatic burette. A very convenient arrangement, using a burette with an automatic zero, is shown in Fig. 16.

Method:

For this determination the soil sample should be ground to pass a 0·5 mm. screen. Transfer a weighed quantity of soil, not exceeding 10 g. and containing about 10–25 mg. of organic carbon to a 500 ml. Erlenmeyer flask. Add 10 ml. of N potassium dichromate followed by 20 ml. of concentrated sulphuric acid. Where large numbers of analyses have to be carried out the dichromate is most conveniently added from a burette with an automatic zero and the sulphuric acid from an automatic pipette. Shake by hand for one minute and leave the flask to stand on a sheet of asbestos for about 30 minutes. Then add about 200 ml. of water, 10 ml. of phosphoric acid and 1 ml. of diphenylamine indicator solution. Titrate by adding ferrous sulphate from the automatic burette until the solution is purple or blue. Continue to add the ferrous sulphate in small lots of about 0·5 ml., until the colour flashes to green. This occurs with little or no warning. Then add 0·5 ml. of N potassium dichromate to restore an excess of dichromate and complete the titration by adding ferrous sulphate drop by drop until the last trace of blue colour disappears. If more than 8 ml. of the 10 ml. of potassium dichromate originally taken have been reduced during the digestion, repeat the determination using a smaller quantity of soil.

The end point can easily be recognized to within one drop of ferrous sulphate. The colour is not always purple on adding the indicator at the beginning of the titration, but the purple

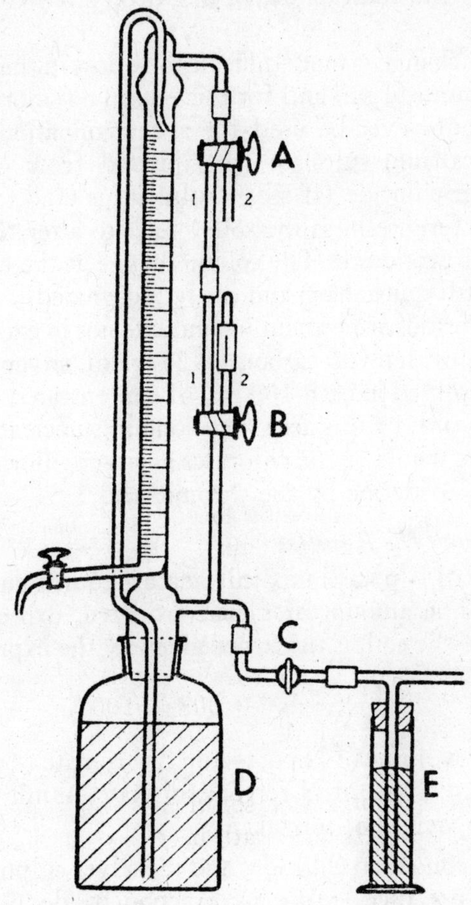

Fig. 16. An automatic burette, with self-adjusting zero, arranged for the preservation of ferrous sulphate solution in an atmosphere of hydrogen.

Suitable pressure from a cylinder of hydrogen is applied through stopcock *C*, the mercury safety valve at *E* preventing excessive pressure. To fill the burette adjust the stopcocks as follows: *A* open to the air (position 2), *B* closed, *C* open. When the burette is nearly full close stopcock *C*. When full, turn stopcock *B* to position *2*, allowing the hydrogen under pressure to escape through the Bunsen valve. The excess of liquid above the zero graduation then siphons back. For use, turn stopcocks *A* and *B* to position *1*, as shown.

or blue colour always appears just before the end point. The original blue colour frequently does not reappear on the addition of 0·5 ml. excess of potassium dichromate but it soon redevelops after the addition of the first drop or two of ferrous sulphate.

The colour change is more difficult to follow in the presence of larger amounts of soil and for this reason not more than 10 g. of soil should ever be used for a determination. Large amounts of calcium sulphate (precipitated from calcareous soils) or silver chloride (if silver sulphate is used to prevent chlorine interference in saline soils) tend to alter the shades of the colours produced. The colour change at the end point, however, is still quite sharp and easily recognized.

Where chlorides are present in amounts not in excess of the molecular equivalent of carbon, 1·25 g. of silver sulphate should be dissolved in each 100 ml. of concentrated sulphuric acid. Twenty ml. of this acid then contain sufficient silver to precipitate the whole of the chlorides as silver chloride and so prevent their oxidation by the chromic acid.

Calculation of the Results:

One ml. of N potassium dichromate is equivalent to 3 mg. of carbon. The amount of carbon oxidized, expressed as a percentage of the soil, is therefore given by the expression:

$$\frac{V_1 - V_2}{W} \times 0{\cdot}003 \times 100$$

where V_1 = volume of N potassium dichromate (10·5 ml.)
V_2 = volume of N ferrous sulphate, in ml.
and W = weight of soil taken.

If the soil contained chlorides and the silver sulphate modification was not used, make a correction by deducting one-twelfth of the percentage of chlorine present from the value calculated above, as explained on p. 214.

Walkley finds that the percentage recovery by this method varies from 60 per cent. for some subsoils to 90 or more per cent. for peat soils, taking values obtained by dry combustion as the standard of comparison. For the majority of agricultural surface soils the mean recovery lies between 75 and 80 per cent. As the recovery factor varies with soil type it is considered undesirable to use a general factor unless it has been correlated with the dry combustion values for the soils under examination. It is preferable to report the results obtained without the use of a recovery factor as "single value" determinations and designate them "Organic Carbon, Walkley and Black values."

The Determination of Humic Acid: Eden's Method.

The procedure used by A. B. Beck in these laboratories is as follows:

Reagents:

50 % Sodium Hydroxide. Dissolve 250 g. of sodium hydroxide in water and dilute to 500 ml.

0·1N Hydrochloric Acid. Dilute 17·5 ml. of concentrated hydrochloric acid to 2 litres. This solution does not require standardization.

0·2N Hydrochloric Acid. Dilute 35 ml. of concentrated hydrochloric acid to 2 litres. This solution does not require standardization.

Standard Acidum Huminicum Solution. Dissolve approximately 0·8 g. of Merck's *acidum huminicum* preparation in 25 ml. of 0·2N sodium hydroxide by warming for a few minutes. Then cool, dilute to about 80 ml., and filter through a collodion membrane filter. Dilute the filtrate to 200 ml. Standardize this solution in the following manner. Pipette 20 ml. into 10 ml. of 0·1N hydrochloric acid, leave to stand overnight and filter through a 9 cm. Whatman No. 41 filter paper, which has been previously dried at 105° C. and weighed. If the filtrate is not clear pass it through the filter a second time. Wash several times with small portions of 0·01N hydrochloric acid. Transfer the filter and precipitate to a weighing bottle, dry overnight at 105° C., cool in a desiccator and weigh. The increase in weight gives the amount of humic acid precipitated from 20 ml. of the standard solution. Make a correction for the small amount of humic acid which escapes precipitation and remains in the filtrate. To do this, make the filtrate alkaline by the addition of 2 ml. of 50 per cent. sodium hydroxide, dilute to 100 ml. and compare the colour with that of a standard prepared by diluting 2 ml. of the original standard solution to 100 ml. after the addition of 2 ml. of 50 per cent. sodium hydroxide. From this comparison calculate the volume of the standard solution equivalent to the colour of the unprecipitated humic acid and deduct this volume from the 20 ml. originally taken.

The colour of standard solutions of humic acid is only stable for about 10 days.

Method:

Transfer 5 g. of soil to a small beaker and add 50 ml. of 0·2N hydrochloric acid or sufficient to decompose all carbonates and humates. After one hour wash the suspension into the centrifuge tube; centrifuge and decant the supernatant liquid. Wash the residue thoroughly with two or three lots of water, centrifuging to clear the suspension each time. Then transfer the soil residue to a large test tube with a graduation mark corresponding to a volume of 100 ml. plus the volume occupied by 5 g. of soil (say 2 ml.). If more than 75 ml. of water are used to transfer the soil from the centrifuge tube to the extraction tube add a little 0·1N hydrochloric acid to flocculate the suspension and, when clear, siphon off the supernatant liquid.

Add 20 ml. of 50 per cent. sodium hydroxide to the soil in the extraction tube. Stir thoroughly to break up all lumps and dilute with water to the graduation mark. Then immerse the tube in a vigorously boiling water bath for 15 minutes, keeping the contents of the tube below the surface of the water in the bath. Stir continuously during the extraction.

After 15 minutes cool, transfer the bulk of the suspension, without dilution, to a dry centrifuge tube and centrifuge until clear (about 10–15 minutes). Transfer some of the clear supernatant liquid to a colorimeter and compare the colour with that of a suitable standard. As a standard use a solution prepared by diluting 2–10 ml. of the freshly standardized *acidum huminicum* solution to 100 ml. with 1 per cent. sodium hydroxide or use an extract prepared from a soil, the humic acid content of which has been accurately standardized.

The percentage of humic acid in the soil is given by the following expression:

$$\frac{D_s}{D_u} \times H \times \frac{100}{5}$$

where D_s = depth of the standard colour solution
D_u = depth of the unknown colour solution
and H = amount of humic acid in 100 ml. of the standard colour solution.

REFERENCES

(1) Arnold, C. W. B. and Page, H. J.: *J. Agric. Sci.*, 1930, *20*, 460–477.

(2) Cornell, H. H. and Leibbrant, F. E. A.: *Imp. Bur. Soil Sci.*, 1933, Monthly Letter No. 25.

(3) Eden, T.: *J. Agric. Sci.*, 1924, *14*, 469–472.

(4) Hosking, J. S.: *J. Agric. Sci.*, 1932, *22*, 91–100.

(5) Joseph, A. F. and Whitfeild, B. W.: *J. Agric. Sci.*, 1927, *17*, 1–11.

(6) Organic Carbon Committee (International Society of Soil Science): *Trans. 3rd Int. Congr. Soil Sci.*, 1935, *1*, 114–127; and *3*, 82–83.

(7) Robinson, G. W. and Jones, J. O.: *J. Agric. Sci.*, 1925, *15*, 26–29.

(8) Robinson, W. O.: *J. Agric. Res.*, 1927, *34*, 339–356.

(9) Schollenberger, C. J.: *Soil Sci.*, 1927, *24*, 65–68.

(10) Waksman, S. A. and Stevens, K. R.: *Soil Sci.*, 1930, *30*, 97–116.

(11) Walkley, A.: *J. Agric. Sci.*, 1935, *25*, 598–609.

(12) Walkley, A. and Black, I. A.: *Soil Sci.*, 1934, *37*, 29–38.

CHAPTER XIII

FREE FERRIC OXIDE

The amount of free iron oxides in the soil is often required, either for ascertaining its distribution in the profile or for studying its contribution to soil colour. Its removal from the soil is beset by many difficulties, for acids sufficiently strong to dissolve ferric oxide also attack and bring about considerable decomposition of silicate minerals. At the present time the best of the available methods yield only approximate, although useful, values.*

In Tamm's method (3) the soil is extracted twice with a solution containing oxalic acid and ammonium oxalate and the iron brought into solution is determined. The extracting solution has a reaction of about pH 3·2. Owing to the attack on the clay minerals by a solution of this reaction the method is only suitable for light sandy soils. With heavy clays it gives unreliable values and it cannot be used for soils containing calcium carbonate.

In Drosdoff and Truog's method (2) hydrogen sulphide is introduced into the system. This rapidly transforms the free iron oxides into the black sulphides, the reaction being the same as that which is used commercially for the removal of hydrogen sulphide from coal gas. The iron sulphides are readily soluble in cold dilute hydrochloric acid. No attack of iron silicates occurs during the treatment with hydrogen sul-

* Since this was written, Allison and Scarseth have published a biological method for the removal of free iron oxides from soils (ALLISON, L. E. and SCARSETH, G. D.: *J. Amer. Soc. Agron.*, 1942, *34*, 616–623). The principle of microbial reduction is employed; a readily available source of energy, such as sucrose, is added to the soil suspension and the latter is incubated under anaerobic conditions. The free iron oxide in the soil is rapidly reduced to the ferrous state and passes into solution. When reduction is complete the supernatant liquid is siphoned off and the residue is leached with a dilute solution of sodium chloride, adjusted to pH 3·0; total iron is then determined in the combined extracts. This method appears to be particularly effective in securing the removal of free iron oxides from the soil. In contrast to the chemical methods, the biological method produces very little attack on alumino-silicate minerals; only small traces of alumina and silica are brought into solution. It should, therefore, prove of considerable value and warrants further investigation on a wider range of soils.

phide but some decomposition takes place during the subsequent extraction with dilute hydrochloric acid.

While Drosdoff and Truog's method is a great advance on previous methods, experience has shown that the whole of the free ferric oxide is not always dissolved in a single treatment. For many soils Beck (1) found it necessary to increase the time during which the hydrogen sulphide is allowed to act on the soil from 30 minutes to 6 hours and to make two extractions in order to remove the colour due to ferric iron, although it was recognized that the extended treatment with hydrogen sulphide and the double extraction led to greater attack and decomposition of complex silicates containing iron. Drosdoff and Truog's method, as used by Beck (1) is given on p. 233.

Truog and his colleagues (4) have recently introduced a further method for the removal and determination of free ferric oxide. In this later method the treatment is somewhat more drastic since the soil is first boiled with a solution of sodium sulphide (pH 12 or thereabouts) and then subjected to the action of nascent hydrogen sulphide, liberated by the addition of sufficient oxalic acid to reduce the pH of the suspension to between 6 and 7. The iron sulphides are finally dissolved at an acidity not greater than pH 3·5. Hydrogen sulphide apparently first converts ferric oxide to a yellow iron sulphide (Fe_2S_3) which breaks down rather easily to the black sulphides, FeS and FeS_2. Iron disulphide is much less soluble than ferrous sulphide in dilute acids, particularly on ageing at a boiling temperature.

Truog's sodium sulphide method is more effective than the earlier method in removing free ferric oxide but it does not eliminate attack on the clay minerals. A single treatment is not always sufficient to remove the whole of the free ferric oxide. The reaction with hydrogen sulphide depends on the state of division and the nature of the iron oxides. If the surface area is not great the attack will be slow and incomplete. Hæmatite, the non-hydrated form of ferric oxide, is much more resistant, even in the colloidal form, than the hydrated forms such as göthite ($Fe_2O_3.H_2O$) and limonite ($Fe_2O_3.3H_2O$). However, hæmatite, when present in soils, appears to be brought into solution satisfactorily by this method.

Truog uses a centrifuge, fitted with special tubes, for separating the soil and solution during the treatment. In these laboratories Hosking has found that this separation can be more conveniently made by filtering, with gentle suction, when many analyses have to be carried out. Details of his procedure are given on p. 234.

In Truog's method all reagents should be added rapidly, except when reducing the reaction of the soil suspension from pH 7 to pH 6. It is within this range that nascent hydrogen sulphide is produced and, if passage through this range is too rapid, the hydrogen sulphide is liberated too quickly for effective action. If, on the other hand, too much time is consumed in this and the later stages, and the temperature is kept at the boiling point, some of the ferric disulphide may become sufficiently aged to resist solution in the acid. For these reasons Truog stresses the necessity for following strictly the details of the procedure. In order to subject the soil to several treatments with nascent hydrogen sulphide, and so secure maximum attack on the ferric oxides, the system is carried through the pH range of 7 to 6 several times. However, passage more than four times through this pH range does not result in any appreciable increase in the solution of ferric oxide. For soils high in iron it is better to make a second extraction on the residue, after removing as much iron as possible in the first extraction. The action between ferric oxide and sodium sulphide is more effective in the presence of oxalic acid than any of the other common acids. However, because of the difficulty in destroying excessive amounts of oxalic acid in the subsequent procedure, hydrochloric acid is used, in part, for all soils except those very high in iron oxides. The latter soils require more drastic treatment.

If the residue after the extraction of free ferric oxide is required for colour determination or mineralogical studies it can be freed from the sulphur, precipitated during the determination, by washing twice with 96 per cent. alcohol to remove water, four or five times with a mixture of one part of carbon bisulphide and two parts of absolute alcohol, and finally several times with absolute alcohol, to remove the carbon bisulphide.

The Determination of Free Ferric Oxide: Drosdoff and Truog's Method.

A. B. Beck's procedure is as follows: Transfer 5 g. of soil to a large centrifuge tube and add sufficient 0·05N hydrochloric acid to decompose all carbonates and humates. If the soil contains much carbonate use 0·1N acid. After a short time decant the acid and wash the residue two or three times with water, centrifuging each time to clear the suspension. Transfer the residue to a 250 ml. beaker using about 60 ml. of water. Add 15 ml. of 30 per cent. hydrogen peroxide, or sufficient to give a concentration of 6 per cent. in the suspension. Place the beaker on a boiling water bath for one hour, or until the dark colour due to organic matter has disappeared. Add more hydrogen peroxide if necessary to complete the destruction of the organic matter. Again separate the residue by centrifuging, washing once with water. If necessary add a little potassium nitrate (about 1–2 g. per 100 ml.) to the water used for washing, to assist coagulation. Rinse the residue in the centrifuge tube back into the beaker and digest for two hours, on a water bath, with 150 ml. of 2 per cent. sodium carbonate to remove colloidal silica. Then cool, centrifuge the suspension and wash several times with the sodium carbonate solution. This leaves the soil residue in the sodium-saturated condition and facilitates its dispersion.

After the removal of the soluble silica, disperse the residue by rubbing with a rubber pestle in a small mortar, using 2 ml. of N ammonia and a little water. Transfer the suspension to a 500 ml. Erlenmeyer flask, dilute to about 200 ml. and shake overnight in an end-over-end shaking machine. Then pass in a slow stream of hydrogen sulphide until the solution is saturated (about 20 minutes), stopper and shake again for six hours. After shaking, warm the suspension to about 50° C. and slowly add 0·1N hydrochloric acid until most of the black colour, due to iron sulphides, disappears. Then add a further 50 ml., to ensure an excess of acid, and leave for 10 minutes. Cool and transfer to a centrifuge tube and collect the supernatant liquid after centrifuging. Wash four times with about 50 ml. portions of 0·05N hydrochloric acid, decanting each time after centrifuging. Combine the solution and washings, boil to remove hydrogen sulphide, add 5 ml. of concentrated nitric acid to oxidize the ferrous salts to the ferric state, and evaporate to dryness. Continue heating on the water bath for 30 minutes after the residue becomes dry, to dehydrate the silica.

Take up the residue in 5 ml. of concentrated hydrochloric acid and 30–40 ml. of water. Filter through a 9 cm. Whatman No. 44 filter paper and wash several times with hot 5 per cent. hydrochloric acid. Collect the filtrate in a 250 ml. beaker. When quite cold precipitate iron in this solution by means of cupferron (see p. 248).

For many soils it is desirable to carry out a second extraction to secure more complete removal of the free ferric oxide. When this second extraction is required, disperse the residue from the first extraction by adding 4 ml. of N ammonia and rubbing with a rubber pestle. Transfer the suspension to a 500 ml. Erlenmeyer flask, dilute to about 200 ml., saturate with hydrogen sulphide as before and continue exactly as in the first extraction. Determine the amounts of iron brought out in each extraction separately.

The Determination of Free Ferric Oxide: Truog's Sodium Sulphide Method.

J. S. Hosking's procedure is as follows:

Reagents:

Sodium Sulphide Solution. Dissolve 25 g. of sodium sulphide ($Na_2S.9H_2O$) in 125 ml. of water.

Oxalic Acid. Dissolve 10 g. of oxalic acid in 100 ml. of water. This solution is approximately saturated.

0·001N Oxalic Acid. Dissolve 0·063 g. of oxalic acid in 1 litre of water.

2N Ammonium Hydroxide. Dilute 150 ml. of concentrated ammonia (S.G. 0·91) to 1 litre.

2N Hydrochloric Acid. Dilute 175 ml. of concentrated hydrochloric acid to 1 litre.

Acidified Sodium Chloride Solution. Dissolve 5 g. of sodium chloride in 1 litre of water and add 1 ml. of 2N hydrochloric acid.

5 % Sodium Chloride Solution. Dissolve 50 g. of sodium chloride in 1 litre of water.

Indicator Solutions. Brom cresol purple, cresol red or phenol red, brom thymol blue, methyl red, brom phenol blue and thymol blue are required. For these solutions see p. 127.

Method:

If the soil is rich in ferric oxide carry out the determination on a 2 g. sample but use 5 g. for light sandy soils or those low in ferric oxide.

Transfer 2 g. or 5 g. of soil to a 100 ml. beaker, add 10 ml. of a 30 per cent. solution of hydrogen peroxide and sufficient 2N hydrochloric acid, drop by drop, to reduce the reaction of the suspension to about pH 5, using brom cresol purple as an external indicator on a spot plate. Cover the beaker with a clock glass and allow to stand overnight. Then place the beaker in a boiling water bath for 20 minutes, stirring occasionally. After this time remove the beaker, add an additional 10 ml. of hydrogen peroxide and heat as before for a further 20 minutes. If organic matter still persists repeat the treatment once more.

When the organic matter has been destroyed, rinse the cover glass and sides of the beaker, dilute to about 25 ml. and boil for 5 minutes. Remove and rinse the cover glass, place the beaker on the water bath and evaporate the contents to a thin paste but do not let it dry out. Dilute with 20 ml. of water and again evaporate to a paste to remove hydrogen peroxide.

Transfer the paste to a 400 ml. beaker, diluting with water and cleaning the sides and bottom of the small beaker with a rubber-tipped stirring rod. Dilute the suspension to about 200 ml., add 2 ml. of sodium sulphide solution, cover the beaker and boil for 5 minutes. Allow to cool slightly and, while stirring, add 2 g. of ammonium chloride. Heat to 80–90° C. and maintain at about this temperature during the whole of the following treatment. Do not boil. Throughout the next stages all operations must be carried out expeditiously except in passing from pH 7 to pH 6, the stage in which nascent hydrogen sulphide is liberated. The liberation of hydrogen sulphide should extend over a minute or two to secure efficient reaction between it and the ferric oxides. All reagents must be added with vigorous stirring and prolonged standing must be avoided. Owing to the colour and turbidity of the suspension all the indicators required must be used as external indicators. To avoid undue delays it is desirable to prepare a spot plate with several drops of each of the indicator solutions required, before commencing the subsequent operations. After completing operations rinse the spot plate back into the suspension to avoid loss which would otherwise occur.

Rapidly reduce the reaction of the suspension to pH 7 by the addition of oxalic acid, either as finely divided crystals or as a saturated solution, drop by drop. Use cresol red or phenol red as an external indicator on a spot plate. Then more slowly, during the course of a minute or two, acidify the solution to pH 6 by the addition of oxalic acid solution, drop by drop (brom thymol blue external indicator). Stir well, add 4 ml. of sodium sulphide solution and again acidify rapidly to pH 7 and more slowly to pH 6 as before. After this second liberation of nascent hydrogen sulphide, rapidly acidify to pH 3·5 by the addition of oxalic acid (crystals or solution), using methyl red or brom phenol blue as the indicator. Stir and allow to stand for five minutes or until the black sulphides have dissolved. If the suspended material assumes a light grey colour, free from red or brown shades, a single further treatment will be sufficient. If any red or brown colour is apparent it denotes undissolved iron oxides and a more prolonged treatment is necessary.

If a single treatment only is required, bring the reaction of the suspension back to pH 7 (brom thymol blue) by the rapid addition of 2N ammonia, drop by drop, stirring vigorously. Using 2N hydrochloric acid, added drop by drop, once more gradually acidify to pH 6 and then rapidly to pH 3·5. Leave for 5 minutes. If the colour of the suspended material is now greyish white the treatment is complete.

If the suspended material still retains a noticeable red or brown colour after the first reduction to pH 3·5 bring the solution back to pH 7 by the addition of 2N ammonia, add 2 ml. of sodium sulphide solution and, using oxalic acid, acidify through the three stages, as before, to pH 7, 6, and 3·5. Add a further 2 ml. of sodium sulphide and repeat the whole process. If any reddish or brownish colour still persists after

5 minutes at pH 3·5 treat once more, using hydrochloric acid for the final reduction to pH 3·5. If colour still persists it is preferable to filter at this stage and repeat the sulphide treatment on the residue.

After final acidification to pH 3·5 wash down the sides of the beaker and cover glass and place the beaker on a water bath. Stir frequently until coagulation takes place. Any black particles or stainings, due to undissolved iron disulphide, may be disregarded as these will go into solution during the subsequent peroxide treatment. Filter through a small Buchner funnel fitted with a 4·5 cm. Whatman No. 50 filter paper, using gentle suction. Return the first part of the filtrate to the funnel if it is noticeably turbid. Wash with a 0·5 per cent. solution of sodium chloride acidified with 1 ml. of 2N hydrochloric acid per litre. Wash three times in all, using about 20 ml. of solution each time. Return the sample to the beaker, washing the funnel and filter paper with the smallest possible amount of 0·001N oxalic acid. Add 3–5 ml. of 30 per cent. hydrogen peroxide and leave for 10 minutes or until any vigorous reaction subsides. With vigorous stirring bring the solution to pH 2 by the addition of a saturated solution of oxalic acid (thymol blue external indicator). If after 5 minutes standing solution of iron disulphide is not complete warm for a time to 60–70° C. but do not boil. If necessary add a few ml. of 5 per cent. sodium chloride to flocculate the suspension.

Filter through the same funnel and filter paper previously used, collecting the filtrate in the flask containing the first filtrate. Wash completely with 10–20 ml. portions of the acidified 0·5 per cent. sodium chloride solution, draining completely between each addition.

Transfer the filtrate to a 250 ml. evaporating basin, adding more as evaporation proceeds until the whole of it has been transferred. Evaporate until the volume is reduced to about 50 ml. Remove from the water bath, cover with a clock glass and add 10 ml. of nitric acid, to destroy the oxalic acid and oxidize any sulphur precipitated during the evaporation. Digest for about 15 minutes on the water bath, remove and rinse the cover glass, and evaporate to dryness. Remove from the bath, add a little water and 5 ml. of nitric acid and again evaporate to dryness. Take up in 10 ml. of water and 5 ml. of hydrochloric acid and evaporate to dryness to convert the nitrates to chlorides. Continue heating for one hour after the hydrochloric acid has been expelled, to dehydrate the silica completely. Take up in dilute hydrochloric acid, filter off the silica and determine iron in an aliquot of the filtrate, exactly as described on pages 246 to 249.

REFERENCES

(1) Beck, A. B.: *J. Coun. Sci. Ind. Res. Australia*, 1939, *12*, 128–136.

(2) Drosdoff, M. and Truog, E.: *J. Amer. Soc. Agron.*, 1935, *27*, 312–317.

(3) Tamm, O.: *Medd. Skogsförsoksanst.*, 1922, *19*, 387–404.

(4) Truog, E., Taylor, J. R., Pearson, R. W., Weeks, M. E. and Simonson, R. W.: *Soil Sci. Soc. Amer. Proc.*, 1937, *1*, 101–112.

CHAPTER XIV

THE SEPARATION AND ANALYSIS OF THE CLAY FRACTION

For characterizing the clay fraction the molecular ratios $\frac{SiO_2}{Al_2O_3}$ (Silica: Alumina Ratio) and $\frac{SiO_2}{Al_2O_3 + Fe_2O_3}$ (Silica: Sesquioxide Ratio) are most useful and little is gained by making a more complete analysis. Since ratios only are required, the presence of organic matter or an excess of flocculating agent in the clay separate is without effect on the values.

Methods for the separation and analysis of the clay fraction were studied by the Soils Sub-Committee of the Agricultural Education Association (1). It concluded that the technique of separation of the clay fraction was of greater importance than the methods used for its analysis, since the variations in results obtained in different laboratories could be largely traced to differences in this separation. The customary methods of silicate analysis were considered satisfactory for the analysis of the clay separate.

Particle size affects the composition of the clay. In general, clays become less siliceous as the particle size decreases. Although the above mentioned Sub-Committee did not make any recommendation regarding an upper limit of particle size for the fraction to be separated for silicate analysis, it is generally considered that the International Clay Fraction covers too wide a range and includes some unweathered silicates. It is now widely accepted that clay separated according to the old British standard, with a maximum particle size corresponding to a settling velocity of 8·6 cm. in 24 hours (0·0001 cm. per sec.), is more suitable. To secure uniformity and enable the direct comparison of results this standard fraction should be separated for all clay analyses.

Details of two convenient methods are given below. In the

first method dispersion is brought about by pre-treatment of the soil with sodium chloride and this method is recommended for most soils. The use of hydrogen peroxide may be necessary for some organic soils but its use inevitably leads to some attack of the silicate material. For strongly acid soils Muir has suggested the use of neutral sodium acetate, instead of sodium chloride, so as to reduce the loss of sesquioxides by leaching, particularly after the use of hydrogen peroxide. Certain ferruginous and lateritic soils are not satisfactorily dispersed by the sodium chloride method and for these Puri's method of dispersion (2), in which the soil is first boiled with ammonium carbonate solution, is recommended. In the method recommended for the analysis of the clay fraction silica is separated as usual by evaporation with hydrochloric acid, iron and aluminium are precipitated and ignited as oxides and iron is determined separately after precipitation by cupferron.

In these laboratories Dixon and Walkley *(priv. comm.)* have developed a micro-method in which the determination of silica and iron can be carried out on 0·1 g. of clay. Such a method enables the direct determination of SiO_2 and Fe_2O_3 on samples obtained by pipetting from a suspension, as in the standard method of mechanical analysis. If a micro-method for the determination of aluminium on the same sample could be developed the ease of obtaining the clay sample and the simplicity of the procedure would enable large numbers of analyses to be carried out rapidly. Silica is determined by obtaining the loss of weight on digesting a clay with hydrofluoric and sulphuric acids, removing all sulphates remaining at the end of the operation by fusion with sodium hexametaphosphate. The method yields results about one per cent. higher than usual due to traces of sulphates originally present in the clay. However this difference is not important as it is only of the same order as the mean error found in the determination of silica in different laboratories during the co-operative work of the Agricultural Education Association already mentioned. After the determination of silica, iron is determined polarographically in an aliquot.

The Separation of the Clay Fraction: Sodium Chloride Method.

The clay is separated directly from soils with moderate amounts of clay (e.g. 20 per cent. or more) but for lighter soils it is preferable to disperse the soil more fully at the start and concentrate the finer portions by eliminating most of the coarse and fine sand fractions before commencing the clay decantation.

Reagents:

5N Sodium Chloride. Dissolve 585 g. of sodium chloride in water and dilute to 2 litres.

0·5N Sodium Carbonate. Dissolve 26·5 g. of anhydrous sodium carbonate in water and dilute to 1 litre.

0·5N Acetic Acid. Dilute 28 ml. of glacial acetic acid to 1 litre.

5N Calcium Chloride. Dissolve 1,085 g. of $CaCl_2.6H_2O$ or 445 g. of dry $CaCl_2$ (75 % $CaCl_2$) in water and dilute to 2 litres.

Dilute Calcium Chloride. Add 5 ml. of 5N calcium chloride to 1 litre of water.

Method:

(a) *For soils with more than 20 per cent. of clay.*

Weigh out sufficient soil to give approximately 5 g. of clay and transfer to a 600 ml. beaker. Add 200 ml. of water and boil for 5–10 minutes. While still hot add 75 ml. of 5N sodium chloride, stir well and leave to stand overnight. Decant the clear supernatant liquid and discard it. Transfer the residue, through a 90 mesh sieve, to a 500 ml. tall shaped beaker, graduated at a height of 8·6 cm. above the bottom. Rub gently and wash all the clay through the sieve. Reject the sandy residue remaining on the sieve. Add 35 ml. of 5N sodium chloride to the contents of the beaker, stir well and place aside to stand. Decant twice at intervals of 24 hours, or after shorter periods of standing, if the supernatant liquid is clear. Refill the beaker each time with water.

Then add 25 ml of 0·5N sodium carbonate solution, fill to the 8·6 cm. graduation mark and allow to stand in a place of reasonably constant temperature and out of direct sunlight.

After 24 hours decant, again discarding the supernatant liquid if it is still clear. If, however, it is turbid collect it in a 3 litre bottle. Refill the beaker with water to the 8·6 cm. mark, stir well and continue the decantations at intervals of 24 hours, collecting the clay suspension in the 3 litre bottle. As the excess of sodium chloride is removed the clay readily deflocculates. It is preferable to let the first two lots of the heavy clay suspensions stand for 48 hours before decantation since the concentrated suspensions obtained at this stage may hold up a proportion of the coarser particles. Lengthening the period of sedimentations avoids this source of error. Moreover, no attempt should be made to decant these early suspensions completely for, owing to their turbidity, it is impossible to see that the sediment at the bottom of the beaker is not disturbed during the decantation. Usually three-quarters of the suspension may be poured off with safety. Incomplete pouring off increases the number of decantations necessary to separate the clay fraction but ensures better separation from coarser particles.

After the first four or five decantations add 2–3 ml. of 0·5N sodium carbonate solution each time the beaker is filled, to assist dispersion. During the course of the decantations the soil should be pestled two or three times with a rubber pestle (p. 65) to hasten the complete dispersion of the clay. Make the last two or three decantations with water only. Continue the decantations until, after rubbing the residue in the beaker with the rubber pestle, only a small amount of material remains in suspension for 24 hours.

When all the clay suspension has been collected in the 3 litre bottle add 25 ml. of 0·5N acetic acid and 25 ml. of 5N calcium chloride to flocculate it. After one to two days carefully siphon off the clear liquid and transfer the residue to a 9 cm. Buchner funnel, fitted with a No. 50 filter paper. Return the first runnings to the funnel as the filtrate is always slightly turbid at the start. Wash the bottle used for collecting the clay separate with one lot of 75 ml. of dilute calcium chloride and transfer to the funnel. Then wash the clay in the funnel with one lot of 50 ml. of 50 per cent. alcohol and finally with 25 ml. of absolute alcohol and allow to dry. When air-dry, remove the clay from the filter paper, dry in an oven at 105° C. and weigh.

Grind in an agate mortar and preserve in a small stoppered bottle.

(b) *For soils with less than 20 per cent. of clay.*

Transfer a quantity of soil sufficient to yield 5–6 g. of clay, to a cylinder, add 500 ml. of water, 75 ml. of 5N sodium chloride and 25 ml. of 0·5N sodium carbonate. Shake the cylinder and contents for two hours and place aside to stand. When the supernatant liquid is clear siphon it off as completely as possible. Add 300 ml. of water to the residue in the cylinder, shake again for 10 minutes, then transfer through a 70–90 mesh sieve into a tall 800 ml. beaker, rubbing the residue thoroughly to ensure mechanical dispersion. Discard the sandy residue. When all the fine material has been washed through the sieve stir the suspension vigorously and leave standing for four minutes, so that the greater part of the fine sand will settle in the beaker. Decant the suspension into a 3 litre beaker. Pestle the residue with a rubber pestle, refill the beaker with water to a depth of 10–12 cm. and again decant after 4 minutes. Repeat the pestling and decantation once or twice more so as to separate the whole of the clay from the sandy residue.

Add 75 ml. of 5N sodium chloride to the suspension in the 3 litre beaker. Stir well and leave to stand until flocculated. Siphon off the supernatant liquid, discarding it if it is clear and free from clay, but collecting it in a 3 litre bottle if turbid. In the latter event flocculate it by adding 25 ml. of 5N calcium chloride and 15 ml. of 0·5N acetic acid, after siphoning. Transfer the residue from the large beaker to a 500 ml. tall shaped beaker, graduated at a height of 8·6 cm. above the bottom, and proceed with the clay decantations exactly as described above for soils with more than 20 per cent. of clay.

The Separation of the Clay Fraction: Puri's Ammonium Carbonate Method.

This procedure is recommended for ferruginous and lateritic soils.

Reagents:

N Ammonium Carbonate. Dissolve 78–80 g. of ammonium carbonate in water and dilute to 1 litre.

0·5N Sodium Hydroxide. Dissolve 20 g. of sodium hydroxide in water and dilute to 1 litre.

Method:

Weigh out sufficient soil to give approximately 5 g. of clay and transfer to a 500 ml. tall beaker. Add 250 ml. of N ammonium carbonate and boil gently until the volume is reduced to half, covering the beaker for the first few minutes if necessary. If much frothing occurs add two or three drops of secondary octyl alcohol.

When the volume has been reduced to half, add 25 ml. of 0·5N sodium hydroxide (or lithium hydroxide), dilute to about 250 ml. with hot water and boil again until the volume is once more reduced to half. When cool pour the suspension through a 70–90 mesh sieve and collect it in a 500 ml. tall shaped beaker graduated at the 8·6 cm. level as before. Proceed with the decantation of the clay as in the sodium chloride method given above.

The Determination of Silica, Iron and Alumina in the Clay Separate.

Transfer about 1·2 g. of the finely ground clay separate to a weighed palau basin (with cover) of about 50 ml. capacity and place the basin and contents in a muffle furnace. Raise the temperature to about 900° C. and ignite for 30 minutes at this temperature. Cool and weigh as ignited clay.

Fusion with sodium carbonate:

Add 7–8 g. of sodium carbonate to the ignited clay and mix gently with a platinum or horn spatula, wiping the spatula on a small portion of the fusion mixture which is held in reserve for this purpose. This mixing does not require to be very thorough. Cover the basin and ignite over a Meker burner, gradually raising the temperature until the melt becomes quiescent and there is no further ebullition of carbon dioxide. This generally requires thirty to forty minutes, but this time may be reduced if the contents of the basin are mixed a little by swilling, after fusion has occurred. When decomposition is complete, remove the basin from the burner and place it aside to cool. Heat the lid in order to melt any sodium carbonate that has splashed on to it and so ensure decomposition of any adhering clay particles. A palau basin is very much better than platinum for this fusion on account of the ease with which the fused cake separates from it when cold.

Place the palau cover in a 250 ml. porcelain basin and add sufficient warm water to dissolve any of the melt that may be adhering to it. When this has dissolved, remove and rinse the cover, wiping it with a small piece of No. 44 filter paper to remove the last traces of silica. Pour the wash water from the porcelain basin on to the fused cake in the palau basin and allow it to stand for about half an hour. Then return the water and the disintegrated cake to the porcelain basin and wash the palau basin as completely as possible, using another small piece of filter paper to assist in the removal of the last traces of the fused cake. Cover the porcelain basin with a clock glass and, by means of a funnel drawn out to a fine tip and bent so as to pass between the lip of the basin and the clock glass, add about 25 ml. of concentrated hydrochloric acid. This acid must be added in small portions at a time, so that loss does not occur through the effervescence becoming too violent. Also add about 5 ml. of the acid to the palau basin, to dissolve the last traces of the fused material in it. So little of the fused residue is left in the palau basin after it has been carefully washed that there is practically no attack of the basin, such as would otherwise occur if appreciable quantities of manganates were left in it.

When the effervescence has ceased, rinse the tip of the funnel and remove from the basin. Rinse the clock glass into the basin and add the portion of acid used for rinsing the palau basin. Finally rinse this latter basin, using another small piece of filter paper to assist in cleaning it. If the fusion has been properly conducted the solution in the porcelain basin will now be clear, except for gelatinous pieces of silica. There will be no sign of any unattacked particles of clay.

Silica:

Place the porcelain basin on a water bath and evaporate the contents to dryness, stirring frequently and breaking up the crust of salts that separate in order to hasten the process.

To ensure dehydration of the silica continue the evaporation for one hour after the last of the hydrochloric acid has been expelled, as is indicated by the salts becoming pale in colour. When thoroughly dry, remove the basin from the water bath and moisten the salts with 6–8 ml. of concentrated hydrochloric acid. Stir gently with the rod, breaking up all lumps, and add about 30 ml. of water. Replace on the water bath for 2–3 minutes to dissolve the salts present.

Filter through a 9 cm. Whatman No. 44 filter paper, collecting the filtrate in a small beaker. Transfer as much as possible of the silica to the filter by washing the basin once with hot water, containing 50 ml. of concentrated hydrochloric acid per litre. Then wash the basin twice more, rubbing well with small pieces of filter paper to ensure complete removal of the silica which adheres tenaciously to the glaze of the basin. Return the filtrate to the basin, rinse the beaker and re-evaporate to dryness, stirring as before, to complete the dehydration of the silica. Continue heating on the bath for one hour after the salts lose their deep colour.

While this evaporation is proceeding, complete the washing of the main portion of the silica with hot 5 per cent. hydrochloric acid, collecting the filtrate in a 500 ml. volumetric flask. About six washings are sufficient. Allow the filter paper to drain and transfer to an unweighed platinum crucible.

When the second evaporation is complete, moisten the salts in the basin with about 5–6 ml. of concentrated hydrochloric acid and add 40–50 ml. of water. Warm for a minute or two on the bath to effect solution and then filter through a fresh 9 cm. Whatman No. 44 filter paper, collecting the filtrate in the volumetric flask containing the washings from the first silica separation. Rinse the basin completely, using two small pieces of filter paper as before to assist in the removal of the last traces of silica. Wash the filter paper as before with hot 5 per cent. hydrochloric acid. Finally add the filter paper to the crucible containing the bulk of the silica already separated.

Transfer this crucible to a muffle furnace and ignite carefully to char the filters. Then raise the temperature and continue the ignition for at least half an hour at the maximum temperature of the muffle. Place the lid on the crucible after transferring it to a desiccator, cool, and weigh as "crucible + SiO_2 + x."

After weighing, carefully moisten the silica in the crucible with a little water, add four or five drops of dilute sulphuric acid (1 + 2) and then 5 ml. of hydrofluoric acid or sufficient to dissolve all the silica. Support the crucible on a triangle over a radiation type of hot plate in a fume hood and evaporate carefully to dryness. Do not permit the liquid in the crucible to boil. Finally ignite for 5 minutes over a Meker burner, cool in a desiccator and weigh again as "crucible + x." The loss in weight represents the amount of silica.

Add about 0·4 g. of potassium bisulphate to the crucible and fuse gently over a small flame in order to dissolve the residue. When cool add 10 drops (0·5 ml.) of dilute sulphuric acid (1 + 2) and sufficient water to dissolve the fused cake. Warm until solution is complete and add to the contents of the flask containing the previous filtrates. Dilute the solution in this flask to the graduation mark and mix well. Use aliquots of this solution for the determination of:

(a) $Fe_2O_3 + Al_2O_3 + TiO_2$
(b) $Fe_2O_3 + TiO_2$ (and TiO_2 separately)

Iron, Aluminium and Titanium:

Pipette a suitable aliquot (usually 200 ml.) of the filtrate from the silica separation into a 400 ml. beaker and add 10 ml. of concentrated hydrochloric acid and a few drops of methyl red or sofnol red indicator. Then add concentrated ammonia, from a dipping pipette, slowly and with constant stirring until the point of precipitation is almost reached. Continue the neutralization with more dilute ammonia (1 + 5) until precipitation occurs and the colour of the indicator just changes to the alkaline shade. Add a further three drops of dilute ammonia and allow to stand for a minute or two to enable the colour of the indicator in the supernatant liquid to be seen. Then heat to boiling and boil for 40 seconds. If the red colour of the indicator returns add a further one or two drops of dilute ammonia, but only sufficient just to change it to the alkaline shade. Allow the precipitate to stand for some minutes until the bulk of it has settled and then filter through a 9 cm. Whatman No. 41 filter paper. Wash the beaker and precipitate three or four times with hot 2 per cent. ammonium chloride solution. Return the filter paper and precipitate to the beaker, add 10 ml. of concentrated hydrochloric acid and macerate the filter paper until solution of the precipitate is complete. Then add about 150 ml. of water and a few drops of methyl red or sofnol red indicator. Carefully neutralize and reprecipitate the iron, aluminium and titanium hydroxides exactly as before.

Filter through an 11 cm. Whatman No. 41 or 541 filter paper and wash completely with hot 2 per cent. ammonium chloride. Finally rinse the filter and precipitate once with warm water and allow to drain. If a 541 paper has been used, a filter pump can be used with advantage to drain the filter and precipitate. Transfer the filter and precipitate to a weighed crucible and ignite in a muffle furnace for 30 minutes after the filter paper has been destroyed. When the ignition is complete, cover the crucible and place it in a desiccator to cool. Weigh as $Al_2O_3 + Fe_2O_3 + TiO_2$. Deduct the amount of $Fe_2O_3 + TiO_2$ found later, in order to obtain the amount of Al_2O_3. If a significant amount of P_2O_5 is present it should be determined and deducted from the value for Al_2O_3.

Iron and Titanium:

Cupferron precipitations are best made at temperatures of about 5° C. Low temperatures tend to yield a more granular precipitate, whereas at higher temperatures it becomes tarry and difficult to wash. However, this precipitate can be handled satisfactorily up to 12–15° C. Precipitation and washing should be carried out without undue delay since the excess of reagent decomposes slowly in acid solution even in the cold.

Pipette a suitable aliquot of the filtrate from the silica determination (usually 200 ml.) into a 400 ml. beaker and add 30 ml. of concentrated hydrochloric acid and 25 ml. of macerated filter paper suspension (5 Whatman ashless filter tablets per litre). Cool to room temperature or lower if possible. With constant stirring add 5–10 ml. of a freshly prepared 6 per cent. solution of cupferron, until on testing a drop on a piece of No. 44 filter paper with a drop of potassium ferrocyanide solution, no blue colour is seen at the margin. To ensure an excess add a further 2 ml. of reagent. Filter after two to three minutes through an 11 cm. Whatman No. 41 or 541 filter paper, returning the first portion of the filtrate if it is not clear. Wash repeatedly with cold dilute hydrochloric acid (1 + 10) containing 20 ml. of 6 per cent. cupferron solution per litre. Finally wash once with water and then four or five times with dilute ammonia (1 + 10) to remove most of the cupferron. If a 541 paper has been used it may be drained very completely by the use of a filter pump and this considerably assists the subsequent ignition. Transfer the filter and precipitate to a weighed crucible, ignite as usual, cool, and weigh as Fe_2O_3 + TiO_2.

Titanium:

After the iron and titanium precipitate has been ignited and weighed transfer it to a 100 ml. silica basin, rinsing the crucible with water. Add 20 ml. of dilute sulphuric acid (1 + 1) and 5 ml. of concentrated hydrochloric acid. Heat over a hot plate or radiation type heater at a low heat, until fumes of sulphuric acid begin to appear. Cool, dilute with 30 ml. of water, warming if necessary to dissolve any salts, transfer to a 100 ml. volumetric flask and dilute to about 90 ml.

When cold add 1 ml. of 30 per cent. hydrogen peroxide and dilute to the mark. Compare the colour so obtained with a suitable standard colour solution derived from a standard titanium sulphate solution (see p. 143).

Calculation of the Results:

Express the values for silica, iron and alumina as percentages of the ignited clay and divide each value by its molecular weight to obtain the gram molecular percentage.

(a) Gram molecular percentage of SiO_2

$$= \frac{\text{Weight of } SiO_2}{\text{Weight of ignited clay}} \times \frac{100}{60 \cdot 0}$$

(b) Gram molecular percentage of Fe_2O_3

$$= \frac{\text{Weight of } Fe_2O_3}{\text{Weight of ignited clay}} \times \frac{\text{Total volume of solution}}{\text{Volume of solution taken}} \times \frac{100}{159 \cdot 7}$$

(c) Gram molecular percentage of Al_2O_3

$$= \frac{\text{Weight of } Al_2O_3}{\text{Weight of ignited clay}} \times \frac{\text{Total volume of solution}}{\text{Volume of solution taken}} \times \frac{100}{102 \cdot 0}$$

From these values calculate the following molecular ratios:

$$\frac{SiO_2}{Al_2O_3} = \frac{a}{c}$$

$$\frac{SiO_2}{Fe_2O_3 + Al_2O_3} = \frac{a}{b + c}$$

$$\frac{Fe_2O_3}{Al_2O_3} = \frac{b}{c}$$

REFERENCES

(1) Agricultural Education Association. (Analysis of Soils Sub-Committee): *Agric. Progress*, 1939, *16*, 55–60.

(2) Puri, A. N.: *Soil Science*, 1935, *39*, 263–270.

PART II

CHAPTER I

THE COLLECTION AND PREPARATION OF PLANT SAMPLES

The proper collection of plant samples is quite as important as the use of reliable methods for their subsequent analysis. Unless a sample is truly typical of the material which it is intended to represent, the analytical results are of little or no value, however carefully they may be determined in the laboratory. The exact details of sampling in the field will depend upon the type of material and the purpose for which the sample is required. In studying the influence of such factors as the stage of growth, cultural practice, manurial treatment, or grazing, on the mineral composition of a crop or pasture, the whole of the plant material from selected areas will be required. In other investigations, designed to compare the composition of different species of plants growing on various soil types, the material sampled should be restricted to pure species, all cut at comparative stages of growth. This is important since the mineral composition of all plants varies with the stage of growth and, in general, decreases as maturity is approached. Very young growth is always much richer than older material in total ash and in the individual mineral constituents. The different plant species also differ markedly in their mineral composition. In mixed pasture samples both botanical composition and stage of growth are variables, and care must be exercised in interpreting the analytical results from such samples. Different species recover at different rates after grazing or cutting and, in many long range experiments, this may bring about big changes in the botanical composition of successive samples of the pasture.

When it is necessary to harvest experimental plots for both

yield and chemical composition the number and size of the areas cut on each plot must be chosen so that the significance of the results can be determined statistically. Material from such experiments is frequently subdivided botanically into subsamples, prior to analysis. When a composite sample is to be collected from a field it is usually taken by cutting small areas at random and combining the material. If it is desired to investigate the effect of soil type on the composition of the plant such samples are preferably restricted to a small area typical of the field and a composite soil sample is also taken to represent the same area.

In connexion with grazing experiments it is desirable to cut the pasture samples so as to simulate the grazing animal as closely as possible. At the Waite Institute the pasture is cut, by means of hand shears or a modified sheep shearing machine, about half to three-quarters of an inch above the level of the ground. Any attempt to cut the herbage closer to ground level destroys the plant as well as increases contamination of the sample by soil. On each plot a number of areas of say 5 x 2·5 links are selected at random and the herbage cut from each is lightly packed into separate open mesh bags and transferred as quickly as possible to the laboratory. The herbage of each sample is thoroughly mixed and divided into a number of subsamples, such that each weighs approximately 100 g. (green weight). After consulting a table of random numbers one of these samples is put aside for botanical analysis. The botanical separation should be carried out as soon as possible after cutting. The remaining subsamples are bulked and dried rapidly at 60° C. in an oven through which a strong current of air is drawn. Drying is completed in another oven at 100° C. and, from the dry weights obtained, the yield is calculated. The botanically separated samples are dried in a similar manner and their weights added to those of the other samples.

After drying, the botanical separates from the individual areas cut on each plot are bulked to form a composite sample for that plot and this sample is used for chemical analysis. To reduce the amount of analytical work involved, the samples from replicate plots are also combined at this stage. If there are many replicates the sample may be inconveniently large for grinding and, in such cases, it is further reduced by quartering (p. 255).

Suitable precautions must always be taken to avoid contamination of the sample by soil. When possible the plants should be cut individually at a height sufficiently above ground level to avoid serious contamination. The height at which the sample can be cut depends on the material being harvested and the purpose for which it is required. Tall crops, such as cereals, can be cut at a somewhat greater height than pasture samples, without introducing serious errors, due to the material left standing. The freshly cut material should be picked over by hand to ensure freedom from soil particles. Any dust or soil adhering to the lower leaves should be brushed off. If, owing to the uneven surface of the soil, cutting is difficult and lumps of soil become mixed with the sample, such badly contaminated parts should be discarded. All samples should be picked over again before grinding, as during drying small amounts of soil become detached from the plant material. Owing to the loss, by solution, of some of the more soluble inorganic constituents it is not permissible to wash plant samples to remove soil contamination.

When it is necessary to cut the sample at ground level, as in many pot experiments, the individual plants should be broken apart at their bases and any adhering soil particles removed by careful brushing with a camel hair brush.

In considering the possible effects of contamination of the sample, the nature of the soil and the purpose for which the analysis is required are important and these will, in general, determine the stringency with which contamination must be avoided. For example, a small amount of contamination with a sandy soil is not likely to affect significantly the values for calcium or magnesium in the plant sample whereas admixture with a similar amount of a calcareous soil would cause a greater error for these elements. Soil contamination will not invalidate values for an element like potassium as much as those for elements such as iron and aluminium, since the former occurs in relatively greater amounts in the plant than in the soil while the reverse is true for the two latter elements. In determinations involving the so-called trace elements, manganese, copper, zinc and boron, rigid precautions must be taken to avoid contamination with soil. Particular care should be exercised in collecting samples after top-dressing with fertilizers containing these elements, since significant amounts of them may be retained as a fine dust on the leaves of the plants. The amounts of the trace elements normally present in plants are so small that the error from this source may, under some conditions, be considerable. In plants such as cereals and grasses the fine dust, from the fertilizers applied, may collect between the leaf sheath and the stem and persist for a considerable time. To avoid all possibility of contamination, plant samples should not be handled at the same time as, nor transported in proximity to, fertilizers containing the element or elements under investigation.

The analytical results sometimes indicate the probability of contamination of the sample with soil. In this connexion Aston (1) has suggested that the amount of aluminium present in the plant sample is a valuable guide to the degree of contamination of the sample. Aluminium is not generally recognized as an essential element for plant growth and only small amounts are present in normal plant materials. High values are indicative of contaminated samples.

All samples should be dried as rapidly as possible after collection so as to reduce chemical and biological changes to a minimum. Considerable loss in dry weight may occur, due to respiration, while proteins are also broken down to simpler nitrogenous compounds, if drying is unduly delayed. When green material has to be transported long distances to the laboratory before drying it should be lightly packed in loosely woven bags. Tight packing may lead to considerable overheating, so increasing respiration losses and other changes.

Green material should be dried in a well ventilated oven so that the water vapour is rapidly removed from the immediate vicinity of the sample. The material to be dried should be very loosely packed in open shallow trays, otherwise free movement of the moisture-laden air is impeded. Failure to secure good ventilation leads to considerable decomposition of some of the organic constituents of the sample. Rapidly dried samples are always a much brighter green colour than those badly dried. Drying is most conveniently carried out at a temperature of 60–70° C. in a large steam heated oven so arranged that a strong current of warm air is continuously forced through the oven. If dry weights are required drying can be finished in another oven at 100° C. Samples for the determination of copper or zinc should not be dried in drying ovens made of copper or brass, unless these ovens are completely protected with a suitable synthetic lacquer, to prevent metallic contamination of the sample. Drying ovens made of stainless steel are particularly suitable for the drying of samples for most trace element determinations.

Before grinding, bulky samples are best reduced by quartering until a convenient amount of material is obtained. The material should be handled carefully to avoid excessive separation of leaf, stem and seed. If the individual parts of the plants become badly separated it makes it more difficult to subsample the material accurately, since the different parts of the sample tend to segregate. When large samples are sufficiently reduced by quartering the subsample is chaffed and further reduced by quartering until an amount of 150–250 g. is obtained. The material is then ready for grinding.

The fineness of grinding depends upon the size of the subsamples to be taken for chemical analysis. For many purposes it is sufficient to grind the material until it passes through a sieve with round holes 1 mm. in diameter. If amounts of only 1–2 g. are to be weighed for the individual chemical determinations it is better to grind the sample somewhat finer so that it will all pass through a sieve with round holes about 0·5 mm. in diameter. This gives a more homogeneous sample.

Two mills can be recommended for grinding plant samples. These are the Wiley Mill and the Christy and Norris Junior (C. & N. Mill). Grinding is best carried out on the oven-dried material but, except in very humid localities, both of these mills will grind air-dry samples very satisfactorily.

In the Wiley mill, grinding is accomplished by the action of a set of steel knives revolving against another set of knives mounted in the circular outer casing of the grinding chamber. One of three interchangeable brass sieves (2 mm., 1 mm. and 0·5 mm. round holes) can be fitted into the lower part of this outer casing. Material to be ground is fed in at the top of the mill and carried around between the blades of the revolving and stationary knives until it is sufficiently fine to drop through the holes of the sieve at the bottom. The finely ground material is collected in a drawer below the sieve. This mill grinds plant material with a minimum amount of heating.

In the C. & N. Junior Mill the grinding is accomplished by a four armed beater-cross which revolves freely and at high speed (12,500 r.p.m.) in a plain circular grinding chamber. At this high speed any plant material, fed through the feed inlet, is shattered or torn until it is sufficiently fine to pass through a small iron screen plate clamped in the periphery of the grinding chamber. Several interchangeable screen plates are supplied with the mill. Those with holes 0·4 mm. and 1 mm. in diameter are most useful. Owing to the centrifugal action of this mill a considerable amount of air is drawn through it when in operation. It is therefore necessary to use a dust tight receptacle to collect the ground sample. For most purposes reinforced cellulose extraction thimbles, as supplied with the mill, are very suitable. These can be readily brushed clean between samples. For larger samples silk collecting sleeves are also supplied. The small size and the simplicity of the grinding chamber and beater make this mill very easy to clean between samples. It is also particularly valuable when only small amounts of material are available for grinding.

Special precautions must be observed when grinding plant materials for certain analytical determinations. It is not permissible to use a steel mill for grinding samples in which iron is to be determined since the amount of iron derived from the mill during grinding is sufficient to give entirely erroneous values. In such cases an aliquot of the original sample should be ground by hand, in an agate or porcelain mortar, and used for the iron determination. Alternatively the material can be ground in a C. & N. mill made entirely in bronze, phosphor bronze or other non-ferrous alloy. Steel mills are permissible when manganese is to be determined. The amount of manganese derived from the steel is insufficient to affect the value significantly. The same remarks apply to cobalt.

When it is necessary to determine copper or zinc, very special care must be exercised to avoid contamination owing to the widespread use of brass, an alloy containing both of these elements. The amounts of copper and zinc normally present in plants are so small that the slightest contamination would invalidate the results. Copper and brass must be rigidly excluded from any part of the grinding mill. The use of a Wiley mill with standard brass sieve may cause the copper content of the finely ground plant sample to be increased several-fold. Providing that steel screens are used and the brass thumb screws and feed tray are replaced by steel or stainless steel, the C. & N. mill is very suitable. The mill should never be used with brass fittings, at any time, if it is to be used for trace element work. The Wiley mill can also be adapted if the brass sieve is replaced by one constructed entirely of iron or steel. Wide-stemmed funnels frequently used for transferring ground samples to the sample bottles should be either glass or aluminium. The latter metal is very serviceable.

The metals available preclude the use of any single grinding mill when both copper and iron are to be determined. The material must either be ground by hand in an agate or porcelain mortar or two separate samples must be ground in appropriate mills.

After grinding, all plant samples are mixed thoroughly, transferred to suitable bottles, labelled clearly and tightly corked. For the individual chemical determinations the finely ground air-dry material is spread out in a thin layer and grab samples of suitable weight are taken. These are oven-dried, in weighing bottles, immediately before weighing for analysis. This procedure is preferable to the determination of the moisture in the air-dry sample and the use of the value so determined for all subsequent calculations to the oven-dry basis, since the moisture in the bulk sample may vary considerably with changes in atmospheric humidity.

REFERENCE

(1) Aston, B. C.: *New Zealand J. Agric.*, 1928, *36*, 22–27.

CHAPTER II

METHODS FOR THE ASHING OF PLANT MATERIAL

Before the inorganic constituents of plant substances can be determined it is generally necessary to destroy the organic matter. The methods used to bring about this result fall into two main groups, which may be described as "wet ashing" and "dry ashing" respectively. Wet ashing includes those methods in which the destruction of the organic matter is brought about by oxidation in a liquid medium, while dry ashing refers to processes in which the sample is ignited. The sample may be ignited alone or after moistening with sulphuric acid, in which case a sulphated ash is obtained. At other times it may be ignited with certain basic substances (e.g. lime), or salts which on ignition yield basic substances (e.g. acetates and nitrates of calcium and magnesium). The most common methods of wet ashing include the digestion of the sample with mixtures of sulphuric and nitric acids, or sulphuric, nitric and perchloric acids.

Dry ashing must always be carried out at as low a temperature as possible and the operation cannot be hurried. To obtain a good ash, free from excessive amounts of carbon, requires considerable experience and patience. Some elements may be lost if the ashing is carried out at too high a temperature. High temperatures also favour the formation of complex silicates which are not readily soluble in hydrochloric acid, even after prolonged digestion. It is probable that many of the apparent losses of the inorganic constituents during ashing are due to this cause (4). Fusion of some of the salts of the ash may occur if the temperature is too high and these fused salts surround particles of unburnt carbon, so excluding the air and preventing their free combustion. If ashing is carried out too quickly, deflagration occurs, producing excessive local heating within the glowing mass. Some analysts prefer to treat the sample with sulphuric acid before ashing so as to obtain a sulphated ash, which is less fusible than ordinary ash. However, this does not prevent the formation of varying amounts of difficultly soluble silicates, particularly those of the trace elements.

No loss of phosphorus occurs during the ashing of most plant materials, provided that the temperature does not exceed 600° C. However, when ashing seeds or other plant materials low in basic constituents, the sample should be thoroughly mixed with calcium or magnesium acetate or nitrate, prior to ashing. The acetates are recommended in preference to the nitrates since the strongly oxidizing action of the latter may cause violent deflagration, with considerable rise in temperature of the ash. In the presence of added magnesium salts Ashton (1) has shown that the temperature may reach 800° C. without loss of phosphorus. Chlorine and sulphur are largely lost during ashing unless a basic substance such as lime or sodium carbonate is intimately incorporated with the material before ashing.

No matter how carefully the ashing is carried out, the siliceous residue, left after digestion of the ash with hydrochloric acid, always retains small amounts of some of the constituents. In particular, significant amounts of the trace elements are strongly retained, even after prolonged digestion with hydrochloric acid and they can only be recovered by alkaline fusion or by solution of the silica in hydrofluoric acid. In many cases more than one-quarter of the total amounts of manganese, copper and zinc have been found in the insoluble residue. While the absolute amounts are not sufficient to affect signficantly the value for silica, the errors in the determinations of the individual elements will be serious unless the adsorbed amounts are recovered. Treatment of the plant sample with sulphuric acid or magnesium salts, prior to ashing, does not prevent this retention by the silica.

Plant ash always contains greater or smaller amounts of residual carbon, depending upon the care and skill with which the ashing has been carried out. Carbonate is also a variable constituent. In some plant ashes it is present in considerable quantity whereas in others it is practically absent and the bases are left as silicates or oxides. Its presence depends on the nature of the ash and it may also be considerably affected by slight differences in the temperature and other conditions during ashing. The value for crude ash as ordinarily determined is therefore somewhat variable and there is little point in its accurate determination. Quantitative values are only of general interest. However, ashing must be carried out carefully to ensure accurate values for each of the individual constituents. When a complete analysis is made the most reliable value for total ash is the sum of the individual constituents.

When a more accurate value for total ash is required than that given by crude ash the carbon-free ash can be determined. A carbon-free ash is obtained by extracting the crude ash with water, igniting the residue until free from carbon, returning the aqueous solution to the basin, evaporating and re-igniting. The value so obtained may still be affected by a variable amount of carbonate. To avoid this the value for sulphated ash is sometimes preferred.

For many determinations methods of wet ashing are preferable to, and more convenient than, dry ashing. Since the oxidation is carried out in solution in an acid medium, the temperature cannot exceed the boiling point of the mixture used; complex insoluble silicates are not formed, and all bases are obtained in solution in the excess of acid. When sulphuric or perchloric acids are used the silica is completely dehydrated and left in a form in which adsorption is at a minimum. Such small traces of any of the other plant constituents are retained by the silica that it is only necessary to dissolve it in hydrofluoric acid for the most exacting work.

In the wet digestion methods the greater part of the oxygen required for the oxidation is supplied by the nitric acid. For the most efficient use of this nitric acid, digestions must be carried out at a low temperature, thus avoiding excessive losses by evaporation in the early stages. Perchloric acid considerably assists the digestion since it appears to break down some of the organic compounds into simpler compounds, which are then more readily oxidized by the nitric acid. A small amount of perchloric acid is sufficient for this purpose; the bulk of the oxidation is carried out by the nitric acid. In the presence of perchloric acid much smaller quantities of nitric acid suffice for the digestion. Perchloric acid also prevents the excessive frothing which so frequently occurs when nitric and sulphuric acids alone are used. Instead of perchloric acid, a concentrated solution of sodium perchlorate may be used (6) provided that it is not necessary to determine sodium in the digest. When sodium perchlorate is used an equivalent amount of extra sulphuric acid is usually desirable. For each digestion 8 ml. of a solution of sodium perchlorate (550 g. dissolved and made to 1 litre) and 1 ml. of sulphuric acid are equivalent to 4 ml. of perchloric acid (S.G. 1·54). Since sodium perchlorate is very much cheaper than perchloric acid its use leads to considerable economies. It also offers special advantages in the determination of the trace elements, since it can be purified from them in such a simple and convenient manner (see p. 331), thus avoiding the necessity for purifying perchloric acid by distillation. In many laboratories perchloric acid is regarded as a dangerous reagent. The anhydrous acid explodes spontaneously on contact with organic matter. Aqueous solutions are, however, very stable and possess no oxidizing properties in the cold. For instance, cold aqueous solutions will not oxidize ferrous salts (3). Provided that reasonable care is taken, and that certain precautions are observed, perchloric acid may be used with complete safety for the digestion of plant materials. According to Kahane (3) these may be digested with nitric and perchloric acids alone but the addition of sulphuric acid increases the convenience and safety of the digestion. One important function of the sulphuric acid is said to be that of a diluent for the perchloric acid, regulating the vigour of the attack and promoting smooth decomposition of the organic matter. When carrying out digestions in its absence, Kahane recommends the use of a big excess of perchloric acid, apparently to prevent overheating. In the author's opinion it is very desirable to carry out the later stages of the digestion in the presence of sulphuric acid since this less volatile acid prevents overheating towards the end of the digestion, when the volume is reduced. Unless sulphuric acid is present at this stage the perchloric acid and ammonium perchlorate, which remain at the end of the digestion, may decompose with explosive violence, either through contact with charred organic matter or the walls of the flask, if these become overheated. To avoid its accidental omission at a later stage, it is preferable to add the sulphuric acid at the commencement of the digestion. When so added the digestions require less supervision. It is desirable to use as much sulphuric acid as the subsequent analytical operations will permit. Five ml. is usually a very suitable quantity but for some determinations smaller quantities only are permissible (e.g. 2 ml. for copper and 3 ml. for phosphorus). Sufficient nitric acid should always be used to effect the greater part of the oxidation of the organic matter originally present. The digestion should, of course, be carried out slowly to avoid wasteful loss of nitric acid by evaporation. If the material to be digested is very rich in fat or oil, this should be largely destroyed by a preliminary digestion with nitric and sulphuric acids before the addition of perchloric acid.

The perchloric acid method of digestion has been used extensively in many different laboratories. In these laboratories many hundreds of digestions have been carried out by the technique described on p. 272. The silica which separates retains only the faintest traces of iron, copper, manganese and the other trace elements. This makes the perchloric acid digestion the most suitable method for the determination of these elements.

Details of the most common methods for the ashing of plant materials are given below. Several factors will govern the choice of the method to be used for the destruction of the organic matter. Among these may be mentioned the constituents to be determined in the plant material, the equipment and facilities available, and the personal preference of the analyst. Dry ashing is tedious and lengthy. If only one muffle furnace is available, the number of samples that can be handled is strictly limited. On the other hand the simple equipment necessary for wet digestion makes it possible to carry out a large number of digestions at one time. For the chemical determination of phosphate and nearly all the metallic cations, digestion with sulphuric, nitric and perchloric acid is strongly recommended.

Drying and Weighing of Samples for Ashing.

Since all results are to be expressed on an oven-dry basis it is preferable to weigh the sample in the oven-dry form immediately prior to ashing or digestion. Owing to the hygroscopic nature of dry plant material, drying and weighing are most conveniently carried out in a weighing bottle.

Spread out the finely ground sample on a piece of glazed paper or sheet rubber, mix thoroughly with a spatula and take several small portions at random, to give a true subsample. In this way transfer a suitable weight of the air-dry sample to a weighing bottle, dry in an oven at 105° C. for 12–16 hours and cool in a desiccator. When cool, weigh the sample and weighing bottle. Then transfer the sample to the appropriate silica dish or digestion flask and weigh the empty weighing bottle. The difference corresponds to the weight of the oven-dry sample taken for ashing or wet digestion. Use this value for all calculations.

Dry Ashing.

This method is recommended for the determination of silica and for the preparation of a solution of the ash for the determination of calcium, magnesium, potassium and sodium, when it is not desired to use a wet digestion method for these latter elements. Phosphorus can be determined in the solution provided that the ash is naturally high in basic constituents and the ashing is carried out at a low temperature. Wet digestion with nitric, sulphuric and perchloric acids is, however, strongly recommended in place of dry ashing for the determination of phosphorus and also for the trace elements. Dry ashing, without the addition of a basic substance, cannot be used for the determination of chlorine.

Transfer a weighed quantity of the oven-dried plant material to a flat shaped silica basin which has been previously weighed. For the determination of single elements 5–10 g. of material are usually sufficient. If, however, an acid solution of the ash is to be made for the determination of several constituents, take 25–50 g. of oven-dried material. The silica basin should be sufficiently large to hold the required amount.

Support the silica basin on a triangle, sufficiently far above a gas burner turned very low, or a hot plate, so that the material slowly chars. This operation must be carried out very slowly, particularly if more than 3–4 g. of dry matter are being ashed. When sufficiently charred, transfer the basin to a muffle furnace at a temperature of about 300° C. Stand the basin on a small triangle on the floor of the muffle to avoid overheating. Do not allow the muffle to become too hot but keep the temperature down so that the charred material only glows faintly when the door of the muffle is partly opened. If the temperature rises too much the sample burns too rapidly and this leads to further increase in the temperature with, perhaps, some local fusion of the ash, loss of volatile constituents or the formation of difficultly soluble silicates. Allow the ashing to proceed slowly and when no more glowing carbon can be seen, gradually raise the temperature of the muffle to a very dull red heat (about 500–550° C.) but do not exceed this temperature. A dull red glow should be just visible in the darkened muffle. When several samples are being ignited at once, occasionally change their positions in the muffle, to secure more uniform conditions of ignition.

If the ashing is carried out carefully the resulting ash is greyish white or grey and contains only small amounts of unburnt carbon. It is lightly sintered together but shows no other signs of fusion. The porous structure should not be disturbed during ashing, since it allows free access of air into the mass.

If the weight of crude ash is required, cool the silica dish and its contents in a desiccator and weigh.

Solution of the Ash and Separation of Silica:

When cold carefully moisten the ash with a little water, cover the basin with a clock glass and cautiously add 40 ml. of dilute hydrochloric acid (1 + 1), pouring the acid into the covered basin so as to avoid any loss by effervescence. Place the basin, still covered, on a water bath and digest for 20–30 minutes. Remove and rinse the cover, add 1 ml. of nitric acid to oxidize any ferrous salts and evaporate the contents to dryness. Continue heating for half to one hour on the bath, to dehydrate the silica. If necessary heat for one hour in an air oven at 110° C. to complete the dehydration.

Moisten the dried salts with 10 ml. of dilute hydrochloric acid (1 + 1), add a further 50 ml. of water and warm on the bath until all soluble salts are in solution. Filter through an 11–12·5 cm. Whatman No. 44 filter paper collecting the filtrate in a volumetric flask of suitable size (usually 500 ml. when ashing 25–50 g. of material). Transfer the insoluble residue from the basin to the filter, using a rubber tipped stirring rod to remove particles of silica adhering to the sides of the basin, and wash completely with hot dilute hydrochloric acid (1 + 19).

When the filter paper containing the insoluble residue has been fully washed transfer it to a weighed platinum basin and carefully ignite it to remove the filter paper and any carbon present in the crude ash. Finish the ignition at a bright red heat, cool and weigh as insoluble residue. This insoluble residue consists essentially of silica and gives a value sufficiently accurate for most purposes. However, it contains small amounts of other elements which, while not significantly affecting the accuracy of the silica determination, should be recovered and added to the hydrochloric acid solution of the ash. Moisten the insoluble residue with water, add 2–3 drops of sulphuric acid and approximately 10 ml. of hydrofluoric acid for each 0·5 g. of silica. Evaporate slowly over a hot plate in a fume hood but do not allow the solution to boil. Continue the evaporation until fumes of sulphuric acid are produced. Do not remove the sulphuric acid completely, otherwise some of the iron and aluminium may be rendered insoluble again. When cold add 2 ml. of dilute hydrochloric acid (1 + 1) and a little warm water. Transfer the solution to the volumetric flask containing the filtrate previously obtained. Dilute the solution to the mark on the flask and use suitable aliquots for the determination of such elements as calcium, magnesium, potassium, sodium and manganese. This solution may also be used for the determination of iron, copper and zinc provided that suitable precautions are taken to avoid the accidental introduction of any of these elements, either during manipulation or in the reagents used (see p. 305).

Ashing with the Addition of Sulphuric Acid.

This method is recommended for the ashing of plant materials when the constituents are to be determined spectrochemically, since sulphates are the most convenient salts to use in the electric arc used to excite the spectra. If properly sulphated prior to ashing, the plant ash so obtained is non-basic. When plant materials are ashed without sulphating a basic ash is obtained and slight attack on the silica basin occurs, no matter how carefully the ashing is carried out. Sulphated ash, being less fusible and non-basic, is practically without attack on the silica basin during ashing. Except for ashes unduly high in potassium the glaze of the basin is not impaired in the slightest, even after repeated use.

After drying in a weighing bottle, transfer a suitable quantity of the finely ground material, usually 5 g., to a flat bottomed silica capsule, add 1 ml. of concentrated sulphuric acid for each gram of dry matter, mix well with a fine glass rod and leave overnight, during which time the mass usually becomes semi-fluid. Place the capsule over an electric hot plate and heat gently at a low heat, watching to avoid any tendency to froth over. Continue the gentle heating until the mass chars and dries out somewhat. When nearly dry break up the lumps with a glass rod and wipe the rod with a scrap of filter paper, to remove any material adhering to it. Now heat the capsule more strongly, to fume off as much of the sulphuric acid as possible. Finally transfer the capsule to a muffle furnace at an incipient dull red heat (about 500–530° C., a red colour being just visible in the darkened muffle) and keep at this temperature for 2 hours or until all the carbon is burnt off. During the operation keep the door of the muffle slightly open, leaving a gap of about 1 cm. at the bottom. When the ashing is completed, cool in a desiccator and weigh to obtain the amount of sulphated ash.

When ashed by this method only a small amount of carbon is left and, after thorough mixing, the ash is suitable for use in the spectrographic arc without further treatment. Most ashes are grey in colour but plant materials high in potassium yield a white ash.

If it is desired to use ash obtained by this method for chemical determinations dissolve it in hydrochloric acid exactly as described on p. 264. The preliminary treatment with sulphuric acid does not prevent the retention by the silica of small amounts of other substances, particularly the trace elements. For accurate work recovery of these, by solution of the silica in hydrofluoric acid, is still necessary.

Ashing with Sulphuric Acid: Alternative Method.

This method is recommended when a solution of the ash is required for chemical determinations, provided that it is not necessary to determine the amount of silica. The silica is dissolved in hydrofluoric acid and volatilized during the ashing. In this way two operations are combined and retention of other elements by the silica is avoided.

After drying in a weighing bottle, transfer a suitable quantity of the finely ground material (generally 1–5 g.) to a platinum basin and add sufficient dilute sulphuric acid (1 + 5) to correspond to 1 ml. of sulphuric acid for each gram of dry matter. Stir well so as to moisten completely and rinse the rod with a little water. Heat gently over a radiation type hot plate until the mass dries out and chars. When thoroughly charred break up the lumps with a glass rod and wipe the rod with a scrap of filter paper, to remove adhering material. Heat the platinum basin more strongly to drive off as much sulphuric acid as possible and then transfer to a muffle furnace at a temperature of about 450° C. Gradually raise the temperature until red is just visible in the darkened muffle. (Incipient dull red heat, 500–530° C.) Continue the ignition at this temperature for about one hour or until most of the carbon is burnt off, keeping the door of the muffle slightly open at the bottom.

When the ashing is completed moisten the cool ash with a little water, add 5 drops of concentrated sulphuric acid and 3–5 ml. of hydrofluoric acid. Volatilize the silica by evaporating gently to dryness over a hot plate, taking care to avoid boiling the solution. When dry, ignite in a muffle furnace for 10 minutes at 500–530° C. to remove the last traces of carbon. If carbon still persists add a few drops of sulphuric acid and heat for a further 5–10 minutes. When all the carbon has been destroyed, take up the residue in 5 drops of concentrated sulphuric acid, 2–3 drops of hydrofluoric acid and 2 ml. of water. Digest gently and fume carefully over a hot plate to remove all hydrofluoric acid but do not fume off the sulphuric acid completely. Take up in 2–10 ml. of constant boiling-point hydrochloric acid and 10–20 ml. of water. Warm on a water bath to ensure complete solution. Use this solution for the determination of the metals required.

Ashing with the Addition of Magnesium Acetate.

During ignition magnesium acetate is converted to magnesia and this prevents the loss of acidic constituents, such as phosphoric anhydride, from plant materials (e.g. seeds), the ash of which is low in bases. Magnesium acetate is preferable to magnesium nitrate since nitrates cause deflagration during the ignition. The ash obtained in the presence of added magnesia is said to be more readily soluble in hydrochloric acid, but the silica may still retain significant amounts of the trace elements. For the accurate determination of these elements, solution of the silica in hydrofluoric acid is usually necessary. The most useful application of this method is for the determination of phosphorus, when it is not desired to use a wet digestion method.

After drying in a weighing bottle transfer a suitable quantity of the finely ground material to a flat shaped silica basin. Add 3–5 ml. of magnesium acetate solution (40 g. of magnesium acetate dissolved and diluted to 100 ml.) for each gram of dry material taken, taking care that all the material is brought into contact with the solution. Heat gently over a radiation type hot plate, slowly increasing the heat until the mass chars. Transfer to a muffle furnace at a temperature of about 300° C., supporting the basin on a small silica triangle on the floor of the muffle. Continue the ashing as described on p. 264. Use the ash so obtained for the determination of phosphorus (p. 295).

Ashing with Lime.

For the determination of chlorine the addition of some alkaline substance, such as lime or sodium carbonate, is necessary before ashing, to prevent the considerable loss of chlorine which may otherwise occur. Ashing is more easily carried out if lime is used. The lime must be free from chlorine and is most conveniently prepared by heating reagent grade calcium carbonate in a muffle furnace.

Transfer a suitable quantity of the oven-dry material (usually 3–5 g.) to a 100 ml. silica basin, mix well with one-quarter of its weight of finely divided calcium oxide and sufficient water to give a thin paste. Dry the mixture on a water bath and ash slowly in a muffle furnace commencing at a low temperature, just sufficient to char the mass. Gradually raise the temperature until a very dull red heat (about 550° C.) is reached but do not exceed this temperature. Use the ash so obtained for the determination of chlorine (p. 297).

Ashing with the Addition of Sodium Hydroxide.

For the determination of boron in plant materials it is usually recommended that sodium hydroxide be added to the material before ashing, to prevent loss of boric acid. Boric acid is volatile with steam and some loss may even occur during oven-drying, unless the sample is distinctly alkaline. For this determination it is therefore necessary to use fresh material. The following procedure is essentially that of Dodd (2) and has been used in these laboratories for the determination of boron in apples and pears.

Transfer about 250 g. of the freshly minced sample to a capacious platinum basin and add 20 ml. of a freshly prepared 10 per cent. solution of sodium hydroxide. Mix the sodium hydroxide into the sample, transferring the mixture to a silica basin and returning it to the platinum basin, to ensure thorough mixing. Dry overnight in an oven at 105° C.

At the same time determine the amount of moisture in the fresh sample by drying duplicate portions of 60–80 g. in an oven at 105° C. for 48 hours.

After drying the sample in the platinum basin, transfer it to a muffle furnace, standing the basin on a small triangle on the floor of the muffle to prevent overheating. Commence ashing at a low temperature so as to char the mass slowly. As the volatile constituents are driven off, slowly increase the temperature so that a very dull red heat (530–550° C.) is reached $1\frac{1}{2}$ to 2 hours after starting the ashing. Maintain the muffle at this temperature for a further 3–4 hours so that the mass is charred completely. The muffle temperature must be controlled carefully and not allowed to rise above a very dull red heat. A considerable amount of carbon remains at this stage of the ashing.

Extract the charred residue with three portions, each of 30–40 ml., of hot water and filter through a 9 cm. Whatman No. 44 filter paper, collecting the filtrate in a 100 ml. volumetric flask (boron-free glass). The filtrate is frequently light brown to straw coloured. A stronger colouration indicates that the preliminary charring has not been sufficiently complete.

After the extraction of the soluble portion transfer the filter paper and the charred residue to the original platinum basin, dry in an oven and re-ash in a muffle furnace at a dull red heat (580–600° C.) until the ash is practically carbon-free. This usually takes about 4 hours.

While this further ashing is proceeding transfer the filtrate to a second platinum basin and evaporate it nearly to dryness on a water bath.

When the ash is as carbon-free as possible, extract it again with three lots of hot water, using portions of 15 ml., 10 ml. and 10 ml. respectively. Filter through another 9 cm. Whatman No. 44 filter paper and collect the filtrate in the original 100 ml. volumetric flask. Add one drop of phenolphthalein and make the filtrate just slightly acid by the addition of about 10 ml. of dilute hydrochloric acid (1 + 3), avoiding any loss by effervescence. Transfer the contents of the flask to the basin containing the evaporated residue from the first filtrate and rinse the flask twice with water. Continue the evaporation to small bulk. The excess of base in the first filtrate restores alkalinity and prevents loss of boric acid.

Return the filter paper and any residue to the first platinum basin and complete the ignition of any small amounts of carbonaceous material remaining. When cool add 5 ml. of dilute hydrochloric acid (1 + 3), cover the basin with a clock glass and digest on the water bath for about 15 minutes. Do not let the acid solution evaporate. By this time the second filtrate will have been evaporated to small volume. Transfer the hydrochloric acid solution of the ash to this basin, covering the latter to prevent loss by effervescence. Then add further amounts of dilute hydrochloric acid (1 + 3) until the solution is just acid to litmus paper. Digest for 15 minutes then remove the clock glass. If necessary evaporate until the volume is about 40 ml. but do not prolong this evaporation, since boric acid may be lost if the solution is allowed to become more concentrated.

Filter through a 9 cm. Whatman No. 44 filter paper and collect the filtrate in the 100 ml. volumetric flask previously used. Wash once with warm water containing 10 drops of dilute hydrochloric acid and complete the washing with small portions of warm water, so that the filtrate does not exceed 80–85 ml. Use this solution for the determination of boric acid (p. 315).

Wet Digestion with Sulphuric and Nitric Acids.

This method is sometimes used for the destruction of organic matter prior to the determination of phosphorus. However, the method has little to recommend it owing to the large amounts of nitric and sulphuric acids used and the excessive amount of sulphuric acid remaining at the end of the digestion. The latter interferes with many determinations and precludes the use of more reliable gravimetric methods for phosphorus. The details given below are essentially those described by Richards and Godden (5).

After drying in a weighing bottle, transfer a suitable quantity of the finely ground material to a 300 ml. flat bottomed Kjeldahl digestion flask. The quantity taken may vary from 1–10 g. If required for the volumetric determination of phosphate it should contain approximately 7–30 mg. of phosphoric anhydride if possible. Add 10 ml. of concentrated sulphuric acid and 10 ml. of nitric acid. If more than 2 g. of plant material are taken, use 20–30 ml. of nitric acid, or sufficient to make the sample fluid. Digest carefully over a low flame or a radiation type hot plate at low heat (120–150 watts), avoiding excessive frothing. The reaction is extremely vigorous in the early stages. The heating must be cautious and be interrupted if necessary to prevent the contents of the flask frothing into the neck. When more than 2–3 g. are taken for digestion, keep a large beaker of cold water at hand and plunge any flasks showing excessive frothing into the water. The cooling effect rapidly brings the frothing under control.

Continue the digestion at low heat until brown fumes cease to be evolved and much of the water is driven off. Allow the flask to cool, add a further 5 ml. or, if digesting large amounts of organic matter, 10 ml. of nitric acid and continue the digestion until white fumes of sulphuric acid are produced. Repeat the addition of nitric acid until a clear and colourless digest is obtained. Complete the digestion by heating at the full heat of the hot plate (500–600 watts) for 3–5 minutes. When cold dilute with about 50 ml. of water. The solution so obtained can be used for any determination with which the sulphuric acid does not interfere.

Wet Digestion with Sulphuric, Nitric and Perchloric Acids.

On account of the rapidity and ease with which this digestion is carried out, the relatively small amounts of reagents used and the non-retention of other substances by the silica, this method of digestion is most valuable for the accurate determination of nearly all of the ash constituents. Since it is not easy to remove the silica quantitatively from the digestion flask this is determined separately when required. If redistilled acids are used, this method is also the most suitable technique for the determination of the trace elements. The digestion is carried out in a flask and risk of contamination by extraneous matter during ashing is thus completely eliminated. The precautions to be taken in the determination of trace elements are discussed on p. 305.

Provided that the sulphuric acid is not omitted there is no danger of explosive decomposition of perchloric acid or ammonium perchlorate at the end of the digestion, when digesting ordinary plant materials including most seeds. With seeds very rich in fats or oils it may, however, be necessary to carry out a preliminary digestion with nitric acid.

After drying in a weighing bottle transfer a suitable quantity of the finely ground material to a 300 ml. flat bottomed Kjeldahl digestion flask. For the determination of a single element 1–5 g. is usually a suitable amount but up to 10 g. may be taken if necessary. Add 4 ml. of perchloric acid (S.G. 1·54) and sufficient nitric acid to ensure complete oxidation of the organic matter (15 ml. for quantities up to 2 g. and 7 ml. for each additional 1 g.). Then add 2–5 ml. of sulphuric acid. The larger amount of sulphuric acid is preferable provided that it does not interfere with the subsequent determinations. For such determinations as copper and phosphoric anhydride smaller amounts only are permissible and these amounts are indicated under the respective determinations. *At least 2 ml. of sulphuric acid should always be present* as this ensures a sufficiency of an acid of high boiling point to prevent overheating of the digest in the later stages, after the nitric acid has been expelled. In the absence of sulphuric acid, local overheating of the digest at this stage may lead to decomposition of the ammonium perchlorate with explosive violence.

After the addition of the perchloric, nitric and sulphuric acids, mix the contents of the flask by swilling and heat gently at a low heat (120–150 watts) on a hot plate of the open radiation type (e.g. Gilmer heater) for 3–5 minutes or until the first appearance of dense brown fumes. If a Bunsen burner is used it must be turned very low. When dense brown fumes appear remove the flask from the heater for about five minutes to allow the initial vigorous reaction to subside. Then replace the flask on the heater and continue the digestion slowly, and at low heat, until the appearance of dense white fumes of sulphuric acid. If there is any tendency to bump remove the flask, add two small glass beads and continue the digestion.

The rate of digestion is important. If the temperature of the heater is too high wasteful loss of nitric acid occurs before oxidation of the organic matter is completed, while too low a temperature unduly prolongs the time required for the digestion. When 25–30 ml. of nitric acid are used the digestion should take approximately 90–100 minutes to reach the fuming stage. If the nitric acid is boiled off too rapidly oxidation is not complete and the liquid in the flask becomes black with charred organic matter. If this occurs remove the flask, add a further 1–2 ml. of nitric acid and continue the digestion at a slower rate.

Continue the digestion, at low heat, for 5–10 minutes after the appearance of the dense white fumes of sulphuric acid then digest for a further 1–2 minutes at the full heat (500–600 watts) of the hot plate. If the digestion is complete the liquid will be colourless at this stage. Any sign of carbonization indicates incomplete digestion. In such a case add 1–2 ml. of nitric acid and digest again to fuming. When cold, dilute the digest with 50–75 ml. of water and use the solution for such determinations as are required.

Micro-Digestion with Sulphuric, Nitric and Perchloric Acids.

This method, used by Walkley (6) for the polarographic determination of zinc and other trace metals in plant materials, is also of value when the small amounts of plant materials available necessitate the use of micro-methods for their analysis. The digestion is very rapid and can be completed in 45–50 minutes. The precautions to be taken in the determination of trace elements are discussed on p. 305.

Take approximately 1 g. of the finely ground air-dry sample, dry overnight at 105° C. in a small weighing bottle, cool in a desiccator and weigh. Transfer the weighed sample to a 50 ml. micro-Kjeldahl digestion flask. Add 10 ml. of nitric acid, 2 ml. of a mixture of equal volumes of sulphuric and perchloric acids and a drop of kerosene to prevent frothing. (Alternatively, 3 ml. of a mixture, consisting of 1 volume of sulphuric acid and 2 volumes of a 50 per cent. aqueous solution of sodium perchlorate, may be used instead of the sulphuric-perchloric acid mixture.) If necessary to prevent bumping, particularly in blank determinations, add one or two small glass beads to the flask. Support the digestion flask on a porcelain ring and heat gently over an open radiation type of hot plate at a low heat. A vigorous reaction, accompanied by the evolution of copious red fumes, generally sets in immediately but the drop of kerosene prevents persistent frothing. This initial reaction becomes quieter after a few minutes and digestion then proceeds smoothly until the nitric acid is nearly completely boiled off. When fumes of perchloric acid begin to appear a second vigorous action sets in. The onset of this vigorous action is fairly sudden and it only lasts for a minute or two. Continue the digestion, still at low heat, for a further 5–10 minutes then raise the temperature to the full heat of the hot plate so that the refluxing of the sulphuric acid takes place at the base of the neck of the flask, or higher. Continue with this vigorous heating for about 5 minutes to dehydrate the silica thoroughly. When completed, the digest consists chiefly of the 1 ml. of sulphuric acid originally added. The hot liquid is yellow-green but becomes colourless on cooling, unless relatively high in manganese. In the latter case a distinct pink colouration persists.

Some charring may occur after most of the nitric acid has been expelled and perchloric acid fumes begin to appear but the carbon which separates is usually oxidized within a minute or two. If it persists, however, add 2–3 ml. of nitric acid and continue with the digestion as usual. Charring is particularly likely to occur if the digestion is carried out too rapidly and the nitric acid is boiled off in the early stages. The temperature of digestion should be such that it takes 35–45 minutes to reach the stage of the second vigorous reaction.

Use the digest for the micro-chemical determination of such elements as are desired.

REFERENCES

(1) Ashton, F. L.: *J. Soc. Chem. Ind.*, 1936, *55*, 106T–108T.

(2) Dodd, A. S.: *Analyst*, 1927, *52*, 459–466.

(3) Kahane, E.: *L'Action de l'Acide Perchlorique. Actualités Scient. et Indust. No. 167 and 168, Paris*, 1934.

(4) Piper, C. S.: *Aust. Chem. Inst. J. and Proc.*, 1939, *6*, 421–427.

(5) Richards, M. B. and Godden, W.: *Analyst*, 1924, *49*, 565–572.

(6) Walkley, A.: *Aust. J. Expt. Biol. and Med. Sci.*, 1942, *20*, 139–147.

CHAPTER III

THE DETERMINATION OF THE MORE COMMON INORGANIC CONSTITUENTS OF PLANTS

SILICA

In the method of ashing given on p. 263 the insoluble residue is determined and this value is generally sufficiently accurate for silica, provided that ashing has been carried out carefully and at a low temperature. It is a little higher than the true value, on account of the small amounts of other elements retained by the silica. Small amounts of silica are also derived from the silica capsules used for ashing, owing to attack by the basic constituents of the ash. If it is necessary to determine the silica more accurately the same method can be followed, ashing a smaller sample of material (2–5 g.), in a porcelain or platinum capsule, weighing the insoluble residue, volatilizing it with hydrofluoric acid and igniting strongly at this stage to expel all sulphuric acid. The loss in weight then corresponds to the silica. In the usual method of ashing, in which the preparation of the solution of the ash is the prime consideration, the final ignition is omitted, to avoid rendering insoluble the iron, aluminium and other substances recovered from the silica.

Silica can also be determined by the nitric-sulphuric-perchloric acid method of wet digestion (p. 272), if the digestion is carried out in a tall beaker so that the silica, which adheres to the walls of the digestion vessel, can be cleaned off and recovered. This method gives silica in a high state of purity, without the need for volatilizing it with hydrofluoric acid. However, digestion in a beaker is not quite as convenient as in a Kjeldahl digestion flask.

CALCIUM AND MAGNESIUM

Owing to the presence of phosphates in plant ash, precautions must be taken to avoid their interference with the determination of calcium and magnesium. If calcium oxalate is precipitated in neutral or alkaline solutions of the ash, calcium and magnesium phosphates will also be precipitated. The oxalate must therefore be precipitated under conditions which preclude the co-precipitation of the alkaline earth phosphates. These requirements are fulfilled by carrying out the precipitation in a weakly acid medium, as in McCrudden's original method. Calcium oxalate is completely precipitated under conditions as acid as pH 4. At pH 4–5 calcium and magnesium phosphates are not precipitated, hence the whole of the calcium separates as oxalate. The precipitate formed at this reaction will include ferric phosphate, from the iron present in the ash, but this does not interfere with the volumetric determination of calcium, by titration of the oxalate with standard potassium permanganate.

To obtain quantitative precipitation of the calcium as oxalate and a good separation from magnesium, the reaction at the time of precipitation must be carefully controlled. Chapman (1) carries out the precipitation at pH 4 using brom cresol green as an internal indicator. The varying proportions of phosphates present in different samples of plant ash affect the buffer capacity of the solutions and R. E. Shapter prefers to control the reaction at the time of precipitation, by varying the amount of sodium acetate present, using methyl red as an internal indicator. If much magnesium is present significant quantities may still be co-precipitated with the calcium oxalate. In accurate work the precipitate should be dissolved and reprecipitated at the same reaction, to avoid this error.

Magnesium is determined by precipitation as magnesium ammonium phosphate in the filtrate from the calcium precipitation, citrate being used to prevent the separation of any traces of iron and aluminium remaining in solution. The method is simple and convenient but is subject to two sources of error. In the first place the magnesium ammonium phosphate precipitated in this method is not constant in composition, owing to the excess of ammonium salts present in the solution. This error can be reduced by carrying out the precipitation in hot solution as described, or it can be eliminated altogether by solution of the precipitate and reprecipitation from cold solution. The other source of error arises from the precipitation of manganese ammonium phosphate with the magnesium precipitate, thus giving high values for magnesium, particularly if much manganese is present in the plant ash. A small part of the manganese is precipitated as oxalate and determined with the calcium. The larger part of the manganese is, however, precipitated with the magnesium. If necessary a correction can be applied for this by determining, colorimetrically, the amount of manganese present in the ignited precipitate.

When the amount of phosphorus in the plant ash is known, it can be quantitatively removed from solution, by precipitation as ferric phosphate. Manganese can also be removed by oxidation to hydrated manganese dioxide and calcium and magnesium then determined in the phosphate-free solution. The method used by R. E. Shapter in these laboratories is given on p. 281. This method is applicable to a wider range of plant materials than the first method and gives more accurate values for calcium and particularly for magnesium. All phosphates present are precipitated as ferric phosphate at pH 5–6 after the addition of a calculated amount of ferric chloride, sufficient to combine with them. The small excess of iron is also precipitated as ferric hydroxide at this reaction. The ferric chloride solution should be measured to the nearest millilitre, so keeping the excess as small as possible and reducing the bulk of the precipitate to be handled. In calculating the amount to be added the small amounts of iron and aluminium, naturally present in the plant ash, are not taken into account. During the separation of the phosphate by this method the solution should not be boiled, nor heated for unduly long periods, since prolonged heating causes the iron precipitate to become slimy and difficult to filter. Before separating the iron precipitate by filtration, manganese is also precipitated as hydrated dioxide, using bromine water and controlling the acidity within narrow limits. The precipitate of iron, aluminium, phosphorus and manganese is then filtered off, redissolved and reprecipitated, to recover the small amounts of calcium and magnesium readily occluded by it. Having obtained a solution free from iron, aluminium, manganese and phosphorus, calcium is precipitated as oxalate and magnesium as magnesium ammonium phosphate or preferably hydroxyquinolate. If magnesium is precipitated as hydroxyquinolate, sodium and potassium can be determined in the filtrate from the magnesium precipitation, since no sodium salts have been used in the method. After the removal of ammonium salts and the destruction of hydroxyquinoline, by treatment with nitric acid and subsequent ignition, the solution is completely free from substances which might interfere with the determination of the alkalies. The determination of calcium, magnesium, sodium and potassium in the one sample is of particular value when only a restricted amount of material is available. In general, however, it is preferable to determine sodium and potassium in separate aliquots of the original solution.

The Determination of Calcium and Magnesium.

When the amount of phosphorus present has not been determined the following method is used by R. E. Shapter in these laboratories.

Reagents:

Oxalic Acid. Dissolve 25 g. of reagent grade oxalic acid in water and dilute to 1 litre.

Ammonium Oxalate. Dissolve 17 g. of ammonium oxalate in 500 ml. of water, to give a solution which is approximately saturated at room temperature.

Sodium Acetate. Dissolve 250 g. of sodium acetate crystals ($NaC_2H_3O_2.3H_2O$) in water and dilute to 500 ml.

Sodium Citrate. Dissolve 50 g. of purest sodium citrate in water and dilute to 500 ml.

Sodium Phosphate. Dissolve 50 g. of sodium phosphate ($Na_2HPO_4.12H_2O$) in water and dilute to 500 ml.

Method:

For this determination it is generally convenient to use an amount of solution corresponding to about 5 g. of the original plant material. Pipette a suitable aliquot of the silica-free solution of the plant ash (p. 264) into a 250 ml. beaker, add 5–10 ml. of concentrated hydrochloric acid and 1 drop of methyl red indicator solution.

Calcium.

Neutralize the solution just to the full yellow colour of the indicator, by the careful addition of dilute ammonia (1 + 4), adding the ammonia slowly, and with constant stirring, so as to avoid the precipitation of calcium and magnesium phosphates through excessive local alkalinity. Carry out this neutralization in the cold. Then add dilute hydrochloric acid (1 + 4) until the indicator just changes to red. After 5–10 minutes add an excess of 10 ml. of dilute hydrochloric acid (1 + 19) followed by 10 ml. of oxalic acid solution. Boil, and by means of a dip pipette, add 10 ml. of a hot solution of ammonium oxalate, adding the reagent drop by drop, with constant stirring, to the boiling solution. Keep boiling for 1–2 minutes so as to assist the formation of a coarse grained precipitate. A precipitate may not always form at this stage. Its formation depends on the reaction and the amount of calcium present. The reaction is somewhat variable, due to the differing buffer capacities of different samples.

Allow the solution to cool and, when cold, neutralize to pH 5 so as to precipitate, or complete the precipitation, of calcium oxalate. To do this add sodium acetate, drop by drop and with constant stirring, until the indicator shows its intermediate orange to pink tint. The colour should have a tendency to red rather than yellow. If the acidity is reduced too much add a few drops of acetic acid until the proper colour is obtained.

After adjusting the reaction to pH 5, allow the solution to stand overnight, then filter through a 9 cm. Whatman No. 44 filter paper. Wash the precipitate with cold water until chlorine free. If much magnesium is present in the original sample, dissolve the precipitate in warm dilute hydrochloric acid and, when cool, neutralize and reprecipitate exactly as in the first precipitation.

Wash the precipitate of calcium oxalate from the filter paper, add a small excess of dilute sulphuric acid and continue the determination, exactly as described on p. 148. For the titration use 0·1N potassium permanganate.

One ml. of 0·1N permanganate corresponds to 0·0020 g. of calcium (Ca) or 0·0028 g. of calcium oxide (CaO).

Magnesium.

Concentrate the filtrate from the calcium precipitation to about 100 ml. and add 5 ml. of hydrochloric acid. Heat to boiling, add 10 ml. of sodium citrate and 10–15 ml. of sodium phosphate and neutralize the solution with dilute ammonia (1 + 1). Then add an excess of ammonia, equal to about one-third of the volume of solution.

Allow the precipitate to stand overnight, filter through a 9 cm. Whatman No. 44 filter paper and wash thoroughly with dilute ammonia (1 + 50) until free from chlorine. Ignite the filter and precipitate and weigh as $Mg_2P_2O_7$. Multiply the weight by 0·2184 to obtain the amount of magnesium (Mg), or by 0·3621 for magnesium oxide (MgO).

If a correction for manganese is desired, determine this element, colorimetrically, in the precipitate.

The Determination of Calcium and Magnesium.

When the amount of phosphorus present is known, R. E. Shapter *(priv. comm.)* recommends the following method for the removal of iron, aluminium, manganese and phosphorus before the determination of calcium and magnesium.

Reagents:

Ferric Chloride. Dissolve 20 g. of anhydrous ferric chloride or 34 g. of the crystalline reagent in water, add 20 ml. of concentrated hydrochloric acid and dilute to 200 ml. Determine the strength of this solution, by precipitation of an aliquot with ammonia. For use, dilute this stock solution so that it contains 0·525 g. of anhydrous ferric chloride per 100 ml. One ml. then corresponds to 1 mg. of phosphorus.

Bromine Water. Prepare a saturated solution of bromine in water.

Ammonium Oxalate. Dissolve 17–18 g. of ammonium oxalate in 500 ml. of water. This solution is approximately saturated at room temperature.

8-Hydroxyquinoline. Dissolve 25 g. of 8-hydroxyquinoline in water containing 50 ml. of acetic acid and dilute the solution to 500 ml.

Method:

For this determination an amount of solution, corresponding to 3–5 g. of plant material, is generally convenient. Transfer a suitable aliquot of the solution of the plant ash (p. 264) to a 400 ml. beaker and add an amount of ferric chloride solution, equivalent to the amount of phosphorus present. Measure the amount of ferric chloride so that the excess does not exceed 1 ml. Add 5–10 ml. of concentrated hydrochloric acid and one drop of methyl red indicator solution. Neutralize carefully with concentrated ammonia, adding the ammonia drop by drop until the indicator just acquires its yellow colour. Now add dilute hydrochloric acid, drop by drop and with constant stirring, until the red colour of the indicator is just restored. Allow the precipitate to stand for 15–30 minutes, stirring occasionally to ensure re-solution of any precipitated calcium and magnesium phosphates. If, on standing, the reaction changes towards the alkaline side add a further drop or two of dilute hydrochloric acid, but do not exceed the acidity indicated by the first appearance of red.

Place the covered beaker, containing the solution and iron precipitate, in a boiling water bath for 5–10 minutes, but not longer. Then remove, and to the hot solution add, drop by drop from a burette, dilute ammonia (1 + 1) until the full yellow colour of methyl red just appears. At this reaction precipitation of iron, aluminium and phosphorus is complete. Now proceed with the precipitation of manganese in the same solution, without filtering off the iron precipitate. Add a further 10 drops of ammonia and, from a second burette, add bromine water, drop by drop, stirring continuously. When the liquid darkens and becomes opaque, test the reaction with a small piece of blue litmus paper, noting the colour before it is bleached by the bromine. Continue the slow addition of bromine, until there is a tendency for the litmus paper to redden. Restore the alkalinity by the addition of a few drops of ammonia and continue adding bromine. Repeat this procedure until 25 ml. of bromine water have been added. Set aside until convenient to filter. Prolonged heating must be avoided as it tends to make the iron precipitate slimy and difficult to handle.

When ready to filter, heat the beaker in a boiling water bath for 5 minutes and, if not definitely alkaline to litmus, add one or two drops of ammonia. Filter through a 9 cm. Whatman No. 41 filter paper, retaining as much of the precipitate as possible in the beaker. Collect the filtrate in a 400 ml. beaker. Wash the precipitate by decantation, using three lots of 2 per cent. ammonium chloride, made just alkaline to litmus with a few drops of ammonia. Then wash the filter paper twice more with the same solution. Return the filter to the beaker in which the precipitation was made, add 5 ml. of concentrated hydrochloric acid and 50 ml. of water, then digest on the water bath until the precipitate is dissolved and the filter paper macerated.

Neutralize the solution as before, but without cooling to room temperature, and add ammonia and bromine, exactly as previously described. Filter through an 11 cm. Whatman No. 41 filter paper and wash the precipitate completely with hot 2 per cent. ammonium nitrate until the filtrate is chlorine-free. Reject the filter and precipitate.

Calcium.

Reduce the volume of the filtrate to about 150–200 ml., by evaporation. Small traces of iron and aluminium that have escaped precipitation may separate at this stage but this will not affect the determination. Add one drop of methyl red to the concentrated filtrate and neutralize it with concentrated ammonia, adding 5 drops in excess. Heat the solution to boiling and add, drop by drop from a dip pipette, an excess of hot ammonium oxalate. An amount of 10–30 ml. is necessary, depending on the amount of calcium present. Continue to boil for 1–2 minutes, or stand the beaker on a warm water bath, to obtain a coarsely grained precipitate. Allow to stand at least four hours, then filter through a 9 cm. Whatman No. 44 filter paper and collect the filtrate in a silica basin. Wash the precipitate three times with hot 0·1 per cent. ammonium oxalate.

Dissolve the precipitate in hydrochloric acid, reprecipitate and determine calcium as in the method given on p. 147. One ml. of 0·1N potassium permanganate corresponds to 0·0020 g. of calcium (Ca) or 0·0028 g. of calcium oxide (CaO).

Magnesium.

Evaporate the filtrate from the calcium precipitation to dryness on the water bath. When dry, turn off the bath and allow it to cool for 5–10 minutes. Then add 14 ml. of water and 20 ml. of concentrated nitric acid, cover the basin with a clock glass and allow the decomposition of the ammonium salts to proceed. After a few minutes slowly raise the bath to boiling and, when the decomposition is complete, rinse the clock glass and remove it from the basin. Evaporate the solution to dryness. If the ammonium salts have not been completely destroyed take up the residue in 5 ml. of concentrated nitric acid and again evaporate to dryness. Dissolve the residue in 5 ml. of water and 10 ml. of concentrated hydrochloric acid and again evaporate to dryness to expel most of the nitric acid. Dissolve the chlorides remaining in 5 ml. of water and 5 ml. of concentrated hydrochloric acid, transfer the solution to a 250 ml. beaker and dilute it, so that it does not contain more than 0·1 g. of magnesium per 100 ml.

Make the solution just slightly alkaline by the addition of concentrated ammonia, heat to 65–80° C. and add 5–10 ml. of the hydroxyquinoline reagent, or sufficient to give a slight excess after the precipitation of all magnesium. Stir well and make distinctly alkaline by the addition of 5–10 ml. of concentrated ammonia, to complete the precipitation. An excess of hydroxyquinoline is indicated by the yellow colour of the solution after the precipitate has settled. If an excess is not present add more reagent, but avoid too large an excess, since hydroxyquinoline is not very soluble in ammoniacal solution and separates from it.

Allow the precipitate to stand until cool, or overnight if desired, then filter through a 9–11 cm. Whatman No. 41 filter paper, washing with warm dilute ammonia (1 + 50). Transfer the precipitate to a crucible, dry and ignite slowly, finishing the ignition at a bright red heat. Weigh as magnesium oxide (MgO). To obtain the weight of magnesium (Mg) multiply by 0·6032.

POTASSIUM

Potassium may be determined directly, in the solution of the plant ash, either by the perchlorate or the volumetric cobaltinitrite methods. Both of these methods possess advantages over the classical determination, which involves the laborious separation of sodium and potassium chlorides and the determination of potassium in the mixed chlorides.

In the perchlorate method sulphates interfere and must first be removed, by precipitation as barium sulphate. Phosphates do not interfere, provided that an excess of perchloric acid is used and the evaporations with this acid are not carried quite as far as in the absence of phosphoric acid. The method given for the determination of potassium in the hydrochloric acid extract of soils (p. 149) is applicable.

Sulphates do not interfere with the volumetric cobaltinitrite method and the procedure outlined on p. 178 for the determination of exchangeable potassium in soils by this method is also applicable to solutions of plant ash. A suitable aliquot, preferably containing less than 40 mg. of potassium, is evaporated to dryness, after the addition of a little calcium chloride, and ignited for a few minutes at an incipient dull red heat. Potassium and other soluble salts are then extracted with hot water and the determination continued as described.

The most convenient and accurate method for the determination of potassium in plant ash is a gravimetric method, based on that originally devised by Krügel and Retter (3) for potassic fertilizers. In this method the advantages of the cobaltinitrite and perchlorate methods are combined, with the mutual elimination of the disadvantages of each. Potassium can be separated quantitatively, as sodium-potassium cobaltinitrite, from weakly acid solutions containing iron, aluminium, calcium, magnesium, phosphates and sulphates, but the composition of the precipitate varies considerably, depending on the nature of the solution, the amount of potassium present, the reagent used and the conditions of precipitation, as well as several other factors (6). Precipitation as sodium-potassium cobaltinitrite is only suitable for the determination of potassium when all these conditions are closely controlled and allowance is made for the variation in the proportion of sodium to potassium in the precipitate (p. 162). However, as a means of separation of potassium, precipitation as cobaltinitrite is excellent. Precipitation is quantitative, if the solutions are not too dilute and an excess of reagent is used. Although the composition of the precipitate is not constant, the potassium in it can be determined by the perchlorate method, since it has been separated from all substances, particularly sulphates, which interfere with this method. The cobalt present in the cobaltinitrite precipitate does not interfere with the perchlorate determination, cobalt perchlorate being readily soluble in alcohol. Potassium perchlorate is constant in composition and accurate values for potassium are readily obtained.

The separation of potassium from sulphates by the combination of the two methods avoids errors due to the retention of potassium by the barium sulphate precipitate. Ammonium salts, if present in the original solution do not interfere, for they are destroyed during the determination. Although ammonium is co-precipitated with potassium as cobaltinitrite, when the precipitate is dissolved in hydrochloric acid and warmed the nitrous acid liberated reacts with any ammonia present, destroying it and setting free elementary nitrogen.

Water saturated with sodium-potassium cobaltinitrite, or 35 per cent. alcohol, are the best wash liquids for washing the cobaltinitrite precipitate. It is least soluble in these solutions. However, a saturated aqueous solution of the salt is not very stable and slowly decomposes within a few hours, while filtration with alcohol tends to be unduly slow. Since washing of the precipitate at this stage need not be very thorough, 5 per cent. acetic acid may be used. Provided that the volume is kept small, the loss of sodium-potassium cobaltinitrite, by solution in this reagent, will not introduce a significant error. Any excess of precipitating reagent remaining in the precipitate does not interfere with the subsequent perchlorate determination.

For the precipitation of sodium-potassium cobaltinitrite, separate solutions of cobalt nitrate and sodium nitrite are preferable to a mixed sodium cobaltinitrite reagent, since the former salts each give stable solutions, whereas sodium cobaltinitrite solution slowly decomposes on keeping.

In dissolving the evaporated residue prior to precipitation with cobalt nitrate and sodium nitrite care should be taken to avoid hydrolysis of iron and aluminium, by adding one or two drops of hydrochloric acid, together with the acetic acid, before the water. Hydrolysis of iron and aluminium may cause the subsequent filtration of the cobaltinitrite precipitate to be unduly slow.

In all methods for the detemination of potassium, precautions must be taken to see that no ammonia is reabsorbed from the laboratory atmosphere, after it has been eliminated from the determination. If it is reabsorbed, as may readily occur during the evaporation of acid solutions, erroneously high values will be obtained for potassium, since ammonia is always partly co-precipitated with potassium.

The Determination of Potassium.

Reagents:

Cobalt Nitrate. Dissolve 200 g. of potassium-free cobalt nitrate in water and dilute to 1 litre.

Sodium Nitrite. Dissolve 350 g. of potassium-free sodium nitrite in water and dilute to 1 litre.

5 % Acetic Acid. Dilute 50 ml. of glacial acetic acid to 1 litre.

Method:

For this method use the solution of plant ash prepared as described on p. 264, taking sufficient to yield 0·2–0·5 g. of potassium perchlorate. Transfer a convenient aliquot of the solution to a pyrex glass evaporating basin and evaporate it to dryness on a water bath. Remove the basin from the bath, add two drops of concentrated hydrochloric acid, 3 ml. of glacial acetic acid and 15 ml. of water. Stir and allow to stand until solution is as complete as possible. Then add 10 ml. of sodium nitrite and 10 ml. of cobalt nitrate, stirring vigorously for about 30 seconds. Cover the basin and leave to stand overnight, to ensure complete precipitation of the potassium.

Filter the solution and precipitate through a 9 cm. Whatman No. 44 filter paper, rinsing the basin and filter paper two or three times with water freshly saturated with sodium-potassium cobaltinitrite, or with cold 5 per cent. acetic acid. It is not necessary to remove the precipitate completely from the basin. Then rinse the upper edge of the filter paper once only, avoiding loss of the finely divided precipitate, through surface tension effects.

When the filter paper has drained place the basin, in which the precipitation was made, beneath the funnel and dissolve the precipitate by adding 5 ml. of warm dilute hydrochloric acid (1 + 4) from a dip pipette. Wash the filter paper alternately with hot water and further 5 ml. portions of hydrochloric acid until the whole of the cobaltinitrite has been dissolved. Complete the washing of the filter paper with hot water alone. To the filtrate add 5 ml. of concentrated hydrochloric acid, cover the basin with a clock glass and heat on a water bath at 60° C. until any ammonia present is destroyed and the solution becomes clear rose red in colour. Remove and rinse the clock glass, then evaporate the solution to dryness. Dissolve the chlorides remaining in 5 ml. of warm water, add 5 ml. of 20 per cent. perchloric acid and evaporate, first on the water bath and finally on a hot plate, until copious fumes of perchloric acid are evolved. An excess of perchloric acid must be present to expel all hydrochloric acid. If an excess of perchloric acid is not present, add more perchloric acid and again evaporate until copious white fumes are produced.

When the white fumes appear add 10–15 ml. of water, to dissolve the perchlorates, and then 1 ml. of perchloric acid and continue the evaporation nearly to dryness, on the water bath. Finish the evaporation over a hot plate until dense white fumes have been produced for some time and the liquid just sets to a pasty crystalline mass when cold.

Continue with the determination, exactly as described on p. 150, washing the perchlorates once with alcohol, before their final evaporation with perchloric acid and filtration through a Gooch crucible charged with asbestos.

Multiply the weight of potassium perchlorate by 0·2823 to obtain the amount of potassium (K) or by 0·3401 to obtain potassium oxide (K_2O).

SODIUM

For the determination of sodium in plant ash a direct method is preferable, particularly when the amounts are small. If determined by difference, after separating the alkalies as mixed chlorides and deducting the amount of potassium chloride present, the whole of the errors fall upon the sodium.

Moreover, complete separation of magnesium from the alkalies is difficult and Shapter (8) has found that small amounts of calcium and magnesium often escape precipitation in the alcoholic ammonium carbonate separation.

Sodium is most conveniently determined in a phosphate-free solution of the plant ash by precipitation as sodium uranyl magnesium acetate [$Na(UO_2)_3Mg(CH_3COO)_9.8H_2O$] according to a modification of Kahane's gravimetric method (5). As the precipitate contains only 1·5 per cent. of sodium the method is very sensitive and small amounts of the element give weighable precipitates. Since sodium uranyl magnesium acetate is not completely insoluble in the reagent, temperature changes during precipitation should be avoided. The volumes of reagents used and the conditions of precipitation must also be rigidly followed. As a large excess of reagent is necessary for the quantitative separation of the sodium salt, not more than 5 mg. of sodium should be present in any one determination, when using the quantities of reagents indicated. By doubling the volumes used throughout, up to 10 mg. may be precipitated. The most suitable amounts of sodium for this determination are 1–5 mg. and, if possible, the aliquot should be chosen to contain an amount within this range. Potassium, if present in relatively large amounts in proportion to the sodium, introduces error, since it is partly co-precipitated with the sodium uranyl magnesium acetate. Precipitation of the potassium uranyl magnesium acetate only occurs slowly and the error increases with the time of standing. Sodium uranyl magnesium acetate is completely precipitated in 30 minutes and filtration should be carried out as soon as possible after this. If the amounts of potassium present are small the error is negligible, even if the precipitate is allowed to stand for several hours. With amounts of 15–20 mg. of potassium, however, the precipitate should not be allowed to stand for more than 1–1½ hours before filtration, to avoid errors from co-precipitation. Amounts of potassium much in excess of 20 mg. should be avoided, by taking smaller aliquots for the determination. So as to prevent errors from this source, the determination of potassium should precede that of sodium. Even when the ratio of potassium to sodium in the plant ash is excessively wide the uranyl acetate method gives much more accurate values than the indirect mixed chloride determination.

Phosphates, unless removed, interfere with this determination, since they give rise to uranyl phosphate which would be precipitated with the sodium uranyl magnesium acetate. As an excess of magnesium does not interfere with the determination, phosphates are most conveniently removed by precipitation as magnesium ammonium phosphate, in ammoniacal solution. The traces of iron and aluminium also present in the ash are separated at the same time.

Calcium only interferes with the determination if sulphates are also present in an amount sufficient to exceed the small solubility of calcium sulphate at the time of precipitation, so leading to the presence of calcium sulphate in the sodium precipitate. In the presence of calcium sulphate smaller than this, correct values are obtained. The method should not be used for the determination of sodium in a sulphated ash without removal of sulphates. However, the trace of sulphuric acid introduced in the removal of silica in the method of ashing described on p. 263 does not yield enough calcium sulphate to interfere with the determination.

The Determination of Sodium.

Reagents:

See p. 176.

Method:

Transfer a suitable aliquot of the solution of the ash, generally corresponding to 1–3 g. of plant material, to a 100 ml. volumetric flask, add 5 ml. of a 5 per cent. solution of magnesium acetate and 30 ml. of concentrated ammonia. Shake vigorously and allow to stand for some time. When quite cold dilute to the graduation mark with water and thoroughly mix the contents of the flask. Leave to stand overnight, to ensure complete separation of all phosphate present.

After standing overnight, filter the solution through a dry 9 cm. Whatman No. 44 filter paper. Discard the first runnings then collect sufficient of the filtrate in a dry flask.

Pipette a suitable volume of the filtrate, sufficient to contain 1–5 mg. of sodium, but not more than 5 mg. nor more than 15–20 mg. of potassium, into a 100 ml. pyrex evaporating basin and evaporate the solution to dryness on the water bath. When dry, add 3 ml. of water and 5 ml. of concentrated nitric acid, cover the basin with a clock glass, and leave on the bath to decompose ammonium salts. When the initial reaction is over, remove the clock glass, rinse it into the basin and re-evaporate to dryness.

When the contents of the basin are dry, remove it from the water bath and allow it to cool. Then dissolve the residue, as completely as possible, in 6 ml. of cold water, stirring with a glass rod. Add 15 ml. of uranyl magnesium acetate reagent and stir vigorously for about 15 seconds or until a precipitate forms. Cover the basin and allow it to stand for 30 minutes, but not longer than 1–1½ hours if much potassium is present. Then filter through a small Gooch crucible, charged with asbestos. Wash the precipitate of sodium uranyl magnesium acetate twice with 2 ml. portions of the reagent and then five times with 96 per cent. alcohol saturated with the triple salt, transferring the precipitate quantitatively to the Gooch crucible. Then dry the crucible in an oven at 105° C. for not more than one hour, cool in a desiccator and weigh. Multiply the weight of precipitate by 0·01500 to obtain the amount of sodium (Na) or by 0·02022 to obtain sodium oxide (Na_2O) in the aliquot taken.

As an alternative to oven drying, ether may be used. In this case wash the crucible and precipitate three times with ether, wipe dry on the outside and allow to stand near the balance for 15 minutes before weighing.

PHOSPHORUS

If the plant material is naturally rich in basic substances phosphorus is not lost during ashing at temperatures up to 600° C. However, during the dry ashing of seeds, in which the proportion of phosphorus to basic constituents is large, some phosphorus may be lost unless an excess of magnesium salts is added to the sample before ashing. In the presence of magnesium salts no loss of phosphorus occurs up to 800° C.

In most plant materials, other than seeds, phosphorus can be determined in the solution of the ash (p. 264) prepared for the determination of calcium, magnesium, potassium and sodium. It can always be determined in the solution of the ash after ashing in the presence of an excess of magnesium salts (p. 268). Some pyrophosphates are formed during all methods of dry ashing and, in order to secure correct values for phosphorus, these must be converted to orthophosphates before precipitation with ammonium molybdate. Conversion of pyrophosphates does not occur immediately the ash is acidified but the acid solution must be boiled vigorously, or heated on a water bath for a considerable time, to ensure complete conversion. This is secured during the preparation of the solution of the ash. Details of a method for the determination of phosphorus in the solution of the ash are given on p. 295.

The most convenient method for the determination of phosphorus is one in which the organic matter is destroyed by digestion with nitric, sulphuric and perchloric acids and phosphorus precipitated in the filtrate from the digest by the Lorenz method. This method is strongly recommended since the digestion is simple and rapid and the phosphorus is determined directly on a separate portion of the original sample. No loss of phosphorus occurs during digestion, either through volatilization or retention in the insoluble residue, and no pyrophosphates are formed. Details of this method are given below.

In most methods phosphorus is precipitated from acid solutions as ammonium phosphomolybdate. As the composition of the precipitate depends upon the conditions of precipitation, it is sometimes dissolved in ammoniacal solution and the phosphorus reprecipitated as magnesium ammonium phosphate. More frequently phosphorus is determined by titrating or weighing the yellow precipitate after precipitating it under controlled conditions. In the usual volumetric methods, based on Pemberton's original titration of the yellow precipitate with standard sodium hydroxide, the factors found by different workers for the phosphorus equivalent of the precipitate show considerable variation and Shapter (7) has also found that the factor varies with the amount of phosphorus precipitated. A different factor is necessary for the determination of very small amounts of phosphorus. Acidity and the temperature at the time of precipitation also affect the composition of the precipitate. Apart from the uncertainty of the values sometimes obtained, the volumetric method is not as convenient as the gravimetric method.

The yellow precipitate, produced under the conditions of precipitation in the Lorenz method, is remarkably constant in composition and the method gives accurate values for phosphorus over a wide range of concentrations. Constancy of composition of the precipitate in the Lorenz method is apparently attained by the higher concentrations of acid and molybdate used for the precipitation. The precipitate contains 1·44 per cent. of phosphorus so the method is capable of determining very small amounts of this element.

After precipitation and filtration the precipitate is washed with dilute ammonium nitrate and moisture removed by washing with acetone (or alcohol and ether). It is then transferred to a desiccator at a reduced pressure, constant weight being attained within 30 minutes. The yellow precipitate obtained by this method filters well, except when only very small amounts of phosphorus are present, as in some cereal straws. In such cases the addition of Celite filter aid (p. 151) assists the filtration.

The most suitable amounts of phosphorus for determination by this method are 1–10 mg., giving precipitates of 0·07–0·7 g., and the amount of sample taken should be judged accordingly. However, amounts down to 0·5 mg. or even less can be determined with considerable precision, while micro-technique can be used for the determination of even smaller amounts. Amounts in excess of 15 mg. should be avoided, since precipitation of this larger amount may unduly reduce the strength of the reagent.

The Determination of Phosphorus (Wet Digestion Method).

Reagents:

See p. 152.

Method:

Digest a suitable amount of plant material, containing about 1–10 mg. of phosphorus, with 3 ml. of sulphuric acid, 4 ml. of perchloric acid and sufficient nitric acid, according to the method described on p. 272. When the organic matter has been completely destroyed and the digest fumed strongly for 2–3 minutes, remove it from the hot plate and allow to cool. Add 30 ml. of hot water, shake until solution is as complete as possible and filter through a 9 cm. Whatman No. 44 filter paper, collecting the filtrate in a tall shaped 150 ml. beaker. Wash the digestion flask and silica residue twice with warm dilute nitric acid (1 + 19) and complete the washing with hot water.

Stand the beaker in a boiling water bath and evaporate the filtrate until the volume is reduced to about 5 ml., or as far as the evaporation will proceed readily. Remove the beaker from the bath, add 15–15·5 ml. of concentrated nitric acid and 34 ml. of water. Heat the contents just to boiling, stirring well to dissolve most of the calcium sulphate. Remove from the flame, stir for 10 seconds to cool the overheated sides of the beaker, rapidly add 50 ml. of Lorenz sulphate-molybdic acid reagent, stir for 30 seconds, cover the beaker and leave to stand for two hours, or overnight if more convenient.

Filter through a Gooch crucible, fitted with a small circle of Whatman No. 42 filter paper, cut so as to cover the holes, but not to touch the edges, or through a Gooch crucible charged with asbestos. Before use, dry the crucible in an oven at 100° C., cool and weigh. If a crucible with a disc of filter paper is used, drying for 15 minutes is sufficient. Wash the precipitate four times with 2 per cent. ammonium nitrate, made just acid to litmus with a drop of nitric acid, if necessary. Then wash three times with acetone and draw air through the crucible for half to one minute. Finally place the crucible in a desiccator, without dehydrating agents, evacuate to about 200 mm. and leave for 30 minutes before weighing.

Multiply the weight of precipitate by 0·0144 to obtain the amount of phosphorus (P) or 0·03295 to obtain phosphoric anhydride (P_2O_5).

The Determination of Phosphorus (Dry Ashing Method).

Reagents:

See p. 152.

Method:

Ash a suitable amount of plant material, containing about 1–10 mg. of phosphorus, by the method described on p. 268. When ashing is complete moisten the residue remaining in the silica basin with about 25 ml. of water. Cover the basin with a clock glass, add 10 ml. of concentrated hydrochloric acid and digest on the water bath for 15–20 minutes. Then remove and rinse the clock glass and evaporate the solution to dryness. Leave the basin on the bath for a further period of half an hour, to render the silica insoluble. Dissolve the dried salts in 20 ml. of warm water and 2 ml. of concentrated hydrochloric acid and again evaporate to dryness and heat on the bath for a further period of half to one hour.

When the silica has been completely dehydrated take up the residue in about 25 ml. of warm water and 5 ml. of concentrated nitric acid. Warm until solution is complete, then filter through a 9 cm. Whatman No. 44 filter paper, collecting the filtrate in a tall shaped 150 ml. beaker. Wash the filter paper and insoluble residue with 3 or 4 lots of hot dilute nitric acid (1 + 39) then continue washing with hot water until the filtrate amounts to about 130 ml. Place the beaker in the water bath and evaporate the filtrate to dryness.

As an alternative to ashing with magnesium acetate the hydrochloric acid solution of the ash, prepared by the method given on p. 264, may be used. In this case transfer a suitable aliquot of the solution to a tall shaped 150 ml. beaker and evaporate the solution nearly to dryness in the water bath. Add 5 ml. of water and 5 ml. of concentrated nitric acid and again evaporate to dryness to remove chlorides.

Whichever method is used to obtain a solution of the phosphoric acid, free from silica and chlorides, proceed from this point in exactly the same manner. Dissolve the residue in the beaker in 15–15·5 ml. of concentrated nitric acid and 34 ml. of water, and add 1 ml. of concentrated sulphuric acid.

Heat the contents just to boiling and precipitate the phosphate in exactly the same way as in the previous method.

CHLORINE

Chlorine is readily lost during the ordinary ashing of many plant materials. For its quantitative retention in the ash the sample must be ignited in the presence of an alkali or an alkaline earth. Husband and Godden (2) have shown that ashing with lime gives the best values. The lime must be thoroughly incorporated with the sample, and the mixture moistened and digested on the water bath to ensure this.

In Husband and Godden's method the ash is extracted with hot dilute nitric acid, a measured volume of standard silver nitrate added, to precipitate all chlorides, and the excess of silver nitrate determined in an aliquot of the solution. Since free nitric acid is present it is necessary to use Volhard's method of titration, in which the excess of silver is titrated with standard ammonium thiocyanate, using ferric alum as an indicator. In this method the precipitated silver chloride is generally removed, by filtration, before carrying out the titration, otherwise the end point is indefinite, owing to the interaction between the silver chloride and the soluble thiocyanate. Instead of filtration, however, some analysts prefer to add 1 ml. of pure nitrobenzene, which prevents this reaction by withdrawing the silver chloride from the aqueous phase and concentrating it at the interface of the two liquids. Volhard's titration is subject to a small error, since the precipitated silver chloride adsorbs about 0·7 per cent. of its weight of silver from the solution. Values for chlorine are therefore slightly high.

The determination of chlorine can be considerably simplified by using an electrometric method for its titration in the plant ash. In the method used by R. J. Best in these laboratories the ash is digested with water, or with a small amount of sulphuric acid, and an aliquot of the filtered digest is titrated with standard silver nitrate, using the silver-silver chloride electrode described on p. 40. The nitric acid digest of the plant ash can also be titrated electrometrically, without filtration, provided that the concentration of the free nitric acid is reduced sufficiently before titration. In this latter case the electrometric titration is most conveniently carried out on the whole of the ash.

Chlorides in plant materials can also be determined very rapidly, and without ashing, by the direct electrometric titration of an aqueous suspension of the sample with standard silver nitrate. The end point is not quite sharp, tending to drift slowly, but the values obtained by this simple method are very useful for many purposes.

The Determination of Chlorine: Husband and Godden's Method.

Reagents:

0·1N Silver Nitrate. Dissolve 16·989 g. of pure silver nitrate in water and dilute to 1 litre.

0·1N Ammonium Thiocyanate. Dissolve about 8 g. of ammonium thiocyanate in water and dilute to 1 litre. Titrate this solution against the standard silver nitrate solution and adjust its volume so that it is 0·1N.

Ferric Alum. To 100 ml. of a saturated solution of ferric ammonium sulphate add 5 ml. of concentrated nitric acid.

Method:

Ash 3–5 g. of plant material with one-quarter of its weight of finely divided lime (chlorine-free) by the method described on p. 268.

Moisten the ash with 25–30 ml. of hot dilute nitric acid (1 + 4), taking care to avoid loss by effervescence. After a few minutes filter the solution through a 9 cm. Whatman No. 44 filter paper, collecting the filtrate in a 250 ml. volumetric flask. Wash the basin and filter paper with warm water.

When the filtrate is cool add an excess of 0·1N silver nitrate, measuring the volume accurately. Shake the solution until the precipitate of silver chloride clots, then dilute to volume and thoroughly mix the contents of the flask. Filter through a dry 11 cm. Whatman No. 44 filter paper, rejecting the first runnings and collecting the rest of the filtrate in a dry flask.

To a suitable aliquot, usually 100 ml., add 2–3 ml. of ferric alum solution and then titrate it with 0·1N ammonium thiocyanate, against a white background, until the first signs of the reddish brown colour of ferric thiocyanate persist. The amount of standard thiocyanate used corresponds to the excess of silver nitrate in the aliquot titrated. From the volume of silver nitrate added calculate the amount of chlorine present in the original sample.

One ml. of 0·1N silver nitrate corresponds to 0·00355 g. of chlorine.

The Determination of Chlorine: Electrometric Method.

Reagents:

N/71 Silver Nitrate. Dissolve 2·393 g. of pure dry silver nitrate in water and dilute to 1 litre.

Equipment: The arrangement of R. J. Best's silver-silver chloride electrode and quinhydrone reference half cell is fully described on p. 41.

Method:

Ash 3–5 g. of plant material with one-quarter of its weight of finely divided lime (chlorine-free) by the method given on p. 268.

Digest the ash with 50 ml. of water or 50 ml. of 0·05N sulphuric acid on a water bath for 10–15 minutes, filter through a 9 cm. Whatman No. 44 filter paper and wash thoroughly with hot water. Collect the filtrate in a 200 ml. volumetric flask and, when cold, dilute to the mark.

To a suitable aliquot, usually 50 ml., add a drop of methyl red and then N sulphuric acid sufficient to give a few drops in excess of the amount necessary to acidify the solution. Introduce the silver-silver chloride electrode and agar connecting tube and titrate the solution with N/71 silver nitrate by the method described on p. 43.

One ml. of N/71 silver nitrate corresponds to 0·0005 g. of chlorine.

Alternative Method:

Ash 1–2 g. as before. To the cold ash add 20 ml. of water, then add dilute nitric acid (1 + 4) just sufficient to acidify the ash. An excess of nitric acid must be avoided. If too much is added nearly neutralize with dilute ammonia. Dilute the solution to about 50 ml., digest for 5 minutes on a water bath and then transfer to a 100 ml. beaker. When cold, introduce the silver-silver chloride electrode and agar connecting tube and titrate, electrometrically, as before.

SULPHUR

Sulphur cannot be determined in the ash obtained by the ordinary methods of ashing. During the ignition of plant materials organic sulphur is oxidized and passes off with the other products of combustion. If these gases are passed through strong oxidizing agents, as in the original Fresenius method, the oxides of sulphur are absorbed and converted to sulphates, which can be precipitated as barium sulphate. Such methods give reliable results but are not always convenient.

Other methods involve oxidation and fusion with sodium peroxide, or oxidation with copper and ammonium nitrates, sulphur being converted to sulphate and precipitated as barium sulphate in each case. Neither method is very convenient and the relatively large amounts of salts present at the time of precipitation of the barium sulphate introduce errors due to occlusion by the precipitate. Values obtained for sulphur in plant materials are frequently unreliable although, in experienced hands, good results can be obtained for many substances by the sodium peroxide method.

Marston (4) finds that, of several methods investigated, determination by combustion with oxygen in a steel bomb is most convenient and gives the most consistent values for sulphur. The method determines both inorganic and organic sulphur. Complete combustion of the organic matter is obtained. The silica and basic constituents become fused into clear glass beads, a large proportion of the nitrogen in the sample is burned to nitric acid, and all of the sulphur is converted to sulphate. Volatile compounds of sulphur are not formed. After combustion the contents of the bomb are dissolved in hydrochloric acid, silica separated and sulphate precipitated as barium sulphate in very dilute acid solution. Throughout the determination electric heaters should be used since sulphur may be derived from the coal gas, if this is used as a source of heat. Details of Marston's procedure are given below.

The Determination of Sulphur (4).

Apparatus:

Emerson Bomb. The standard steel bomb from an Emerson bomb calorimeter is used for this determination. The lead washer of the bomb must be completely covered with a pure tin gasket to prevent loss of sulphur, due to the attack of the lead by the nitric and sulphuric acids produced during combustion of the sample.

Method:

Compress about 2 g. of the finely ground material into a hard tablet around an iron fuse wire, by means of a suitable die. Place the compressed tablet on the capsule, on which it is to be ignited in the bomb, and weigh to obtain the amount of material taken. Make an allowance for the weight of fuse wire used. Set up the capsule and its contents in the bomb and fix the fuse wire in position. Assemble the bomb and screw down tightly. Then connect to an oxygen cylinder, introduce oxygen to a pressure of 20 atmospheres, close the valve and immerse the bomb in water to detect any leak of oxygen. Then fire the bomb electrically and leave for a few minutes to cool. When cool, carefully open the valve to relieve the pressure.

Transfer the contents of the bomb by washing into a pyrex evaporating basin. Acidify by adding 10 ml. of concentrated hydrochloric acid and evaporate to dryness to remove nitric acid and to dehydrate the silica. Repeat the evaporation using a further 5 ml. of hydrochloric acid and 15 ml. of water. When the silica has been dehydrated, dissolve the residue in 3–5 ml. of hydrochloric acid and 15 ml. of warm water, filter through a 9 cm. Whatman No. 44 filter paper and wash completely with warm water.

Adjust the acidity of the filtrate to approximately 0·02N, by just neutralizing with dilute ammonia, using methyl red as indicator, then adding 1 ml. of hydrochloric acid (1 + 1) and diluting to approximately 250 ml. Heat the solution to incipient boiling, add an excess of barium chloride and cover the beaker with a clock glass. Maintain at a temperature of about 100° C. for two hours, then allow to cool slowly overnight. By means of a bent capillary, withdraw the major part of the clear supernatant liquid, then collect the precipitate by filtering the rest of the liquid through a 9 cm. Whatman No. 44 filter paper. Wash thoroughly with cold water. Dry in a platinum crucible and ignite in a muffle furnace. Then treat the residue with one drop of nitric acid, fume off carefully and re-ignite at 500° C. Place the lid on the crucible and leave in the balance room to come into equilibrium with the air.

Multiply the weight of precipitate obtained by 0·1373 to obtain the amount of sulphur.

REFERENCES

(1) Chapman, H. D.: *Soil Sci.*, 1928, *26*, 479–486.

(2) Husband, A. D. and Godden, W.: *Analyst*, 1927, *52*, 72–75.

(3) Krugel, C. and Retter, A.: *Z. Anal. Chem.*, 1934, *96*, 314–319.

(4) Marston, H. R.: *J. Agric. Sci.*, 1938, *28*, 679–687.

(5) Piper, C. S.: *J. Agric. Sci.*, 1932, *22*, 676–687.

(6) Piper, C. S.: *J. Soc. Chem. Ind.*, 1934, *53*, 392T–396T.

(7) Shapter, R. E.: *Aust. Chem. Inst. J. and Proc.*, 1937, *4*, 413–425.

(8) Shapter, R. E.: *Aust. Chem. Inst. J. and Proc.*, 1942, *9*, 42–44.

CHAPTER IV

THE DETERMINATION OF THE TRACE ELEMENTS

In this section methods are given for the determination of boron, cobalt, copper, iron, manganese, molybdenum and zinc in plant materials. Since these elements occur in such small quantities in plants, in some cases to the extent of less than 0·1 mg. per kg., the utmost precautions must be taken to avoid their accidental introduction during the course of the analysis. Amounts of impurities which would pass without notice in the determination of the major constituents, such as calcium, magnesium or potassium, must be scrupulously excluded. In the determination of any of the trace elements all likely sources of their introduction as impurities should be carefully considered. This applies particularly to iron, copper and zinc, since so much laboratory equipment is made in steel, copper or brass. In the determination of micro-amounts of copper and zinc sufficient of these metals may be introduced by the handling of unprotected brass water taps, etc., to invalidate the analysis.

The precautions to be observed during the collection and grinding of the plant sample have already been detailed (pp. 253 and 256). Commonsense precautions should be taken in the laboratory in which the determinations are to be carried out. Dusty operations should not be performed in the same laboratory and salts of the elements to be determined should be handled as little as possible. If a few such simple precautions are taken contamination with boron, cobalt, manganese and molybdenum during analysis will be practically non-existent, except for the small amounts derived from the distilled water and the reagents used. This will be dealt with later. However, greater care is necessary for the exclusion of traces of iron, copper and zinc. When it is not possible to replace metal laboratory fittings with those of aluminium, synthetic plastics, or other materials, they should be protected with a coat of one of the synthetic resin lacquers (colourless or black). This protection is quite effective for door handles, water and gas taps, etc. Any copper or brass parts of the electric oven used for drying, particularly the interior, should be protected with a heat resisting synthetic lacquer. Porcelain Bunsen or Teclu burners should replace gas burners of iron or brass. Copper water baths and water heaters should be rigidly excluded from laboratories in which this element is to be determined. Apart from direct contamination, the basic salts formed on the outside of copper water baths are too readily diffused as a fine dust in the laboratory atmosphere. Stainless steel water baths are very satisfactory and can even be used with reasonable safety in the determination of iron. When low temperature ignitions are necessary, as in the micro-determination of zinc, a sheet of aluminium, bent into a semi-cylindrical shape and placed in the muffle furnace, will prevent contamination from particles falling from the wall of the muffle.

For trace element determinations the composition of the glass used for all chemical apparatus is very important. Borosilicate glass yields measurable amounts of boron to solutions, particularly alkaline solutions. Beakers and flasks made from it must on no account be used during the determination of small quantities of boron. Many resistance glasses, including some of the borosilicate glasses, contain zinc as one of their constituents and such glassware must not be used for zinc determinations. Considerable amounts of zinc are taken up by both acid and alkaline solutions from such glass (17). Of the commercially available glasses suitable for chemical apparatus, Pyrex and Duran are the only ones which are practically zinc-free. Both may contain traces of zinc derived from normal impurities in the raw materials used in their manufacture. Pyrex glass is very satisfactory for use in the determination of all the trace elements except boron. Strong acids can be re-distilled from pyrex glass apparatus and have been stored in pyrex reagent bottles for three years without appreciable contamination with heavy metals. However, alkaline solutions attack it slightly and increase appreciably in zinc content after long standing.

As already mentioned, distilled water may be one of the sources from which impurities of some of the trace elements are derived. Ordinary distilled water contains appreciable amounts of iron, copper, zinc and tin. Owing to the widespread use of tinned copper condensers, it may sometimes contain more copper than the original town supply. Ordinary distilled water is generally entirely suitable for the determinations of boron and manganese. However, for determinations of all other trace elements water of high quality, free from heavy metals, must be used. This is most conveniently obtained by the redistillation of ordinary distilled water from a pyrex glass boiling flask and pyrex condenser.

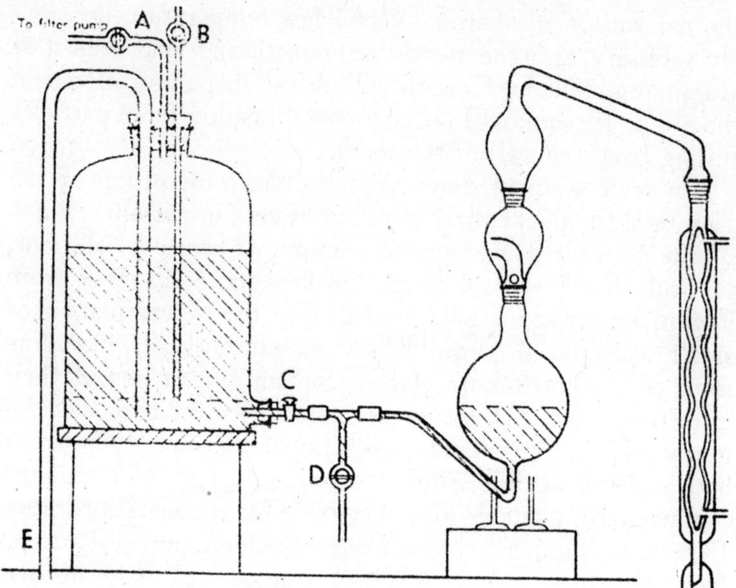

Fig. 17. Pyrex glass still for the redistillation of water.

A simple arrangement for the continuous production of redistilled water at a rate of 1–2 litres per hour is shown in Fig. 17. The supply of ordinary distilled water in the large Mariotte bottle (25 l.) maintains the constant level of water in the boiling flask (2-3 l. capacity). The steam is condensed in a double surface pyrex condenser, the wide bore of which minimizes back pressure in the boiling flask. The latter is heated by two porcelain Bunsen burners. When necessary, the boiling flask can be completely drained by opening the lower stopcock, and the whole apparatus can be arranged for refilling and draining without dismantling it in any way. If an electrical heater is used, automatic relays can be incorporated, to protect the still from boiling dry.

All glass apparatus for the determination of the trace metals must be rinsed with dilute acid and with glass distilled water, after the usual washing with tap water. Repeated rinsings with redistilled water alone are not sufficient to remove completely the last traces of heavy metals, since small amounts are adsorptively bound to the glass surface and can only be removed by replacement with other ions, such as hydrogen ions. Rinsing with half normal acetic or hydrochloric acid, followed by redistilled water, is quite effective in removing these adsorbed heavy metal ions. The reaction is reversible and a clean glass surface will adsorb further quantities of metal ions from tap water if it again comes in contact with it.

To avoid losses during any determination, by adsorption on the glass walls of vessels, transference of solutions from one vessel to another should be carried out at an acid reaction, as far as possible.

Significant amounts of many of the trace elements occur, even in the best grades of analytical reagents. The purification from the last traces of these elements presents difficulties on the commercial scale, but can be easily carried out in the laboratory.

For crystalline salts a single recrystallization of a good quality reagent from glass distilled water generally gives a product of high purity. If freedom from boron is required, the operation must be carried out in boron-free glassware. A single recrystallization reduces the amounts of all of the heavy metals sufficiently to give a very low value in a blank determination. However, it is generally much simpler and more convenient to purify salts from certain heavy metals by an extractive method similar to that used in the actual determination. By extracting a concentrated solution of the salt, at a suitable reaction, with a chloroform or carbon tetrachloride solution of dithizone the whole of the copper, zinc, cobalt and other dithizone metals can be removed. If the excess of dithizone does not interfere, the extracted solution can be diluted to a suitable strength and used directly in the analytical determinations. Sodium diethyldithiocarbamate and amyl alcohol can be similarly used for the removal of copper from certain reagents.

The strong acids are most easily purified from heavy metals by redistillation. Nitric acid can be obtained by the distillation of ordinary nitric acid (S.G. 1·42) in a pyrex glass flask fitted with a pyrex glass condenser. If the first 50–100 ml. of distillate are rejected and the distillate collected until only 10–15 per cent. remains in the boiling flask, a product of high purity is obtained. It should be stored in pyrex glass reagent bottles. Hydrochloric acid is also redistilled at the constant boiling point from a similar all-glass distillation apparatus. Dilute 1 litre of ordinary concentrated hydrochloric acid (S.G. 1·18) with about 650 ml. of water and distil as before, rejecting the first portion of the distillate. This gives an acid of approximately 6N strength. Sulphuric acid is best redistilled from a pyrex glass or silica retort, rejecting the first and last fractions as before. Provided that one or two small glass beads are added, all of these acids distil quietly without bumping. According to Kahane (6), perchloric acid can be distilled at ordinary pressures with complete safety, from an all-glass retort, only a small amount of loss occurring from decomposition. Since only limited amounts of perchloric acid are used in most of the determinations, the amounts of impurities introduced are generally small, provided that a good quality reagent is selected. For many determinations the use of a solution of sodium perchlorate, purified by dithizone extraction, is a simple alternative to the purification of perchloric acid. Likewise, sodium fluoride, purified by dithizone extraction, can with advantage replace hydrofluoric acid for many purposes.

In all determinations, blanks must be carried out and, where necessary, a suitable correction made for the amounts introduced by the reagents. Experience has shown that with careful working it is easily possible to keep the impurities introduced from all sources below 0·5 microgram in the usual copper determination, and below 0·1 microgram in the micro-determination of zinc.

In many determinations the risk of contamination can be decreased, and much time can be saved, by using separate dip pipettes for the addition of each reagent which does not require accurate measurement. When not in use, each dip pipette is kept in a small beaker in a covered glass jar, so protecting it from dust. Since each pipette is kept for one specific reagent, it is thus always ready for use without the necessity for careful washing and drying between determinations. Their use in measuring the amounts of nitric, sulphuric and perchloric acids in the wet digestion method is not only a great convenience but also leads to considerable savings in time.

In the determination of cobalt, copper and zinc, use is made of diphenylthiocarbazone as an extractive reagent. The introduction by H. Fischer of diphenylthiocarbazone, and the extractive methods of analysis based on its use, have opened up an entirely new field for the determination of many of the heavy metal elements, which occur in such minute quantities in plant material. Diphenylthiocarbazone (C_6H_5.NH.NH.CS.N:N.C_6H_5), or dithizone as it is more generally called, forms complexes with many of the heavy metals. Like dithizone itself, these dithizonates are insoluble in water but readily soluble in many organic solvents, such as chloroform and carbon tetrachloride. Since these solvents are inmiscible with water there is a very favourable partition coefficient and, if the hydrogen ion concentration of the aqueous phase is suitably controlled, the heavy metals can be quantitatively extracted. This simple process of extraction enables small quantities of the heavy metals to be separated from complex mixtures of large amounts of other inorganic salts. Thus from a solution of plant ash the traces of cobalt, copper, zinc and any other metal, extractable by dithizone, can be easily separated from the excessive amounts of calcium, magnesium, potassium, sodium and other constituents. If a wet digestion method is used for the destruction of organic matter, the extraction can be carried out without separation of the silica. Filtration, with the possibility of introducing positive or negative errors by contamination from or adsorption by the filter paper, is eliminated and a very clean separation is obtained. Since chloroform and carbon tetrachloride are heavier than water, successive extractions can be carried out in a separating funnel, without transferring the aqueous phase from the funnel.

Many factors govern the partition of the dithizone metals between the aqueous and organic phases. Some metals are readily extracted from acid solutions, others only from neutral or alkaline solutions. Wichmann (18) gives the approximate order of extraction of some of the metals as follows, commencing with those extracted from the most acid solutions: silver, mercury, tin (stannous), bismuth, copper, zinc, lead, thallium and cadmium. The relative positions of nickel and cobalt are not given in the above list, but both are extractable from neutral to alkaline solutions.

There exists a favourable range of hydrogen ion concentration for each metal. From solutions more acid than this, extraction does not take place, or is incomplete. In solutions more alkaline, the metal-dithizone complex is not stable and the metal may pass back to the aqueous phase. The favourable range of hydrogen ion concentration is itself governed by several factors, including the solvent employed, the nature of the salts present in the aqueous solution and the concentrations of dithizone and heavy metals. In the acid and neutral range a tetrachloride solution of dithizone will extract most metals from solutions at about 1 pH unit more acid than will a chloroform solution. However, for the extraction of zinc at pH 9·8 in Walkley's method, a chloroform solution gives a better partition, since at this alkalinity dithizone shows a greater tendency to pass into the aqueous phase from tetrachloride solutions.

The presence of acetates, citrates or phosphates in the aqueous phase affects the range of hydrogen ion concentration, within which extraction of any particular metal ion takes place. In the presence of citrates, so frequently added in the analysis of plant materials to prevent the precipitation of phosphates from alkaline solutions, extraction is not so complete at any given hydrogen ion concentration as in acetate solutions. For complete extraction, the hydrogen ion concentration must be suitably reduced. Cyanides, iodides and thiosulphates, as well as certain organic compounds, can form complexes (competitive complexes) with some of the heavy metals. These complexes may be more stable than the dithizonate, and so change the range of hydrogen ion concentration favourable for the extraction of the metal, or prevent its extraction altogether. Thus, although dithizone will extract some ten or eleven metals from an alkaline solution, the addition of an excess of potassium cyanide to such a solution will prevent the extraction of all metals but tin, bismuth, thallium and lead. Larger concentrations of dithizone also favour the partition of the heavy metals into the organic dithizone phase.

By the suitable choice of conditions numerous differential extractions are possible. Many useful separations have been worked out empirically, but a considerable amount of work remains to be done in the detailed study of the equilibria involved, to give a more fundamental basis to the separations. So many factors are involved that the most favourable conditions of extraction have not been precisely defined for all the metals.

Dithizone, as purchased, is seldom sufficiently pure for use. A suitable method for its purification from oxidation products and heavy metal impurities is given on p. 332. Dithizone is insoluble in water but readily soluble in dilute solutions of ammonia. On making such a solution slightly acid with hydrochloric acid the pure dithizone can be extracted with carbon tetrachloride or chloroform. It is more soluble in the latter solvent than in the former, but for many purposes the tetrachloride solution possesses slight advantages over that in chloroform. Tetrachloride is also a cheaper reagent than chloroform.

Dithizone solutions are reasonably stable if kept in a cool, dark place. They can be kept for one to two years without any protective reagent, if stored in a refrigerator at 3–5° C. The solutions slowly oxidize to a yellow diphenylthiocarbadiazone which is not extractable by dilute ammonia. While the presence of this oxidation product in the reagent is objectionable for the colorimetric determination of dithizonates, it does not interfere with any of the extractive methods described later. Its formation only represents a decrease in the strength of the dithizone. Dithizone solutions may be tested for purity by shaking a portion with aqueous ammonia. In the absence of heavy metal impurities the organic phase will become water-white, if the dithizone solution is fresh, or yellowish to straw coloured, if older and contaminated with oxidation products. In either case residual red to violet colours, due to heavy metal impurities, indicate that the dithizone must be further purified before use.

In transmitted light dilute solutions of dithizone in chloroform or carbon tetrachloride appear blue-green and green respectively, but as the concentration is increased the colour changes abruptly from green to red. This gives an approximate method, sufficient for most practical purposes, for determining the strength of dithizone solutions (2). To obtain the strength of a solution, determine the depth of a column of liquid which just begins to impart a red colour when viewed in transmitted light. For chloroform solutions the concentration of dithizone, in mg. per litre, corresponds to a constant (380) divided by the depth of the column in millimetres. For tetrachloride solutions the value of the corresponding constant is approximately 950.

In addition to chemical methods for the determination of the trace elements, spectrochemical and polarographic methods are proving of increasing value, particularly in the determination of certain elements which are difficult to determine chemically. In spectrochemical methods the determination is usually carried out on the ash of the plant. For the best results in quantitative work, the ash should be nearly carbon-free. For some of the trace elements, e.g. molybdenum and zinc, determinations by the Lundegardh or Ramage flame methods are impracticable, owing to the lack of sensitivity. It is generally preferable to use the electric arc for the excitation of the spectra. The latter method is very convenient since the plant ash is arced directly. The need for the preparation of a solution of the ash, with the possibility of the loss of some of the elements by adsorption on the insoluble residue, is eliminated. The most stable arc is given by sulphates, hence a sulphated ash is generally preferred. Quantitative methods are based on the arcing of known amounts of sample, under standardized conditions, and determining the intensity of certain characteristic lines for each element. The relationship betweeen line intensity and concentration is usually determined by means of synthetic samples of known composition.

Specially purified graphite electrodes are used to hold the sample of plant ash during arcing. Except for small amounts of boron and iron, these electrodes do not contain any impurities which interfere with the determination of the elements included in this section. Copper electrodes are sometimes used, particularly for the determination of very small amounts of boron.

Spectrochemical methods possess the great advantage that several elements can be determined at the same time and in the same sample. Moreover once a plate has been taken and filed there is a permanent record of the elements present in the sample. The minimum amounts of the various elements that can be detected, expressed in parts per million parts of ash, are given later under each element. These limits can only be stated approximately, since they vary somewhat with the type of equipment available, details of the method of excitation and the nature of the other ash constituents. If present in an amount below its threshold value the element in question escapes detection. Concentrations in the ash above these threshold values can, however, be determined with the usual spectrochemical accuracy, namely ± 10–25 per cent. for visual methods or ± 5 per cent., when using a microphotometer to measure the intensity of the lines.

As the concentration of the element in the material being arced is the practical factor limiting the sensitivity of the spectrochemical method, the sensitivity can be greatly increased by carrying out a preliminary chemical concentration, either by precipitation or extractive enrichment of the element or elements required. In this way extremely small concentrations of an element in the original material can be determined.

Polarographic methods, originally introduced by Heyrovsky, are proving of considerable value in the determination of certain of the trace elements in plants. On account of the small quantities present it is usually necessary to carry out a preliminary chemical concentration of the element before its polarographic determination.

In the polarographic method of analysis, the solution to be analysed is placed in a small glass electrolysis vessel. A pool of mercury on the bottom of the vessel forms the anode and mercury in a glass capillary tube dipping just below the surface of the liquid forms the cathode. This capillary tube is so arranged that small drops of mercury continually form at its tip and drop off slowly at a constant rate, every 3–4 seconds. The mercury surface of the cathode is thus repeatedly renewed. When a solution containing an electrolyte is electrolyzed between such electrodes, and the current that flows plotted against the voltage applied, characteristic step-like curves are obtained. The position of each step on the voltage scale is specific for each particular element and represents its deposition potential in the solution under investigation. The height of the step, i.e. the increase in the current flowing, is a function of the concentration of the element in the solution under investigation. Thus, measurements of both voltage and current enable the specific determination of an element and its concentration in the solution being electrolyzed.

In the case of zinc in a basal solution of ammonium chloride and potassium thiocyanate the deposition potential is $-1 \cdot 02$ volts and measurement of the current for an applied voltage range of $0 \cdot 8 - 1 \cdot 2$ volts enables the increase, corresponding to the amount of zinc in the electrolyte, to be determined. The usual polarograph is so arranged that the voltage applied to the electrodes can be continuously and automatically increased by means of a potentiometer. Geared to the potentiometer is a drum holding a sheet of sensitized photographic paper, and a spot of light, reflected from the galvanometer mirror, automatically produces a continuous record of the current flowing between the electrodes for each applied voltage. In some of the most recent instruments a pen recorder takes the place of the galvanometer and sensitized paper. From the magnitude of the increase in current at any defined voltage, the concentration of the element in the solution can be determined. More detailed explanations of the polarographic method of analysis have been published by Walkley (15) and by Lingane and Kolthoff (8). The accuracy attainable depends on several factors, including the amount of the element present and the nature and amount of the other constituents. Under favourable conditions an accuracy of ± 2 per cent. is attainable.

BORON

The amounts of boron occurring in plant materials are very variable and values from 2–100 mg. per kg. of dry matter have been reported for normal plants. In cases of boron injury much higher amounts are often found.

The usual chemical methods for the determination of boron are based either on the titration of boric acid, or the development of a blue colour with quinalizarin in concentrated sulphuric acid. Boric acid is a very weak acid and cannot be titrated directly in dilute solution. However, in the presence of multivalent alcohols, a complex acid is formed, which is much more highly ionized and can be titrated directly. Mannitol is better than glycerol for this purpose, since it can readily be obtained pure, a smaller amount is needed, and the change at the end point is sharper. Invert sugar is also very suitable and considerably cheaper than mannitol for this titration.

Carbon dioxide must first be removed by boiling the acidified solution containing the boric acid to be determined. The solution is then neutralized to methyl red, invert sugar or mannitol added, and titrated with carbon dioxide-free alkali, until the phenolphthalein end point is reached. The amount of alkali used between the methyl red and phenolphthalein end points, corrected for that used in a blank determination, corresponds to the boric acid present. Carbonates, phosphates and other substances, buffering the solution between pH 4·5 and 8·4, must be absent, otherwise they will affect the titration between the end points of the two indicators used.

The volumetric method is not very sensitive. The titration corresponds to the following equation:

$$H_3BO_3 + NaOH = NaBO_2 + 2H_2O$$

Thus 1 ml. of the most dilute standard alkali that can be conveniently used for this titration (0·04N) corresponds to 0·43 mg. of boron. It is therefore generally necessary to use rather large samples of material to obtain sufficient boric acid for the final titration. However, useful values can be obtained. Dodd's method (4), details of which are given on p. 315, has been used successfully for the determination of boron in fruit and leaves. It is desirable to determine phosphate colorimetrically on an aliquot of the solution used for the titration, as a check on the completeness of the preliminary separation of phosphate. The trace of phosphate remaining should not exceed an amount equivalent to one drop of the standard alkali.

For the micro-determination of boron the quinalizarin colorimetric method offers a useful alternative to the volumetric method. In concentrated sulphuric acid solutions quinalizarin changes to a deep blue colour in the presence of extremely small amounts of boron, and the change in colour is related to the amount of boron present. The colour depends on several factors. It is developed rapidly at first, but continues to increase slowly, until it reaches full intensity in about 24 hours. Small differences in the amount of quinalizarin present affect its intensity. However, its intensity is most strongly affected by small differences in the concentration of the sulphuric acid, present in the final solution. The colour is most intense in concentrated sulphuric acid and decreases as the concentration falls. In sulphuric acid solution of about 93 per cent. the method is most sensitive with amounts of 2 micrograms of boron. At this level differences of 0·1 microgram can be readily detected. Using sulphuric acid of 89 per cent. strength the method is not so sensitive, but larger amounts of boron can be determined.

The method suffers from the inconvenience necessitated by working with colour solutions in concentrated sulphuric acid and the difficulty of preventing reagents containing acid of this strength from absorbing moisture from the atmosphere. Since slight changes in the strength of the acid in the colour solution have a profound influence on the shade of colour it is necessary to protect the sulphuric acid solutions from moisture during storage and to measure accurately the volumes of acid and aqueous solution used in each determination. A difference in strength of 0·5 per cent. between the sulphuric acid in the standard and unknown can give rise to an error of 8–15 per cent. Details of Maunsell's method are given on p. 316.

The spectrochemical determination of boron in plant ash, although not very sensitive, is sufficient for its determination in the amounts present in plant materials. The lower limit for its detection is about one part in 100,000 of ash, corresponding to 1 mg. per kg. of original sample. In amounts greater than this it can be readily determined. Its presence has been noted in all plant samples analyzed spectrochemically in these laboratories.

The Determination of Boron: Dodd's Method (4).

Reagents:

N Sodium Hydroxide. Dissolve 40 g. of reagent grade sodium hydroxide in water and dilute to 1 litre.

0·04N Calcium Hydroxide. Dilute lime water to this strength and standardize by titration against potassium biiodate.

Calcium Chloride. Dissolve 20 g. of reagent grade calcium chloride ($CaCl_2.6H_2O$) or 13 g. of calcium chloride (dry) in water and dilute to 100 ml.

Invert Sugar. Dissolve 500 g. of reagent grade sucrose in 170 ml. of hot water, boil and add 10 ml. of N sulphuric acid. After 30 seconds add 250 ml. of water containing 10 ml. of N sodium hydroxide. Mix well. If not quite neutral to phenolphthalein adjust the reaction accordingly.

Method:

Ash the sample in the presence of sodium hydroxide and obtain the boric acid in solution, free from silica by the method given on p. 269. To this solution, in the 100 ml. volumetric flask, add 5 ml. of calcium chloride solution. Make just alkaline to phenolphthalein, by the cautious addition of N sodium hydroxide from a burette, shaking the flask continuously during the addition. Add sodium hydroxide until a faint pink colour persists. If made too strongly alkaline some borate is precipitated with the phosphates. Dilute the solution to the graduation mark, mix well and allow to stand for a few minutes. Then filter through a dry 9 cm. Whatman No. 44 filter paper, discarding the first runnings and collecting the filtrate in a dry flask. Transfer 75 ml. of the clear filtrate to a 250 ml. Erlenmeyer flask, add 3 drops of methyl red and make just acid by the addition of one or two drops of N sulphuric acid. Then add an excess of 4–5 drops of this acid and boil gently for 2 minutes to expel carbon dioxide, using two glass beads to promote even boiling. Cover with a clock glass, to prevent re-absorption of carbon dioxide, and cool rapidly in running water.

When cold, titrate with 0·04N calcium hydroxide, to the neutral point of methyl red. Read the burette at this stage as this is the starting point of the boric acid titration. Add 5 ml. of invert sugar solution (or 3–5 g. of mannitol) and continue the titration to the phenolphthalein colour change. The amount of alkali between the methyl red and phenolphthalein end points corresponds to the boric acid present in the 75 ml. aliquot. One ml. of 0·04N calcium hydroxide corresponds to 0·432 mg. of boron.

Carry out a blank determination, using all the reagents including the sodium hydroxide used for the ashing. Include 0·05 g. of potassium hydrogen phosphate in the blank determination so as to allow for incomplete separation of the last traces of phosphate. Correct the titration for the amount of standard calcium hydroxide used in the blank.

The Determination of Boron: Maunsell's Quinalizarin Method (11).

Reagents:

Concentrated Sulphuric Acid. 99·4 per cent. This acid should be kept in a bottle fitted with an automatic pipette to deliver 9 ml. The bottle and pipette must be protected, by guard tubes, to prevent ingress of moisture. In withdrawing 9 ml. portions, the draining time of the pipette must be closely standardized.

Quinalizarin Solution. Dissolve 0·01 g. of quinalizarin in a mixture of 90 ml. of 98·5 per cent. sulphuric acid and 10 ml. of water.

Dilute Sulphuric Acid. Dilute 5 ml. of concentrated sulphuric acid to 500 ml.

Calcium Hydroxide. Prepare a saturated solution.

Standard Boric Acid. Dissolve 2·857 g. of boric acid in water and dilute to 1 litre. One ml. contains 0·5 mg. of boron. Dilute 20 ml. of this solution to 1 litre to obtain a solution containing 10 micrograms of boron per millilitre. For a working standard dilute this tenfold, so that 1 ml. = 1 microgram of boron.

Method:

Ignite 1 g. of the plant sample in a platinum basin in a muffle furnace at approximately 450° C. for a short time. Cool, add 2 ml. of saturated calcium hydroxide and evaporate to dryness. Re-ignite the material at approximately 600° C. When cool take up the residue in 6–15 ml. of dilute sulphuric acid, depending on the amount of boron present. Stir thoroughly with a glass rod, transfer to a centrifuge tube and centrifuge until the supernatant liquid is clear. By means of a pipette, transfer a 1 ml. aliquot to a dry test tube, add 9 ml. of concentrated sulphuric acid and 0·5 ml. of quinalizarin solution. Stir with a dry glass rod, place in a desiccator and leave overnight. Next day determine the colour (red units) in a tintometer, or compare the colour with that of a series of standards developed at the same time. To prepare the standards, take the required amount of boric acid solution, dilute to 1 ml. with water and add sulphuric acid and quinalizarin exactly as in the unknown determination.

When using a tintometer prepare a standard graph correlating the red units with the amount of boron present. From this graph read off the amounts of boron in the unknown solution. The tintometer is most suitable for matching the colours produced by quantities of 1–4 micrograms of boron. Very small quantities (0·1–1 microgram) are best judged by comparing the colour visually rather than by the tintometer.

COBALT

On account of its importance in the nutrition of ruminants it is often necessary to determine cobalt in plant materials. The amounts present are usually extremely small and very variable. Few figures are available except for pasture samples. Cobalt deficient pastures may contain as little as 0·01 mg. per kg. of dry matter and probably average about 0·04 mg. per kg. Normal pastures generally contain 0·05 to 0·5 mg. of cobalt per kg., 0·1–0·2 mg. per kg. being common values. There is some seasonal variation in the amount of cobalt present. Much higher values may be found, particularly in the young growth in pastures which have been treated with cobalt fertilizers.

The spectrochemical and polarographic determinations of cobalt present many difficulties, owing to the small amounts present. The lower limit for the spectrochemical determination is about 5 parts of cobalt per million parts of plant ash, corresponding to about 0·5 mg. per kg. of the original sample. It can therefore seldom be determined spectrochemically, unless a preliminary separation is carried out and the cobalt concentrated by precipitation, or by one of the methods of extractive enrichment. The proximity of its deposition potential to that of zinc, and the small amount present in relation to the amount of this latter element, makes the polarographic determination of cobalt uncertain.

Cobalt forms an intensely coloured soluble compound with nitroso-R-salt (sodium salt of 1-nitroso 2-napthol 3: 6-disulphonic acid). This cobalt compound is stable in nitric acid and forms the basis of an extremely sensitive colorimetric method for the determination of small amounts of the element. Copper and iron if present in excessive quantities may interfere with the determination, but, according to McNaught, it is seldom necessary to separate the small amounts present in plant materials, before the determination of cobalt.

In Kidson and Askew's method (7) details of which are given on p. 321, the plant material is ashed with nitric acid, particular care being taken to ensure decomposition of the last traces of organic matter, which might otherwise give rise to a yellow colour in the final solution and lead to fictitiously high values for cobalt. After solution of the ash and removal of silica, nitroso-R-salt is added and the cobalt colour developed under controlled conditions. The conditions prescribed for the development of the cobalt derivative of nitroso-R-salt must be closely followed. An unduly prolonged time of boiling in the acid condition, after the addition of the reagent, leads to definite bleaching of the colour while the addition of an excessive amount of sodium acetate causes a greenish yellow colour to appear in the solution. After the formation of the cobalt complex, nitric acid is added to stabilize this complex, to destroy the corresponding iron and copper compounds and to reduce the intensity of colour of the excess reagent. The colour is then compared against standard colours, prepared under exactly the same conditions. Satisfactory determinations of cobalt in plant samples can be made by this method in the presence of 6 mg. of iron, 6 mg. of manganese and 1·5 mg. of titanium.

McNaught (12) has developed a similar method for the determination of cobalt and finds that quantities of cobalt from 0·05 to 20 micrograms can be accurately determined. The colour is easier to match if the amount does not exceed 5 micrograms. The colour is most conveniently matched in long, narrow tubes. Owing to the small amounts of cobalt present, and the colour of the excess of reagent, matching by means of a colorimeter is impracticable. Iron, up to 1,000 micrograms, and copper up to 100 micrograms, do not interfere with the determination of 1 microgram of cobalt in plant ash by McNaught's method. If, however, iron and copper are present in excessive amounts they should be removed from the solution of the ash, by extraction with ether and precipitation with hydrogen sulphide respectively. Excessive amounts of iron in plant ash usually denote gross soil contamination.

After developing the cobalt colour it should be protected from light as much as possible; it must on no account be exposed to direct sunlight. At low cobalt levels there is an increase in the intensity of the apparent cobalt colour on exposure to light, due to the action of light on the 1-nitro 2-napthol 3:6-disulphonic acid (resulting from the action of nitric acid on the excess of nitroso-R-salt). Especially when determining amounts of cobalt below 1 microgram, the tubes used for matching should be shielded from light. They can be conveniently protected by enclosing them in a length of rubber tubing.

Marston and Dewey (10) have made an exhaustive study of the conditions governing the extraction of cobalt by dithizone and its subsequent determination as the coloured complex with nitroso-R-salt. Details of their method for cobalt in plant materials are given on p. 323. In this method the organic matter is oxidized by nitric, sulphuric and perchloric acids and silica is separated by filtration. Most of the sulphuric acid is removed by evaporation and volatilization, since the large amounts of alkali required to neutralize it may introduce extraneous cobalt and the high concentration of the neutral salts interferes with the formation and partition of the dithizone metal complexes. Copper is then removed by extraction with dithizone at a reaction of about pH 3–4, in the presence of citrate, and can be determined in this extract. The reaction of the copper-free solution is then adjusted to pH 8·3, using a buffered soda solution to avoid local excessive alkalinity and precipitation of iron and calcium phosphates. At this reaction cobalt is quantitatively extracted by dithizone. The dithizone in the extract is then destroyed and the cobalt converted to sulphate. After removal of the excess of sulphuric acid the cobalt sulphate is dissolved in a buffer solution of pH 8·0 and the colour developed with an excess of nitroso-R-salt. The colour is then stabilized with nitric acid and the excess of reagent destroyed by bromination. This last step destroys all of the light yellow colour due to the 1-nitro 2-napthol 3:6-disulphonic acid, produced by the action of nitric acid on the excess of nitroso-R-salt. After removing the excess of bromine, the pure red colour of the cobalt complex remains. The colour of the cobalt complex is very stable and is unaffected by bromine in relatively strong acid solution, so that, after discharging the colour of the excess of reagent, the red colour remaining is directly proportional to the concentration of cobalt. It can then be compared directly in a colorimeter, or determined in a photo-electric colorimeter. By this method 1 microgram of cobalt can be estimated with considerable precision (± 5 per cent.) while amounts exceeding 5 micrograms can be determined with the precision usually attainable in direct colorimetry (± 1 per cent.).

Owing to the difficulties which sometimes arise from the precipitation of iron and calcium phosphates, leading to low values for cobalt, many analysts prefer to separate cobalt from an acid solution. S. T. Evans *(priv. comm.)* precipitates cobalt, together with copper and iron, by means of α-nitroso β-napthol in dilute acid solution, dissolves the precipitate and extracts copper and cobalt from the solution by means of dithizone. R. W. Pickering *(priv. comm.)* extracts cobalt and copper as thiocyanates in a mixture of amyl alcohol and ether, and then converts the copper and cobalt to sulphates. In both methods the cobalt is determined colorimetrically, in the presence of copper, as the complex with nitroso-R-salt. Copper does not interfere with the cobalt colour provided that an excess of nitroso-R-salt is used and the colour is developed at pH 8 in the presence of citrate. Pickering has simplified the final operations involved in developing the colour of cobalt with nitroso-R-salt, retaining the bromination to decolourize the excess of reagent. Details of this method for the development of the cobalt colour are given on p. 326.

Cobalt occurs in soils in concentrations of 20–100 times as great as that in plants. The utmost precautions must be taken to avoid contamination of pasture samples with soil. One per cent. of soil in a pasture sample may easily double the apparent amount of cobalt in the plant material. Most reagents are reasonably free from cobalt. However, for critical work it is desirable to use recrystallized salts and redistilled acids as far as possible.

The Determination of Cobalt: Kidson and Askew's Method (7).

Reagents:

Hydrochloric Acid. Dilute one volume of concentrated hydrochloric acid with two volumes of water.

0·5N Hydrochloric Acid. Dilute 45 ml. of concentrated hydrochloric acid to 1 litre.

Nitric Acid. Dilute concentrated nitric acid with an equal volume of water.

Potassium Hydroxide. Dissolve 10 g. of potassium hydroxide in water and dilute to 100 ml.

Phenolphthalein Solution. Dissolve 0·2 g. of phenolphthalein in 100 ml. of alcohol.

Nitroso-R-Salt. Dissolve 0·1 g. of the pure crystals in water and dilute to 100 ml.

Standard Cobalt Solution. Dissolve 0·4037 g. of pure cobalt chloride ($CoCl_2.6H_2O$) in water, add 10 ml. of hydrochloric acid and dilute to 1 litre. One ml. of this solution corresponds to 100 micrograms of cobalt. From this solution prepare a working standard containing 0·5 micrograms of cobalt per millilitre.

Method:

Transfer 10 g. of the oven-dry sample to a silica basin, add 30 ml. of dilute nitric acid (1 + 1) and mix thoroughly. Ignite over an Argand burner and then in a muffle furnace, as completely as possible, at a very dull red heat. Moisten the ash with a further 5–10 ml. of dilute nitric acid, dry over a hot plate at low heat, and again ignite in the muffle furnace at a very dull red heat. Repeat the treatment with nitric acid, drying and igniting once more. These repeated treatments are necessary to destroy some form of organic compound which otherwise is liable to persist and interfere with the final cobalt colour.

Dissolve the ash in 30 ml. of dilute hydrochloric acid (1 + 2), heating gently on the water bath or hot plate for about 20 minutes. Filter through an 11 cm. Whatman No. 44 filter paper collecting the filtrate in a 100 ml. pyrex basin. Wash the filter and residue three times with hot water. Then evaporate the filtrate nearly to dryness, take up in 10 ml. of water, add 1 ml. of dilute hydrochloric acid (1 + 2) and 8 drops of dilute nitric acid (1 + 1). Transfer the solution to a 100 ml. Erlenmeyer flask, keeping the volume of solution at approximately 15 ml. Boil for a few minutes to oxidize any reducing substances, cool and add exactly 2 ml. of nitroso-R-salt solution, followed by 2 g. of sodium acetate. Heat to about 70° C., add 6 drops of phenolphthalein solution (not more) and then potassium hydroxide solution, drop by drop, with constant shaking, until a pink colour just appears. Immediately add 0·5N hydrochloric acid until the pink colour is just discharged. Precise adjustment of the reaction at this stage is important. Boil for exactly two minutes, keeping the solution acid by the addition of a drop or two of 0·5N hydrochloric acid if it shows a tendency to become alkaline to phenolphthalein. After boiling for two minutes add 5 ml. of dilute nitric acid (1 + 1) and boil for a further two minutes. Cool quickly by means of running water. Dilute to a volume of 25 ml. and filter, in the dark, into a colorimetric tube. Compare the colour against that of a standard, freshly prepared as described below. When very small amounts of cobalt are present keep the solutions in the dark as much as possible and make the colour comparison without undue delay, for the colour of the solution deepens on exposure to light. This effect is not so important when the amount of cobalt present exceeds 5 micrograms.

To prepare the standard colour solutions dilute suitable aliquots of the cobalt chloride solution to 15 ml. and add 8 drops of dilute nitric acid (1 + 1) and 1 ml. of dilute hydrochloric acid (1 + 2) to each. Then add exactly 2 ml. of nitroso-R-salt followed by 2 g. of sodium acetate and develop the colour exactly as described above.

The Determination of Cobalt: Marston and Dewey's Method (10).

Reagents:

Concentrated Sulphuric Acid. Redistil from pyrex glass (p. 306).

Concentrated Nitric Acid. Redistil from pyrex glass (p. 306).

Perchloric Acid S.G. 1·54. Highest grade analytical reagent.

0·2M Citric Acid. Dissolve the purest citric acid in water and standardize by titration of an aliquot with a standard solution of carbonate-free sodium hydroxide, using phenolphthalein as indicator.

N Sodium Hydroxide. Prepare this solution from a concentrated (carbonate-free) solution of sodium hydroxide (p. 120).

Buffered Soda. Dissolve 6·184 g. of boric acid and 35·62 g. of sodium phosphate ($Na_2HPO_4.2H_2O$) in water, add 500 ml. of N sodium hydroxide and dilute to 1 litre.

Dithizone in Carbon Tetrachloride. (0·05 per cent. dithizone). Dissolve 0·25 g. of dithizone in carbon tetrachloride and purify by the method given on p. 332. Dilute the final carbon tetrachloride extract to a volume of 1 litre and keep in a pyrex glass bottle in a refrigerator at 2–5° C.

Dithizone in Chloroform. (0·2 per cent. dithizone). Dissolve dithizone in chloroform containing one per cent. of alcohol and purify from oxidation products and heavy metals in the same manner as the tetrachloride solution.

Brom Phenol Blue. See p. 127.

Cresol Red. See p. 127.

Methyl Red. Dissolve 0·025 g. in 100 ml. of 60 per cent. alcohol.

Phenolphthalein. Dissolve 0·05 g. in 100 ml. of 50 per cent. alcohol.

Nitroso-R-Salt. Dissolve 1 g. of the pure salt in water and dilute to 500 ml. This solution is quite stable. (Nitroso-R-salt may be prepared as follows: Dissolve 35 g. of the refined sodium salt of 2-napthol 3:6-disulphonic acid (R salt) in about 400 ml. of water, acidify with 10 ml. of concentrated hydrochloric acid, cool to 10° C. and add drop by drop, over a period of 30 minutes, 7 g. of sodium nitrite dissolved in 25 ml. of water. Keep the reaction mixture cooled with ice. Filter off the yellow crystals, wash with ice-cold water and finally with cold alcohol.)

Bromine Solution. Dilute a saturated solution of bromine in water until it is 0·2N, determined by titration with standard thiosulphate after the addition of potassium iodide. The solution is not stable.

Standard Cobalt Solution. From a solution of cobalt nitrate, containing 1 mg. of cobalt per millilitre, prepare fresh solutions as required containing 1 microgram of cobalt per millilitre.

Method:

Digest 10 g. of plant material (or sufficient to contain 1–5 micrograms of cobalt) in a 500 ml. Kjeldahl digestion flask, using the method described on p. 272. For this digestion use 80 ml. of nitric acid, 3 ml. of sulphuric acid and 5 ml. of perchloric acid. When digestion is complete, cool and dilute the contents of the flask with 10 ml. of water. Boil for a few minutes and then filter through a 9–11 cm. Whatman No. 44 or 544 filter paper, which has been previously soaked in 10 per cent. hydrochloric acid and thoroughly washed. Collect the filtrate and washings in a silica basin. Cautiously evaporate and fume off, over a carefully regulated hot plate, until only a trace of sulphuric acid remains. A trace of sulphuric acid must be left at this stage to prevent the formation of insoluble iron oxides by decomposition of ferric sulphate. Precipitation at any stage of the analysis is to be avoided, owing to the tendency of cobalt to be adsorbed.

After fuming off all but the last drop of sulphuric acid, take up the residue in 7·5 ml. of 0·2N citric acid and dilute to about 30 ml. Add 5 drops of brom phenol blue and run in N sodium hydroxide, drop by drop, until a distinct greenish-blue colour appears through the yellowish tint, imparted by the ferric citrate. The solution must still be acid to methyl red. Transfer the solution to a 100 ml. separating funnel, rinse the basin with water and dilute to about 50 ml. so that the citrate concentration is 0·03M. Extract this solution by shaking vigorously with three or more lots, each of 20 ml., of dithizone in chloroform, separating the chloroform layer each time and collecting it in a 50 ml. micro-Kjeldahl digestion flask. Three separations are usually sufficient, but the extraction must be continued until the dithizone is no longer exhausted by combination with heavy metals from the aqueous phase. Reserve the chloroform extract for the determination of copper (p. 335).

After the removal of copper from the aqueous phase, wash this solution by shaking once with 20 ml. of chloroform, discarding the chloroform. Then add a few drops of phenolphthalein and adjust the reaction of the aqueous phase to pH 8·3, by cautiously titrating with buffered soda until the first sign of purplish-pink colour appears. About 6 ml. of buffered soda are required. To detect this colour change, the titration must be carried out in a good light and against a white background. After adjusting the reaction to pH 8·3 extract the cobalt by shaking vigorously with three lots, each of 10 ml. of the tetrachloride solution of dithizone, separating the tetrachloride layer each time and collecting it in a pyrex boiling tube (about 2·5 cm. x 10 cm.). Distil off the solvent and digest the residue with 1 ml. of nitric acid, 0·2 ml. of sulphuric acid and 0·5 ml. of perchloric acid, until quite colourless. Rinse the contents of the tube into a silica basin, remove the water by evaporation and fume off the sulphuric acid completely by heating in a muffle furnace for five minutes at 350° C.

Dissolve the residue in 1 ml. of 0·2M citric acid and rinse back into the boiling tube, using a minimum amount of water. If the volume exceeds 5 ml. reduce to this volume by evaporation of the acid solution. Cautiously adjust the reaction to pH 8·0 by adding 1·2 ml. of buffered soda. Check the reaction by withdrawing a small drop and testing with cresol red. Then develop the cobalt nitroso-R-salt complex by introducing 1 ml. of 0·2 per cent. nitroso-R-salt, by means of a pipette with a fine jet, shaking vigorously during the addition. Boil the solution for one minute, add 1 ml. of concentrated nitric acid and boil for a further minute. Then discharge the yellow colour of the excess reagent by adding 0·5 ml. of 0·2N bromine to the warm solution, allowing the reaction to proceed for five minutes before boiling for one minute to expel the excess of bromine. Even boiling can be assured by the use of a glass boiling stick or one or two glass beads.

After removing the excess of bromine, cool the solution, dilute to 10 ml. and compare the colour against a standard prepared by adding the required amount of cobalt to 1 ml. of 0·2M citric acid, adjusting the reaction with 1·2 ml. of buffered soda and developing the cobalt colour exactly as described for the unknown solution above.

The Determination of Cobalt: Development of Cobalt Nitroso-R-Salt Complex according to Bayliss and Pickering (1).

This modification of Marston and Dewey's method for the development of the cobalt colour is as follows:

Reagents:

Concentrated Nitric Acid. Redistil from pyrex glass (p. 306).

Concentrated Ammonia. Dissolve ammonia gas in glass distilled water.

M Ammonium Citrate. Dissolve 210 g. of reagent grade citric acid in about 400 ml. of water, add 200 ml. of concentrated ammonia and cool the solution. Then add further amounts of ammonia until the solution is just alkaline to phenolphthalein, used as an external indicator. Dilute to 1 litre and remove metal impurities by extracting the solution with successive small portions of dithizone dissolved in chloroform. Remove the excess of dithizone by shaking with one or two lots of chloroform alone. Finally remove the excess of chloroform by means of amyl alcohol. Store the purified reagent in a stoppered reagent bottle.

Nitroso-R-Salt. Dissolve 1 g. of pure nitroso-R-salt in water and dilute to 100 ml.

Bromine Water. Prepare a saturated solution in water.

Standard Cobalt Solution. See p. 321.

Method:

After separation of the cobalt from the plant ash it is obtained in solution in a small excess of sulphuric acid. Copper does not interfere unless present in an excessive amount.

Add 5 ml. of concentrated ammonia to the solution, to neutralize the sulphuric acid, and evaporate on a water bath to remove the excess of ammonia. Take up the residue in 5 ml. of M ammonium citrate, 5 ml. of water and 0·5 ml. of nitroso-R-salt. Heat for about 10 minutes on the water bath and then add 5 ml. of concentrated nitric acid. Heat for a further 5–10 minutes, transfer the solution to a boiling tube and dilute to 15 ml. While still warm add 2·5 ml. of bromine water, allow to react for 15 minutes and then boil to expel the excess of bromine. When cold, dilute to 15 ml. and compare the colour with that obtained from a known amount of cobalt treated similarly.

COPPER

The most common range of copper in plant materials is covered by 1–20 mg. per kg. of dry matter. These small amounts can be determined accurately by chemical and spectrochemical methods.

Copper is extractable by dithizone in carbon tetrachloride from dilute mineral acid solutions as acid as pH 1. At pH 3, however, equilibrium between the tetrachloride and the aqueous phases is much more rapidly established and the whole of the copper is quickly extracted. This gives a rapid and quantitative separation of copper from most of the other ash constituents. Apart from the fact that the colour is not specific for copper under these conditions of extraction, copper dithizonate is not very suitable for the exact quantitative determination of copper. Having extracted the copper as dithizonate it is preferable to make use of a second reaction in which copper forms a strongly coloured complex with diethyldithiocarbamate. This compound can be extracted by amyl alcohol and its colour is proportional to the amount of copper. Of the metals either wholly or partly extractable by dithizone at pH 3, bismuth is the only other one which also gives a coloured compound with sodium diethyldithiocarbamate. If it is present it, therefore, interferes with the determination of copper. It is practically never detectable in plants in significant quantities. A confirmatory test can always be applied, either to prove its absence or to correct for the amount of colour due to it. This is done by adding potassium cyanide to the solution remaining after the colour comparison. On the addition of cyanide the whole of the copper is converted to colourless cuprocyanide, since potassium cuprocyanide is more stable than copper diethyldithiocarbamate, and the copper passes out of the amyl alcohol to the aqueous phase. Bismuth does not form a corresponding double cyanide. Hence any colour remaining in the amyl alcohol layer indicates the presence of bismuth. By comparing its intensity a suitable correction can be applied to the amount of copper previously determined.

The method which has been used in these laboratories for some hundreds of copper determinations is based essentially on that of Sylvester and Lampitt (14), modified to ensure that the whole of the copper is extracted from the sample. Details are given on p. 331. The method involves no filtrations. Organic matter is destroyed by digestion with nitric, sulphuric and perchloric acids and copper separated from most of the other inorganic constituents, by extraction of the solution at pH 3, with dithizone in carbon tetrachloride. The tetrachloride is then removed by evaporation and the dithizone destroyed by digestion with a few drops of sulphuric and perchloric acids. This leaves the copper, and any other metals extracted by dithizone at pH 3, in solution as sulphates, in an excess of sulphuric acid. After dilution and the addition of an excess of ammonia, copper is precipitated as copper diethyldithiocarbamate. This is extracted by shaking with a measured volume of amyl alcohol, which dissolves the copper compound to give a strongly coloured solution. The intensity of this yellowish brown colour of the amyl alcohol phase is then compared, in a colorimeter, with that of a standard containing a known amount of copper. If necessary, a correction is made for any bismuth present.

This method is extremely sensitive since the whole of the copper is separated from the other ash constituents and concentrated in the few millilitres of amyl alcohol used. It is superior to other methods which use either dithizone or diethyldithiocarbamate alone, particularly when only very small amounts of copper are present. In methods using dithizone alone, copper is extracted from a dilute mineral acid solution, the excess of dithizone removed from the tetrachloride phase and the colour of the copper dithizonate determined against a standard. Although the method is very sensitive the colour is not very stable and its shade and intensity are very susceptible to slight changes in the conditions prevailing during the extraction and the removal of the excess of dithizone.

The method described below is very straightforward and there is little trouble from interference, due to other ash constituents. The digestion with nitric, sulphuric and perchloric acids is preferable, but copper can also be determined after dry ashing with sulphuric acid, provided that the silica is removed by hydrofluoric acid treatment (p. 337). After the wet digestion small amounts of nitrosyl-sulphuric acid remain and, unless these are destroyed, they oxidize some of the dithizone during the extraction. This oxidation can be prevented by diluting the digest and boiling it for 10–15 seconds, before its neutralization and extraction.

Some plant materials contain larger amounts of calcium than usual and, in such cases, it may be difficult to keep all of the calcium in solution at the time of the extraction. For such samples, rich in calcium, it is necessary to take a smaller weight of material. Some copper may be adsorbed on the calcium sulphate, unless it is entirely in solution at the time of the extraction.

Solutions of sodium diethyldithiocarbamate are stable for several months, if kept in the dark. In ammoniacal solutions, in the presence of a small excess of the reagent, the colour of copper diethyldithiocarbamate is also quite stable. It will remain unchanged for at least 10 days in the dark or in diffused daylight. It fades more rapidly in bright sunlight. The colour can always be restored to its full intensity by the addition of two or three drops of fresh reagent. The intensity of the colour is not affected by the amount of ammonia used to neutralize the solution.

In making colour comparisons the standard should not be left for unduly long periods in the cup of the colorimeter, otherwise evaporation of the amyl alcohol occurs and the concentration of the copper diethyldithiocarbamate increases above the theoretical value. This increase in concentration of the standard colour solution leads to erroneously low values for copper in solutions compared against it. In cool weather only a small amount of evaporation occurs in one hour, but in warm weather the error is not negligible.

In determining copper in some substances high in iron, e.g. root materials, the colour of the amyl alcohol layer is greyish-brown, rather than the true colour of copper diethyldithiocarbamate, and it is difficult to match it with that of the standard. The small amounts of iron usually present in plant materials do not affect the colour. The interference due to excessive amounts of iron can be prevented by the addition of sodium pyrophosphate, before developing the colour with sodium diethyldithiocarbamate, since ferric pyrophosphate is a more stable complex than the carbamate.

The optimum amount of copper for determination by this method is 5 to 10 or 15 micrograms, but quantities of 2–35 micrograms can be conveniently determined, using the standard amounts of reagents given below. Larger amounts may be determined by taking a suitable aliquot, or by increasing the amount of amyl alcohol used. Quantities of 1–5 g. of plant material are usually very convenient when using either the nitric-sulphuric-perchloric acid digestion or the method of dry ashing.

The precautions to be observed in preventing contamination

with copper during analysis have already been indicated. Since it is not possible to remove completely the last traces of copper from the reagents a blank determination must always be carried out, using the same quantities of reagents as in an actual determination. The amount of copper found in the blank determination should not exceed 0·5–0·6 micrograms.

The spectrochemical determination of copper is very sensitive, since copper is one of the elements most easily excited in the electric arc. Flame methods of excitation are equally sensitive for this element. Its lines are clearly recognizable when present in amounts even less than one part per million parts of ash. On the basis of an ash content of 10 per cent., this corresponds to a sensitivity better than 0·1 part of copper per million parts of plant material.

The Determination of Copper (After Wet Digestion)

Reagents:

Concentrated Sulphuric Acid. Redistil from pyrex glass (p. 306).

Concentrated Nitric Acid. Redistil from pyrex glass (p. 306).

Perchloric Acid S.G. 1·54. Specially select a reagent quality perchloric acid which is low in copper or redistil from pyrex glass (p. 306).

Sodium Perchlorate. Dissolve 550 g. of pure sodium perchlorate in water and dilute to 1 litre. Add a few drops of a dilute solution of sodium hydroxide until the solution is alkaline, when tested with phenol red as an external indicator. Filter, if necessary, and transfer to a separating funnel. Shake with successive 20–30 ml. portions of dithizone solution (see below) until it is seen that no more heavy metals are being extracted. Then shake with a final lot of 50 ml. of carbon tetrachloride alone, separating as completely as possible. Acidify with a few drops of sulphuric acid and store the purified solution of sodium perchlorate in a pyrex reagent bottle. The excess of dithizone remaining in it does not interfere with its use.

Ammonia S.G. 0·91. Dissolve pure ammonia gas in glass distilled water.

Ammonium Citrate. Dissolve 200 g. of reagent grade citric acid in about 1 litre of water and neutralize to about pH 7–8 with concentrated ammonia (about 225 ml.), using brom thymol blue as an external indicator. Filter, if necessary, and transfer the solution to a large separating funnel. When cool extract by shaking with successive portions of dithizone solution, as described above for sodium perchlorate, until no further heavy metals can be extracted. At this stage the carbon tetrachloride layer shows a pure green colour. Then shake with 50 ml. of carbon tetrachloride and separate as completely as possible. Dilute to 2 litres and store in a pyrex reagent bottle. The excess of dithizone remaining in the aqueous phase does not interfere.

Brom Phenol Blue Indicator Solution. Dissolve 0·1 g. in 50 ml. of redistilled alcohol and dilute to 100 ml. with water.

Dithizone in Carbon Tetrachloride. Dissolve 0·25 g. of dithizone in 600 ml. of carbon tetrachloride, warming to about 50° C. to assist solution. Transfer to a large separating funnel and shake with 350 ml. of water containing 3–4 ml. of concentrated ammonia, to extract the dithizone as the soluble ammonium salt. Reject the carbon tetrachloride layer and wash the aqueous solution by shaking with three separate lots, each of 75 ml. of redistilled carbon tetrachloride, rejecting the tetrachloride layer each time. Then add 600 ml. of redistilled carbon tetrachloride and make the aqueous layer slightly acid with redistilled hydrochloric acid. Shake thoroughly. The purified dithizone passes into the tetrachloride phase. Transfer this to a second separating funnel and wash by shaking with three separate lots, each of 150 ml. of water, discarding the aqueous layer each time. Transfer the purified solution of dithizone to a stoppered pyrex reagent bottle, dilute with a further 750 ml. of carbon tetrachloride and store in a refrigerator at 2–5° C.

For preparing the reagent technical carbon tetrachloride (sulphur-free) is quite satisfactory, provided that it is redistilled from glass.

Sodium Diethyldithiocarbamate. Dissolve 3 g. of sodium diethyldithiocarbamate in 100 ml. of water and filter.

Keep this solution in the dark in a pyrex glass dropping bottle, fitted with a ground-in pipette.

Amyl Alcohol. Reagent grade.

Standard Copper Solution. Dissolve 0·3930 g. of crystalline copper sulphate ($CuSO_4.5H_2O$) in water, add 5 ml. of concentrated sulphuric acid and dilute to 1 litre. Dilute 10 ml. of this standard solution to 1 litre, also including 5 ml. of sulphuric acid to stabilize it. Each millilitre of the diluted standard solution corresponds to 1 microgram of copper.

Sodium Pyrophosphate. Heat 20 g. of anhydrous sodium phosphate (Na_2HPO_4), in a silica basin, in a muffle furnace for 3 hours at 370–450° C. When cool, dissolve in 400 ml. of water, transfer to a separating funnel, add 5 ml. of ammonia and 30–40 ml. of amyl alcohol. Then add 10 drops of sodium diethyldithiocarbamate solution and shake vigorously so as to extract the copper from the aqueous phase. Allow to stand for 1 hour and transfer the aqueous solution to a pyrex reagent bottle.

Method:

Digest a suitable amount of material (generally 1–5 g.), sufficient to yield 5–15 micrograms of copper, with 2 ml. of sulphuric acid, 4 ml. of perchloric acid and sufficient nitric acid, according to the method described on p. 272. If desired the perchloric acid in the above digestion can be replaced by 8 ml. of the purified sodium perchlorate solution. In this case use an additional millilitre of sulphuric acid.

When the digestion is complete, remove the flask from the heater, cool and dilute with 60 ml. of water and boil vigorously for 10–15 seconds, to decompose any nitric-sulphuric acid compound which may destroy the dithizone by oxidation. This boiling is most conveniently carried out over a Bunsen flame but the flask must be constantly agitated to prevent violent bumping.

While still warm, add 10 ml. of ammonium citrate solution, to dissolve any hydrolyzed manganese compounds and to prevent the precipitation of phosphates on neutralization. Allow to cool, add 5 drops of brom phenol blue and carefully neutralize with strong ammonia solution adding the ammonia by means of a dip pipette and shaking the flask during the operation. The ammonia should be added carefully and only sufficient to produce a blueish-green colour in the solution; excessive alkalinity is to be avoided since it may cause precipitation of phosphates. Rinse the neck of the flask and just restore the acid (yellow) colour of the indicator by the addition of one or more drops of dilute sulphuric acid (1 + 2), contained in a dropping bottle fitted with a ground-in pipette. This gives a solution of approximately pH 3·0.

Cool to room temperature and transfer the contents of the digestion flask, without filtration, to a separating funnel of sufficient capacity (160–200 ml.). Rinse the Kjeldahl flask twice, using about 25 ml. of water in all. At this stage the volume of the solution in the separating funnel amounts to about 90–100 ml. It must be sufficient to keep all calcium sulphate in solution.

Add 10 ml. of dithizone reagent to the solution in the separating funnel and shake vigorously for 30–40 seconds. After standing for a minute or two, separate the tetrachloride layer as completely as possible, but do not allow any of the aqueous phase to enter the bore of the stopcock. Collect the separated dithizone solution in a 50 ml. micro-Kjeldahl flask, or suitable boiling tube. Repeat the extraction with two further portions of dithizone, each of 5–6 ml., shaking for 30 seconds each time and collecting the tetrachloride layer as before. The whole of the copper is usually obtained in the first extraction if sufficient dithizone is present, since at this reaction copper dithizonate is formed in preference to the dithizonates of the other metals normally present in plants. An excess of dithizone is indicated by the greenish colour of of the reagent persisting during the second and third extractions. The second and third extractions serve to ensure an excess of dithizone and to remove the small amounts of copper left in the separating funnel owing to the incomplete removal of the tetrachloride phase at each separation. If a greenish shade does not persist during the second and third extraction slightly larger volumes of dithizone should be used or a fourth extraction made. Sometimes, however, the dithizone layer becomes a cherry red colour. This may occur when extractions are made at reactions less acid than pH 3, due to the extraction of other metals. This does not interfere with the determination, but prevents the positive recognition of the complete extraction of copper. It may be avoided by the further addition of one of two drops of sulphuric acid, after the first or second extraction.

When the combined dithizone extracts have been collected, heat the micro-Kjeldahl flask over a hot plate until the carbon tetrachloride is just boiled off. Remove the flask from the hot plate and add 30 drops (approximately 0·75 ml.) of concentrated sulphuric acid and 2 drops of perchloric acid from dropping bottles with ground-in pipettes. Digest over a porcelain Bunsen burner or the full heat of a hot plate until a clear digest is obtained (5–7 minutes). Cool, dilute with about 4 ml. of water and transfer to a 25 ml. stoppered graduated test mixer, rinsing the micro-Kjeldahl flask with three lots, each of about 2 ml., of water. Make distinctly ammoniacal by adding 4 ml. of concentrated ammonia, shake and allow to cool. If several determinations are being carried out, dilute all the solutions to the same volume, generally about 20 ml., add 3 drops of sodium diethyldithiocarbamate reagent and shake. Extract the yellow copper complex by shaking with 4–10 ml. of amyl alcohol, accurately measured by means of a pipette or burette. A quantity of 5 ml. is convenient for most analyses. Allow the solutions to stand for about half an hour, or until the amyl alcohol layer is quite clear, then withdraw some of it by means of a dry pipette and compare the colour in a micro-colorimeter (cups holding 1 ml.) with the colour of a standard solution prepared as described below. The colour is quite stable, except in direct sunlight, and can be compared at any convenient time after its development. The standard for comparison can vary from 0·5 to 2 times that of the unknown without introducing a detectable error, so long as the volumes of the aqueous and amyl alcohol phases are identical in each.

If excessive amounts of iron are present, as for example, in plant roots, add 2 ml. of sodium pyrophosphate before adding the ammonia and sodium diethyldithiocarbamate. This will prevent interference by the brownish colour of ferric diethyldithiocarbamate.

The amyl alcohol colour solutions are most conveniently

removed from the stoppered cylinders by means of a 10–15 ml. pipette supported in a clamp and connected (through rubber tubing and a length of capillary tubing) to a two-way stopcock, one arm of which is open to the atmosphere and the other to a very slowly running filter pump (Fig. 18). Sufficient of the amyl alcohol layer is then drawn off into the pipette, without any of the aqueous phase, and held in the pipette to enable the colorimeter cups to be rinsed and refilled. The pipette is rinsed with ether between successive determinations, the last traces of ether being removed by drawing a stream of air through the pipette. The ether must be removed completely as traces of it cause fading of the copper colour.

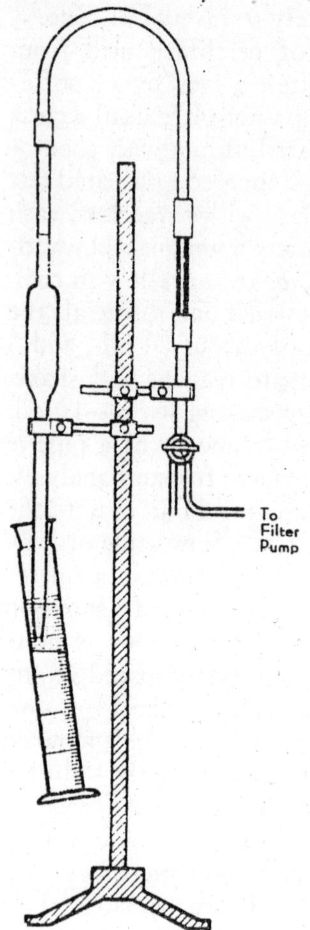

Fig. 18. Pipette arranged for withdrawing the amyl alcohol layer in the determination of copper.

Carry out a blank determination exactly as for an ordinary determination but place one or two glass beads in the Kjeldahl flask during the digestion, to promote even boiling. Deduct the amount of copper found from that determined above.

Standard Copper Colour Solution.

Pipette 1–15 ml. of the dilute standard copper solution, equivalent to 1–15 micrograms of copper, into a 25 ml. graduated test mixer, add 20 drops of concentrated sulphuric acid and 4 ml. of concentrated ammonia. Dilute with water to the same volume as the unknown solution and, when cool, add 3 drops of sodium diethyldithiocarbamate. Then add the same volume of amyl alcohol as that added to the unknown solution, shake vigorously and allow to stand until the amyl alcohol layer is clear. Standard colour solutions containing 1 microgram of copper, for comparison with the blank determination, and 5, 10, and 15 micrograms are most suitable for general use.

Test for the Absence of Bismuth:

After making the colour comparison return some of the amyl alcohol phase to the test mixer from which it was originally taken. Add 2 ml. of a freshly prepared 10 per cent. solution of potassium cyanide and shake vigorously. If the colour is completely removed from the amyl alcohol layer it indicates that bismuth is absent. If, however, some colour persists, due to the presence of bismuth, determine its intensity in terms of the standard copper colour and make a suitable correction for it.

The Determination of Copper (After Dry Ashing).

Reagents:

In addition to most of the reagents mentioned for the previous method the following are required.

Hydrochloric Acid. Constant boiling point acid redistilled from pyrex glass (p. 306).

Hydrofluoric Acid. Reagent grade, selected for its relative freedom from copper. It is preferable to replace it with sodium fluoride.

Sodium Fluoride. Dissolve 40 g. of sodium fluoride in water and dilute to 1 litre. Add a few drops of sodium hydroxide until the solution is alkaline to phenol red (external indicator). Extract with successive portions of dithizone in carbon tetrachloride, exactly as described for sodium perchlorate (p. 331). When free from heavy metals store in a pyrex glass reagent bottle.

Method:

Destroy organic matter by ashing a suitable quantity of material, sufficient to give 5–15 micrograms of copper, with sulphuric acid in a platinum basin (p. 267) and remove the silica by treatment with hydrofluoric acid as detailed. Owing to the practical difficulties in obtaining hydrofluoric acid free from copper it is preferable to use 10–15 ml. of purified sodium fluoride and 1 ml. of concentrated sulphuric acid.

When the excess of sulphuric acid has been removed from the silica-free ash, by ignition in a muffle furnace at 500° C., take up the residue in 5 drops of sulphuric acid, 5 ml. of hydrochloric acid and 10 ml. of water. Warm on a water bath to ensure complete solution, then dilute with a further 30 ml. of water and cool.

When cold, transfer the solution to a separating funnel, rinsing the basin with three small portions of water. Add 10 ml. of ammonium citrate and 5 drops of brom phenol blue. Adjust the solution to pH 3 and complete the determination of copper as described under the previous method.

IRON

The amounts of iron in plant materials are very variable. While 10–250 mg. per kg. of dry matter is not an uncommon range the leafy parts of plants may contain up to 1,000 mg. per kg. The latter quantities hardly justify the inclusion of iron among the trace elements. Its determination is considered here, however, since all the remarks regarding the special precautions to be observed to avoid contamination during sampling and analysis apply with equal force for iron.

Owing to the considerable amount of iron in normal soils, care must be exercised to exclude iron from this source during the taking of the sample. Grinding of samples in a steel mill can also lead to erroneously high values. Grinding in a Wiley mill, in one example reported, increased the iron content by 30 mg. per kg. Small amounts of iron are of widespread occurrence in ordinary distilled water, concentrated acids and all commercial reagents. Suitable precautions must be taken to avoid errors from these sources (p. 305).

Iron in plant materials is most conveniently determined colorimetrically, using either the red colour produced when a ferric salt reacts with thiocyanate or that produced by ferrous salts with αα′-dipyridyl. Both methods are very sensitive and can detect amounts of 1–10 micrograms of the element.

In the thiocyanate method ferric thiocyanate is formed by the addition of potassium thiocyanate to a ferric salt in dilute acid solution. The colour of the ferric thiocyanate is proportional to the amount of iron present. The colour comparison is facilitated by extracting the ferric thiocyanate with amyl alcohol. The colour of the amyl alcohol layer is then compared with that of standards prepared under the same conditions.

The colour of ferric thiocyanate is affected by the hydrogen ion concentration so that it is important that both the standard and the unknown colours be developed under comparable conditions. Orthophosphates, in the amounts present in plant ash, do not interfere with the determination but small amounts of pyrophosphate seriously reduce the intensity of the colour produced, due to the formation of slightly ionized ferric pyrophosphate. Pyrophosphates are not formed during methods of wet digestion, but after all methods of dry ashing they must be completely hydrolyzed, by boiling in dilute acid solution. The colour of ferric thiocyanate is also reduced by excessive amounts of sulphates. Nitric and hydrochloric acids do not interfere with the determination but the nitric acid must be completely free from nitrous acid.

In the dipyridyl method, originally proposed by Hill (5) for the determination of iron in biological materials, use is made of the intense red colour which is produced when ferrous salts react with αα′-dipyridyl. The colour is almost specific for ferrous iron and is very stable. Standard colour solutions remain unchanged for at least 48 hours; if protected from oxidation in sealed tubes they can be kept indefinitely. The complex formed by ferrous salts with αα′-dipyridyl is much more stable than ordinary ferrous salts; it is only oxidized by relatively powerful oxidizing agents. The colour is produced by iron in the ferrous condition. If ferric salts are present, they must first be reduced to the ferrous state by the addition of a suitable reducing agent. For this purpose Hill used sodium hyposulphite (hydrosulphite) but Parker and Griffin (13) prefer *p*-hydroxyphenylglycine. This latter reducing agent is readily available as the photographic developer "Glycin."

In the presence of an excess of dipyridyl, the colour of the ferrous dipyridyl complex is constant over the pH range 3·5–8·5. In solutions more acid than pH 3·5 it is partly or wholly dissociated so that the full colour is not developed.

The determination is most conveniently carried out in an acetate buffer solution, since ferrous iron can be readily kept in solution in such a solution. Details of Parker and Griffin's method are given on p. 342.

The Determination of Iron: Thiocyanate Method.

Reagents:

Concentrated Sulphuric Acid. Redistil from pyrex glass (p. 306).

Concentrated Nitric Acid. Redistil from pyrex glass (p. 306).

Perchloric Acid. S.G. 1·54. Redistil from pyrex glass (p. 306), or select a high grade reagent sufficiently low in iron.

30 per cent. Hydrogen Peroxide. High quality reagent grade.

6N Hydrochloric Acid. Redistil from pyrex glass (p. 306).

Potassium Thiocyanate. Dissolve 20 g. of the purest potassium thiocyanate in water and dilute to 100 ml.

Amyl Alcohol. Reagent grade.

Standard Iron Solution. Dissolve 0·7022 g. of pure ferrous ammonium sulphate [$FeSO_4.(NH_4)_2SO_4.6H_2O$] in about 100 ml. of water and 10 ml. of concentrated sulphuric acid, in a silica basin. Add 5 ml. of concentrated nitric acid, to oxidize the ferrous salt to ferric sulphate. Heat gently, to expel oxides of nitrogen, transfer to a 1 litre volumetric flask and, when cold, dilute to the mark. This solution is stable indefinitely and contains 100 micrograms of iron per millilitre. From this solution prepare a more dilute standard, containing 1 microgram of iron per millilitre, by diluting 10 ml. to 1 litre and adding 15 ml. of concentrated sulphuric acid before adjusting to volume.

Method:

Digest 1 g. of plant material with nitric, sulphuric, and perchloric acids by the method given on p. 272. Use 3 ml. of perchloric acid and 2 ml. of sulphuric acid. After the sulphuric acid digest has been fumed strongly for 2–3 minutes remove from the heater and when sufficiently cool add, drop by drop, 10 drops of hydrogen peroxide. Again heat to fuming for about one minute. Dilute the digest with 50 ml. of water, add 5 ml. of constant boiling point hydrochloric acid and boil for 30 seconds, keeping the flask agitated to prevent bumping. Without filtering off the silica, transfer the diluted digest to a 100 ml. volumetric flask and, when cold, dilute to volume.

By means of a pipette transfer a suitable aliquot, usually 10 ml., to a 50 ml. stoppered test mixer. Add one drop of nitric acid (free from nitrous acid) and dilute to 45 ml. Add 10 ml. of amyl alcohol, accurately measured, and 5 ml. of potassium thiocyanate solution. Shake the cylinder vigorously for 30–40 seconds to extract the ferric thiocyanate into the amyl alcohol phase, then set aside to separate.

When the supernatant amyl alcohol layer is clear compare its colour, by means of a colorimeter, with that of a suitable standard, prepared as described below. For withdrawing the amyl alcohol layer the pipette shown in Fig. 18 is most convenient.

To prepare standard colour solutions for comparison, pipette suitable amounts of the standard iron solution, corresponding to 3–10 micrograms of iron, into 50 ml. stoppered test mixers. If the volume taken is less than that of the aliquot of the unknown, dilute to this volume with dilute sulphuric acid (15 ml. of sulphuric acid per litre) so that the acid concentration of the standard and unknown will be comparable. Add one drop of nitric acid, dilute to 45 ml. then add 10 ml. of amyl alcohol and 5 ml. of potassium thiocyanate, exactly as in the unknown. Shake vigorously and, when clear, use the amyl alcohol layer. Prepare the standard solutions at the same time as the unknowns.

Alternative Method:

Ash 1 g. of plant material and remove the silica with hydrofluoric acid by the method described on p. 267, using the smallest amount of hydrofluoric acid possible. Dilute the solution of the ash (in 10 ml. of hydrochloric acid and 20 ml. of water) to 60–70 ml. and digest, under a clock glass, on the water bath for at least one hour, to ensure hydrolysis of any pyrophosphates which would otherwise interfere. Transfer the solution to a 100 ml. flask and when cold dilute to volume. Using a suitable aliquot, continue with the determination exactly as described for the first method.

For the standard solutions used for the colour comparison hydrochloric acid must be used in place of most of the sulphuric acid. Prepare the diluted iron standard solution by diluting 10 ml. of the stronger stock solution to 1 litre using 100 ml. of constant boiling-point hydrochloric acid (6N) instead of 15 ml. of sulphuric acid. Likewise, when preparing the standard colour solutions, adjust the volume of the standard iron solution to that of the aliquot of the unknown with dilute hydrochloric acid (100 ml. of 6N hydrochloric acid per litre) so as to maintain comparable acidities throughout.

The Determination of Iron: Dipyridyl Method.

Reagents:

Concentrated Sulphuric Acid. Redistil from pyrex glass (p. 306).

Concentrated Nitric Acid. Redistil from pyrex glass (p. 306).

Perchloric Acid S.G. 1·54. Redistil from pyrex glass (p. 306), or select a high grade reagent sufficiently low in iron.

Concentrated Ammonia. Dissolve ammonia gas in water.

p-Hydroxyphenylglycine. Dissolve 0·1 g. of *p*-hydroxyphenylglycine (photographic "Glycin") in 100 ml. of 0·4N sulphuric acid.

αα′-Dipyridyl. Dissolve 0·2 g. of αα′-dipyridyl in 100 ml. of 10 per cent. acetic acid.

N Ammonium Acetate. Dissolve about 70–80 g. of reagent grade ammonium acetate in water or prepare a solution from redistilled acetic acid and concentrated ammonia (above). The solution should have a reaction of about pH 6·4.

Standard Iron Solution. Dissolve 0·7022 g. of pure ferrous ammonium sulphate in water, add 10 ml. of concentrated sulphuric acid and dilute to 1 litre. This solution contains 100 micrograms of iron per millilitre. From this solution prepare a more dilute standard, containing 5 micrograms per millilitre, by diluting 50 ml. to 1 litre and adding 10 ml. of concentrated sulphuric acid before adjusting to volume.

Method:

Digest 1–2 g. of plant material with nitric, sulphuric, and perchloric acids, by the method given on p. 272. Use 3 ml. of perchloric acid and 2 ml. of sulphuric acid. After the sulphuric acid digest has fumed strongly for 2–3 minutes, remove it from the heater and, when cool, dilute it with about 50 ml. of water. Boil for 15–30 seconds, keeping the flask agitated to prevent bumping. Transfer the diluted digest to a 100 ml. volumetric flask and, when cold, dilute to volume.

Filter the solution through a dry 9 cm. Whatman No. 44 filter paper, collecting the filtrate in a dry 100 ml. Erlenmeyer flask. Discard the first runnings. By means of a pipette, transfer a suitable aliquot, containing 10–100 micrograms of iron, to a 50 ml. volumetric flask. Add 0·5 ml. of *p*-hydroxyphenylglycine and 1·0 ml. of αα′-dipyridyl. Then add concentrated ammonia, carefully, until the red colour of ferrous dipyridyl just appears. Dilute nearly to volume with N ammonium acetate solution and, when cold, adjust exactly to volume with this solution.

By means of a colorimeter compare the colour developed with that of a standard colour solution containing approximately the same amount of iron. To prepare suitable colour solutions pipette 2–20 ml. portions of the standard iron solution into 50 ml. volumetric flasks, add 0·5–1 ml. of concentrated sulphuric acid and dilute to about 30 ml. Then add *p*-hydroxyphenylglycine, αα′-dipyridyl and concentrated ammonia exactly as in the actual determination. Finally dilute to volume with N ammonium acetate and mix well.

Instead of an ordinary colorimeter the colour may be measured by means of a photo-electric instrument. In this case prepare a calibration curve, connecting percentage light transmission with amount of iron present, from the readings obtained from several standard colour solutions.

MANGANESE

Plant materials contain manganese in amounts usually varying from 10–150 mg. per kg. It is frequently present in sufficient quantity to give a pink colour to the plant ash on acidifying it, or to the sulphuric-perchloric acid digest of plant material. In general plants growing on acid soils contain more manganese than those on neutral or alkaline soils.

By spectrochemical methods manganese can be determined in plant ash when present to the extent of one part per 100,000. Since this corresponds to an amount of the order of 1 mg. per kg. of original dry matter, this determination is sufficiently sensitive to give accurate values.

The chemical determination of manganese in plant material is based on its oxidation to permanganate, the colour of which is proportional to the amount of manganese present. The method is both sensitive and accurate. As little as 20 micrograms of manganese can be determined, if Nessler tubes are used for the colour comparison. When using a colorimeter, about 0·25 mg. of manganese in a 50 ml. flask gives a very suitable colour, but amounts from 0·1 mg. to 1 mg. can be satisfactorily determined. For larger amounts the colour should be developed in a larger volume of solution.

Significant amounts of manganese are frequently retained in the insoluble siliceous residue of the ash, if organic matter is destroyed by dry ashing. These amounts must be recovered by solution in hydrofluoric acid. No such retention occurs in the wet digestion with sulphuric, nitric and perchloric acids and this method is strongly recommended for the destruction of organic matter.

Manganese is oxidized to permanganate in acid solution, either by means of ammonium persulphate, in the presence of a silver salt, or by potassium periodate as in Willard and Greathouse's method (19). If ammonium persulphate is used the true permanganate shade is not always developed and this makes colour matching with that of the standard difficult. Pure colours are always produced in the periodate method. When developed by this method the permanganate colour is very stable. If not exposed to direct sunlight or to reducing vapours, its intensity will remain unchanged for three months in the presence of a small excess of periodate.

Very few substances interfere with the accuracy of the determination. Reducing substances must, of course, be completely absent otherwise iodine will be liberated from the periodate. Chlorides, if present, should be removed by evaporation with nitric acid to avoid the use of excessive amounts of potassium periodate in their oxidation to chlorine. Small amounts of iron interfere, since the coloured ferric salts affect the shade of the permanganate and make exact colour matching difficult. This source of interference can, however, be overcome by the addition of phosphoric acid since ferric phosphate is almost colourless in cold solution.

A sufficient concentration of free acid must be maintained during the development of the colour, otherwise some of the manganese may be precipitated as hydrated dioxide, and so escape oxidation to permanganate, particularly if much manganese is present. Small amounts of phosphoric acid are very effective in preventing this precipitation. The total acidity should be kept within the limits prescribed in the method. In the presence of too little or too much acid the colour is slow in appearing. However, once the colour has started to appear it develops rapidly and further boiling for 1–2 minutes ensures its maximum intensity. During this time water should be added to prevent excessive increase of concentration of the acid by evaporation of the solution.

If much calcium is present it may not be possible to keep all the calcium sulphate in solution in the final colour solution. If this is so it should be centrifuged or filtered through a dry sintered glass funnel, so as to obtain a clear solution for the colour comparison. The substitution of phosphoric acid for sulphuric acid, used in the dry ashing method given on p. 263, eliminates the formation of sparingly soluble sulphates and clear solutions are obtained.

Partial fading of the permanganate colour sometimes occurs when the nitric-sulphuric-perchloric acid digestion has been used but this can be completely prevented by the use of ammonium persulphate in the final stages of the digestion. The addition of hydrogen peroxide at this stage will also prevent this source of error.

The amounts of manganese occurring in distilled water and good quality reagents are very small. It is not necessary to redistil the acids or recrystallize other reagents before use.

The heating of all solutions in the water bath after nearly full dilution eliminates possible errors from any traces of reducing substances in the distilled water.

The Determination of Manganese: Periodate Method.

Reagents:

Concentrated Sulphuric Acid.

Concentrated Nitric Acid.

Perchloric Acid S.G. 1·54.

Phosphoric Acid S.G. 1·70.

Ammonium Persulphate. Dissolve 20 g. of ammonium persulphate in warm water and dilute to 100 ml. Prepare this solution immediately before use.

Potassium Periodate.

Standard Manganese Sulphate Solution. Dissolve 0·5756 g. of pure, dry potassium permanganate in about 500 ml. of water in a 2 litre volumetric flask. Add 40 ml. of concentrated sulphuric acid and reduce the permanganate, by the careful addition of sodium metabisulphite solution, until the manganese solution just becomes colourless. Oxidize the excess of sulphurous acid by the addition of a little nitric acid. When cool dilute to 2,000 ml. and store in a stoppered reagent bottle. This solution contains 0·1 mg. of manganese per millilitre.

Method:

Digest 2–10 g. (usually 5 g.) of material with nitric, sulphuric and perchloric acids as described on p. 272. For this digestion use 5 ml. of sulphuric acid and 4 ml. of perchloric acid. When the organic matter has been destroyed and the digest has been fumed for 2–3 minutes at the full heat of the hot plate remove the flask and allow to cool. When cool add 5 ml. of freshly prepared ammonium persulphate solution, return the digest to the hot plate, heat again to fuming and fume strongly for five minutes.

When cool add 2 ml. of phosphoric acid and 35–50 ml. of warm water. Filter through a 9 or 11 cm. Whatman No. 44 filter paper and wash well with hot water. Collect the filtrate in a

200 ml. silica basin and evaporate it on the water bath until the volume is reduced to about 25 ml. Then add one or two glass beads to promote even boiling and about 0·3 g. of potassium periodate crystals. Boil cautiously over a low flame until the colour of permanganate appears. Dilute with 25–35 ml. of water at this stage and boil for a further two minutes to ensure maximum colour development. Transfer the solution to a suitable sized volumetric flask, dilute nearly to volume and place in a boiling water bath for 15 minutes. When cold dilute to the mark and compare the colour against that of a standard solution.

To prepare standard manganese colour solutions pipette appropriate amounts of the standard manganese sulphate into 100–150 ml. volumetric flasks, add 10 ml. of concentrated sulphuric acid and 2 ml. of phosphoric acid and dilute each solution to about 60 ml. Add about 0·3 g. of potassium periodate and heat in a boiling water bath for about five minutes after the appearance of the permanganate colour. Dilute nearly to volume and leave in the water bath for a further 15 minutes. When cold, dilute to volume.

Alternative Method:

Ash 2–10 g. (usually 10 g.) of oven-dry material, following the details of the method given on p. 263 as far as the solution of the ash in hydrochloric acid and the evaporation to render the silica insoluble. From this point proceed as follows:

To the dried salts in the silica basin, add 5 ml. of concentrated nitric acid and 25 ml. of water. Warm until all the salts are in solution and filter through a 9 cm. Whatman No. 44 filter paper, collecting the filtrate in a 100 ml. silica basin. Wash well with warm water. Evaporate the filtrate to dryness on the water bath to remove chlorides.

Transfer the filter paper containing the insoluble residue to a platinum basin and ignite it carefully, to remove the filter paper and any carbonaceous residues remaining. Moisten the ignited residue with water and add sufficient hydrofluoric acid (3–10 ml.) to dissolve the silica. Heat very gently, avoiding boiling, until nearly all the hydrofluoric acid has been volatilized. Then add 5 ml. of phosphoric acid and heat more strongly for a few minutes to expel the last traces of hydrofluoric acid. When cool, dilute with 20 ml. of water and wash the solution into the silica basin containing the evaporated residue from the filtration. Add 0·3 g. of potassium periodate and proceed to develop the colour exactly as described in the previous method. As sulphuric acid is not used, sparingly soluble calcium sulphate is not formed and clear solutions are obtained.

Prepare the standard colour solution as previously described, using, however, 5 ml. of phosphoric acid to acidify the solution, instead of the mixture of sulphuric and phosphoric acids.

MOLYBDENUM

According to D. Bertrand molybdenum occurs in the aerial portions of plants to the extent of 0·5–5 mg. per kg., although some fruits and seeds may contain up to 50 mg. per kg. Few determinations of these very small amounts of molybdenum have been made. At the present time it would appear that Marmoy's method (9) is the most suitable for general use.

In this method, details of which are given below, molybdenum is determined as thiocyanate, after reduction of any hexavalent molybdenum by stannous chloride. The stannous chloride also reduces ferric iron and so prevents interference by it. The molybdenum thiocyanate is extracted with ether from an acid solution of the plant ash and the colour of the ethereal solution is compared with a suitable standard. The method will detect 1 microgram of molybdenum.

Molybdenum can be detected and determined by spectrochemical methods when it occurs in the plant ash to the extent of one part per 100,000. In a large proportion of the samples examined the amounts of molybdenum do not reach this value and it remains undetected. For its spectrochemical determination in these cases it is necessary to use some means of preliminary chemical concentration. For this purpose precipitation with 8-hydroxyquinoline, in the presence of iron and aluminium as carriers, is sometimes advocated. Flame excitation, by Lundegardh's method, is probably more sensitive than the arc method but its application is not so convenient.

The Determination of Molybdenum: Marmoy's Method (9).

Reagents:

Concentrated Hydrochloric Acid. S.G. 1·16.

Potassium Thiocyanate. Dissolve 10 g. of potassium thiocyanate in water and dilute to 100 ml.

Stannous Chloride. Dissolve 10 g. of stannous chloride ($SnCl_2.2H_2O$) in dilute hydrochloric acid (1 + 9). Prepare this solution each day that it is required.

Ethyl Ether. Shake pure ether with one-tenth of its volume of a mixture of potassium thiocyanate and stannous chloride reagents, immediately before use.

Standard Molybdenum Solution. Dissolve 0·552 g. of ammonium molybdate, $(NH_4)_6Mo_7O_{24}.4H_2O$, in water and dilute to 1 litre. Dilute 20 ml. of this solution to 200 ml. to obtain a solution containing 30 micrograms of molybdenum per millilitre.

Method:

Transfer 2 g. of plant material to a silica basin and ignite in a muffle furnace at 450–500° C. When cool, moisten the ash with water, add 10 ml. of hydrochloric acid, warm and transfer to a 100 ml. beaker, diluting to about 40 ml. Boil for a few minutes and filter through a 9 cm. Whatman No. 44 filter paper into a 100 ml. graduated flask. Wash with hot water until the volume is about 80 ml., cool and dilute to the mark. Transfer an aliquot containing not more than 20 micrograms of molybdenum (usually 50 ml.) to a separating funnel, diluting to 50 ml. if necessary. Add sufficient concentrated hydrochloric acid to make the concentration 7·0 ml. of acid in the 50 ml. of solution. Mix thoroughly and add 3·0 ml. of potassium thiocyanate and 3·0 ml. of stannous chloride in the order named, shaking between each addition. After one minute add 10·0 ml. of pure ether from a burette and shake the mixture vigorously for about 30 seconds, taking care to release internal pressure, without losing any of the solution. When the layers have separated, drain the aqueous layer into a beaker and collect the ether in a specimen tube of approximately 10 mm. diameter. Return the aqueous phase to the separating funnel and extract again with 5·0 ml. of ether. Collect the ether layer with the first extract. If necessary repeat the extraction until the ether layer is colourless. Two extractions are, however, usually sufficient.

Prepare a standard molybdenum thiocyanate solution by transferring 5 ml. of molybdenum solution, containing 30 micrograms of molybdenum per ml., to a separating funnel. Dilute with water to 50 ml., add 7 ml. of concentrated hydrochloric acid, mix thoroughly and add 3 ml. of potassium thiocyanate and 3 ml. of stannous chloride, exactly as previously described. Extract with 10 ml. of ether and four further portions each of about 5 ml., collecting the ether extracts in a 25 ml. volumetric flask. Dilute the combined extract to 25 ml., using ether that has been shaken with thiocyanate and stannous chloride. The solution so prepared contains 6 micrograms of molybdenum per millilitre. Protect it to prevent changes in concentration as a result of evaporation. This solution cannot be kept for longer than a day.

Compare the unknown solution against the standard in the following manner. Introduce a suitable amount of ether, previously shaken with thiocyanate and stannous chloride, into a second specimen tube identical with that containing the unknown solution. From a micro-burette add the standard molybdenum thiocyanate solution, until the colour matches that of the unknown, when viewed from above against a white background. The amount of molybdenum added then corresponds to that present in the original aliquot taken.

Carry out a blank determination using all the reagents and correct for any molybdenum found.

ZINC

Zinc commonly occurs in plants to the extent of 5–80 mg. per kg. of dry matter. Amounts below 20–25 mg. per kg. are probably most frequent.

It has been found in these laboratories that the polarographic method is most convenient for the determination of the small amounts of zinc that normally occur in plant materials. The determination is rapid and accurate. The amounts are nearly always too small to make a direct determination and it is usually necessary to carry out a preliminary chemical separation of zinc, from the bulk of the other inorganic constituents, by extraction with a chloroform solution of dithizone at pH 9·8. In the procedure recommended, organic matter is destroyed by digestion with nitric, sulphuric and perchloric acids and the zinc extracted, without removal of the suspended silica. The dithizone in the chloroform extract is then destroyed by means of a further wet digestion with a few drops of sulphuric and perchloric acids and the excess of acid expelled by volatilization. The residue is dissolved in a measured volume of a basal solution of ammonium chloride and potassium thiocyanate and polarized between potentials of 0·8 and 1·2 volts at the dropping mercury electrode. The current flowing is recorded photographically by the polarograph. Its increase is proportional to the concentration of zinc in the solution being polarized. Copper, lead, nickel, and other metals extractable by dithizone may also be present in the final solution but they do not interfere with the determination.

The whole determination is carried out on the micro-scale. Using the quantities of reagents specified in the method (p. 354) the optimum amount of zinc for this determination is about 3–300 micrograms. Over this very wide range the accuracy attainable is ± 2 per cent. Amounts smaller than 3 micrograms can be determined but the accuracy falls off and is about ± 10 per cent. for one microgram. Amounts as small as 0·1 microgram of zinc can be detected. One g. of plant material is suitable for all determinations, since 3 micrograms of zinc in this amount correspond to 3 mg. per kg. of dry matter.

Most of the precautions to be observed in preventing contamination with zinc during the determination of this element have already been outlined. The avoidance of glassware containing zinc and the necessity for the use of specially purified reagents are again stressed.

Although zinc is readily extracted from many solutions at about pH 8 by a chloroform solution of dithizone, extraction is often incomplete at this reaction in the presence of suspended silica, such as that derived from the wet digestion of plant materials. Under these conditions Walkley (16) has shown that a reaction of pH 9·5–10 is favourable for complete extraction, zinc dithizonate being stable up to pH 10·5 in the presence of an excess of dithizone.

The spectrochemical method for the determination of zinc in plant ash is not very sensitive unless a line (2138·6 *A*) in the ultra-violet region of the spectrum is used. This necessitates the use of special photographic plates, sensitized to radiation of this wavelength. Under these conditions zinc can be determined when present in the ash to the extent of about 10 parts per million. This corresponds to an amount of the order of 1 mg. of zinc per kg. of plant material. However, when ordinary photographic plates only are available, the most sensitive line that can be used is in the near ultra-violet (3345 *A*) and, unless zinc is present in moderately large amounts, it then escapes detection altogether. Using this line, it is necessary to have about 100 parts of zinc per million parts of ash before it can be detected. This corresponds to about 10 mg. per kg. in the original dry matter of the plant. In amounts above this threshold value zinc can be detected with the usual spectrochemical accuracy (p. 311).

Methods for the chemical determination of small amounts of zinc fail to approach the polarographic method, either in convenience or accuracy. They require considerably larger samples or involve much more manipulation. The usual methods of extraction with dithizone lack specificity, since more than a dozen other metals form coloured complexes with this reagent under the conditions favourable for the extraction of zinc. Zinc can be determined, after the separation of copper, by extraction with dithizone from an ammoniacal solution, and conversion of the zinc dithizonate to sulphate, by digestion with nitric, sulphuric and perchloric acids. The zinc can then be determined micro-volumetrically, titrating the iodine liberated from potassium iodide by zinc sulphate in the presence of ferricyanide according to the reaction:

$$3ZnSO_4 + 2KI + 2K_3Fe(CN)_6 = Zn_3K_2(Fe(CN)_6)_2 + 3K_2SO_4 + I_2$$

Sodium thiosulphate is used to titrate the liberated iodine. Since 1 ml. of 0·002N thiosulphate corresponds to 196 micrograms of zinc the method is not capable of determining small amounts of zinc with great accuracy.

The method of Cowling and Miller (3) is much more promising since it enables the determination of 5–30 micrograms of zinc. It is a photometric mixed colour method based on the extraction of all "dithizone" metals from a solution of the plant ash, separation of zinc and other metals from copper, by shaking with 0·02N hydrochloric acid, and re-extraction of the zinc from this solution with dithizone, under controlled conditions, and in the presence of sodium diethyldithiocarbamate. The latter compound forms stable competitive complexes with all the metals present except zinc so that, in the second extraction with dithizone, zinc is the only metal which passes into the dithizone-carbon tetrachloride phase. However, the whole of the zinc is not extracted; a little remains in the aqueous phase as carbamate. The conditions of extraction, particularly the hydrogen ion concentration, concentration of carbamate and the volumes of the aqueous and tetrachloride phases, must therefore be accurately controlled and reproduced in the standard used for comparison. The percentage light transmission of the zinc dithizonate is determined, without removal of the excess of dithizone, in a photo-electric colorimeter, using a suitable light filter. The amount of zinc is obtained from a standard curve connecting light transmission and the amount of zinc present. Cadmium is the only metal likely to interfere with the determination, since it is partly extracted in the dithizone phase. It may, however, be present in amounts up to half that of the zinc, before serious error occurs.

The method can be modified to a one colour method by removing the excess of dithizone from the final extract, by shaking it with dilute ammonia. When this step is included an ordinary colorimeter can be used to determine the intensity of the colour of the zinc dithizonate remaining. However, this modification makes the method less specific for zinc. High values are obtained, owing to partial decomposition of lead diethyldithiocarbamate in ammoniacal solution and conversion to dithizonate.

The method, as outlined by Cowling and Miller, does not ensure complete extraction of the zinc from the insoluble siliceous residue of the plant ash. Provided that the method is modified to secure this, using, for example, a wet digestion for the destruction of the organic matter and making the first extraction of all "dithizone" metals at pH 9·8 according to Walkley's procedure (below), the method should give satisfactory values for the amounts of zinc in plant materials. Details for this procedure are given on p. 359.

The Determination of Zinc: Walkley's Polarographic Method (17).

Reagents:

Concentrated Sulphuric Acid. Redistil from pyrex glass (p. 306).

Concentrated Nitric Acid. Redistil from pyrex glass (p. 306).

Perchloric Acid S.G. 1·54. Redistil from pyrex glass (p. 306).

Sodium Perchlorate. For the purification of this reagent see p. 331.

Chloroform. Redistil the commercial grade reagent from pyrex glass and add one per cent. of its volume of redistilled absolute alcohol.

Dithizone in Chloroform. Dissolve 5 g. of commercial dithizone in 500 ml. of chloroform and store in a refrigerator at 2–5° C. This is purified for use during the course of the preparation of the next reagent.

Ammonium Citrate Buffer. Dissolve 5 g. of citric acid in 50 ml. of water and 200 ml. of 4N ammonia and transfer to a separating funnel. Add 10 ml. of the above chloroform solution of dithizone and shake vigorously. Discard the chloroform layer, which contains all the impurities originally present in the dithizone, citric acid and ammonia. Shake twice more with 10 ml. lots of chloroform only, again discarding the chloroform each time. The solution has a pH value of 10·7 and is orange in colour due to the purified dithizone which has dissolved in it. Since dithizone oxidizes fairly rapidly in aqueous solutions it is advisable to prepare a stock mixture of citric acid and ammonia and purify sufficient on each day that it is required. Freshly prepared in this way the reagent is free from zinc and 25 ml. of it contains sufficient dithizone for the quantitative extraction of at least 350 micrograms of zinc from a plant digest, by the method given below.

Basal Solution of 0·1M Ammonium Chloride, 0·02M Potassium Thiocyanate and 0·0002 per cent. Methyl Red. Dissolve 5·35 g. of ammonium chloride and 1·94 g. of potassium thiocyanate in water, add 5 ml. of a 0·04 per cent. aqueous solution of methyl red and dilute to 1 litre. Good quality reagent grade salts can generally be used for this solution without further purification.

Standard Zinc Sulphate Solution. Dissolve 2·875 g. of crystalline zinc sulphate ($ZnSO_4.7H_2O$) in water and dilute to 1 litre to give a 0·01M solution. Check the zinc concentration, by precipitation as zinc ammonium phosphate. From this solution prepare a more dilute standard, 0·001M in zinc sulphate in a basal solution of the same composition as that stated above.

Mercury. Shake the best reagent grade of mercury with redistilled nitric acid (1 + 20), wash with water and dry. Treat residues from the electrolysis cell in the same manner. Mercury purified by this treatment is sufficiently pure for the anode pool in the electrolysis cell. Purify all mercury required for the cathode by distillation at reduced pressure after the preliminary acid treatment.

Hydrogen. This must have a low oxygen content. Commercial supplies which have been prepared electrolytically are generally suitable.

Apparatus:

Polarograph. This should be either a photographic or a pen-recording type. A current-compensator, for annulling the effect of the condenser current, is advisable if the instrument is to be used at its maximum sensitivity.

It is convenient to work in a room, thermostatically controlled at about 20° C., or to enclose the electrolytic cell in some form of air thermostat. If the internal-standard method is to be employed, however, no temperature control is necessary.

Glassware. In the course of each determination only four vessels are required. These are:

(1) A 50 ml. micro-Kjeldahl digestion flask, preferably flat bottomed.

(2) A 100 ml. pear-shaped separating funnel with short stem. No lubricant is used on the stopcock.

(3) A small vessel for collecting the chloroform extracts. For this a straight sided weighing bottle (pyrex glass), 30 mm. diameter, 45 mm. high and 30–35 ml. capacity, with an interchangeable ground glass stopper, has proved most convenient. It is sufficiently wide-mouthed to facilitate the evaporation of the chloroform extract and the acids used for digesting the latter; it also serves to contain the final 1 ml. of solution before transference to the polarizing cell.

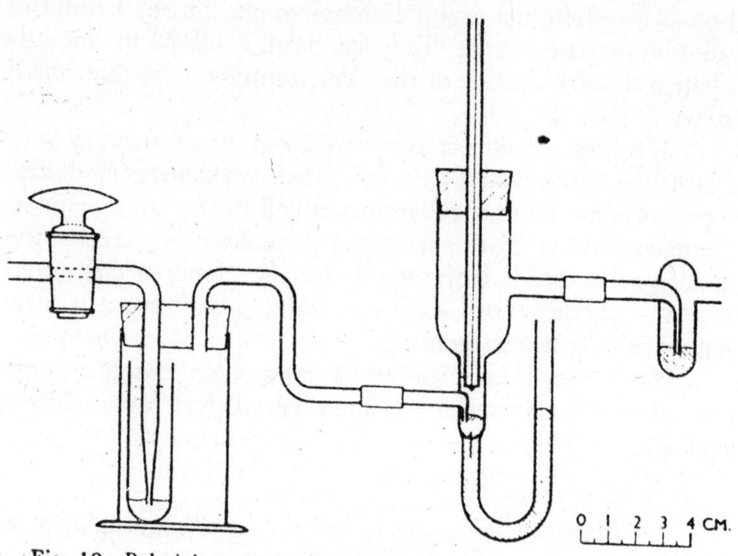

Fig. 19. Polarizing vessel and gas train used in the polarographic determination of zinc. (After A. Walkley (11)).

(4) An electrolysis or polarizing cell, with accompanying gas train (Fig. 19). The narrowest part of the cell has an internal diameter of 8 mm. When it contains 0·4 ml. of solution the depth of the liquid is about 8 mm., which is quite sufficient for adequate immersion of the cathode. For routine work it is advisable to have two gas trains and four to six electrolysis vessels.

The cathode can be prepared by drawing out a length of capillary tubing (0·5 mm. bore) until the bore is reduced to about 0·05 mm., but such capillaries are apt to be somewhat fragile. Capillary tubing of fine and uniform bore (about 0·05 mm.) can now be purchased and it is preferable to use this for the cathode. Such tubing does not require to be drawn out further. After several trials it is comparatively simple to cut off a length which will give the desired drop rate.

Method:

Digest 1 g. of plant material with nitric, sulphuric and perchloric acids by the method described on p. 274. If preferred, 2 ml. of purified sodium perchlorate may be used, instead of the perchloric acid. There is no need to increase the amount of sulphuric acid, as 1 ml. is ample.

When the digestion is completed, cool, add 15 ml. of water, and bring to the boil over a porcelain Bunsen burner, keeping the flask agitated to prevent bumping. When cool add 25 ml. of the ammonium citrate buffer solution and cool again. This gives a reaction of approximately pH 9·8.

Add about 5 ml. of chloroform to a 100 ml. separating funnel, pour in the neutralized digest and rinse the digestion flask two or three times, using about 10 ml. of water in all. Shake the contents of the separating funnel vigorously for about one minute and then allow to stand while the next samples are being shaken. Run off as much of the chloroform layer as is possible, without allowing any of the aqueous layer, or the silica which collects at the interface, to enter the bore of the stopcock. Repeat the extraction with two further lots, each of 5 ml. of chloroform and collect all the separated fractions in the small straight-sided bottle. If the whole of the zinc has been extracted, the final extract will be green. If it is any shade of blue, purple or red, continue the extractions until a pure green is obtained. For all plant samples examined, three extractions have always been found sufficient.

Evaporate the chloroform extracts to dryness over a hot plate at low heat, preferably with the assistance of some form of forced draught. Avoid boiling, since this always leads to bumping. When dry, add 2·5 ml. of nitric acid, 0·5 ml. of perchloric acid and 2 drops of sulphuric acid and heat again.

Keep the temperature as high as possible, but avoid boiling, and evaporate until the sulphuric acid is fuming strongly. Since it is difficult to expel all the acid, without overheating the bottom of the vessel, it is preferable to finish the evaporation by placing the bottle in a muffle furnace, fitted with a sleeve of sheet aluminium, to avoid contamination from any particles dropping from the top of the muffle. After heating for 2–3 minutes at about 300–350° C., cool the bottle to room temperature. When cool, only a very small, dry, white residue should be visible on the bottom.

By means of a pipette add 1 ml. of the basal solution (ammonium chloride, potassium thiocyanate and methyl red), paying strict attention to the draining time of the pipette, and stopper the bottle. This solution dissolves all of the zinc within a minute or two.

To a small electrolysis cell, which has been thoroughly dried, add sufficient mercury (approximately 0·3 ml.) to form the anode pool and then pipette in approximately 0·4 ml. of the solution to be polarized. Put the capillary (cathode) into position and pass a slow stream of hydrogen through the liquid for 5 minutes. Then turn off the hydrogen, note the temperature near the electrode and apply a polarizing potential of 0·8 volts. Record the current-voltage curve over the range 0·8–1·2 volts.

The relation between step-height and concentration of zinc in the solution being polarized is linear over the range 3–300 p.p.m. To obtain the calibrating factor, take several measurements of the step-height of the standard 0·001M zinc solution at 20° C. and add or subtract 1·7 per cent. for each degree C. below or above 20° C. Correct the step-height of the polarogram of the unknown zinc solution for temperature in the same way. Then calculate the zinc concentration from the corrected step-height by simple proportion.

Carry out a blank determination on the same amounts of reagents normally used and correct for the amount of zinc found. With purified reagents the blank should not exceed 0·10 micrograms of zinc.

When no temperature control is possible it is of advantage to use a basal solution containing a known amount of cadmium and then to calculate the zinc content from the ratio of the zinc and cadmium steps. This ratio is independent of temperature change over the range 10–35° C. When using this internal-standard method make up the basal solution with potassium thiocyanate and ammonium chloride as before but add cadmium sulphate to give a cadmium concentration of 0·00025M. This concentration is suitable for determining zinc contents ranging from 5–75 p.p.m.

The Determination of Zinc: Photometric Method.

This method is essentially that of Cowling and Miller (3), modified to ensure complete extraction of the zinc from the siliceous residue.

Reagents:

In addition to the concentrated redistilled acids and the ammonium citrate buffer solution used in the previous method, the following reagents are required:

N Ammonium Hydroxide. Dissolve ammonia gas in glass distilled water and dilute to the proper strength.

0·02N Hydrochloric Acid. Dilute redistilled hydrochloric acid with water and standardize to within one per cent.

Carbon Tetrachloride. Redistil a good quality commercial grade carbon tetrachloride (sulphur free), using a pyrex glass distillation apparatus with fractionating head. Keep the redistilled product in the dark.

Dithizone in Carbon Tetrachloride. Prepare this solution exactly as described on p. 332 but dilute the final tetrachloride solution of purified dithizone to a volume of 2,500 ml. Store in a pyrex glass reagent bottle in a refrigerator at 2–5° C.

0·5M Ammonium Citrate Solution. Dissolve 210 g. of citric acid in about 1,500 ml. of water. Add concentrated ammonia until the solution has a pH of about 8·5–8·7, when a small portion is tested in a comparator, using cresol red as indicator. Dilute to approximately 2 litres, filter, if necessary, and transfer to a large separating funnel. Shake with an excess of dithizone reagent and separate the tetrachloride phase. An excess of dithizone is indicated by a yellow to orange colour in the aqueous phase after shaking. Wash the aqueous phase by shaking with two or three portions, each of 100 ml., of carbon tetrachloride, separating and rejecting the tetrachloride phase each time. Repeat this washing until the extract is a pure green colour, indicating complete extraction of all heavy metal impurities. Store the purified ammonium citrate in a pyrex reagent bottle.

Sodium Diethyldithiocarbamate. Prepare this solution immediately before use. Dissolve 0·25 g. of sodium diethyldithiocarbamate in water and dilute to 100 ml.

Mixed Reagent A. Dilute 1 litre of 0·5M ammonium citrate and 300 ml. of N ammonium hydroxide with 3,200 ml. of water. Immediately before use dilute 9 volumes of this ammonia-ammonium citrate solution with one volume of freshly prepared carbamate reagent, to give a sufficient quantity of "Mixed Reagent A" for one series of determinations.

Standard Zinc Sulphate. Place 0·100 g. of pure zinc foil in a litre volumetric flask, add about 200 ml. of water and 10 ml. of concentrated sulphuric acid. Heat on a water bath until the zinc is completely dissolved, cool and dilute to 1 litre. Store in a stoppered pyrex reagent bottle. This solution contains 100 micrograms of zinc per millilitre. To prepare a working standard dilute 10 ml. of this solution to 1 litre and store in a pyrex reagent bottle. One ml. of this diluted standard corresponds to 1 microgram of zinc.

Method:

Digest 1 g. of plant material and extract the metal dithizonates with chloroform, at pH 9·8, exactly as described in the last method (p. 357). In order to reduce the concentration of the ammonia dissolved in the chloroform phase, thoroughly wash each extract as it is obtained by shaking it with 50 ml. of water, in a second separating funnel. Use the same portion of water for washing all three extracts but run off the chloroform phase, each time, before adding the next extract. Collect and combine the chloroform extracts in a third separating funnel, into which 50 ml. of 0·02N hydrochloric acid have been previously pipetted. Shake for 1½ minutes and allow to separate. Then run off the chloroform phase as completely as possible, shaking down the drop of chloroform from the surface of the liquid. Do not let any of the aqueous phase enter the bore of the stopcock. The chloroform phase contains the copper and dithizone and is discarded. Rinse the aqueous phase by shaking with one lot of 5 ml. and one lot of 2 ml. of carbon tetrachloride to remove the last of the chloroform and dithizone from the separating funnel. Shake down the drop from the surface of the liquid each time and run out the tetrachloride as completely as possible. Remove the stopper from the separating funnel and allow the small amount of tetrachloride on the surface of the liquid to evaporate.

To the aqueous solution in the separating funnel, which contains the zinc and other metals free from copper, add, by means of pipettes, 50 ml. of mixed reagent A and 10 ml. of dithizone-tetrachloride reagent. Shake vigorously for one minute and allow to separate. When the tetrachloride phase is clear, flush out the stopcock and stem with 1 ml. or so of it and collect the remainder in a dry test tube.

By means of a pipette, transfer 5 ml. of the carbon tetrachloride extract to a 25 ml. volumetric flask, dilute to volume with clear carbon tetrachloride and determine the percentage light transmission of the diluted solution with a photo-electric colorimeter, using a Corning No. 401 filter, or a spectrophotometer using a wavelength of about 535 millimicrons. From the standard calibration curve read off the amount of zinc corresponding to the light transmission found. Correct for any zinc found in the blank determination. Make the photometric readings within two hours of the final extraction and protect the dithizone solutions from light, as far as possible.

To prepare the calibration curve, connecting percentage light transmission with the amount of zinc, proceed as follows:

Add 5 ml. of chloroform to a separating funnel. Pipette in 0–30 ml. of the standard solution of zinc sulphate so as to give a range of about six amounts of zinc, say 0, 5, 10, 15, 25 and 30 micrograms, and if necessary add water to bring the volume to 30 ml. Then add 30 drops (about 0·75 ml.) of concentrated sulphuric acid and 25 ml. of the ammonium citrate buffer solution (containing dithizone, p. 354) to bring the reaction to about 9·8.

Extract the zinc, and carry out the steps for the removal of copper and the re-extraction of the zinc in exactly the same manner as in an actual determination, paying strict attention to the volumes of solutions used throughout. Finally measure the light transmission of the diluted dithizone extracts and construct the calibration curve. Once this curve has been prepared it can be used for all subsequent determinations, carried out under precisely the same conditions.

REFERENCES

(1) Bayliss, N. S. and Pickering, R. W.: *Thesis*, University of Western Australia, 1941.

(2) Clifford, P. A. and Wichmann, H. J.: *J. Assoc. Off. Agric. Chem.*, 1936, *19*, 130–156.

(3) Cowling, H. and Miller, E. J.: *Ind. Eng. Chem. (Anal. Ed.)*, 1941, *13*, 145–149.

(4) Dodd, A. S.: *Analyst*, 1927, *52*, 459–466.

(5) Hill, R.: *Proc. Roy. Soc. (London)*, 1930, *107*, 205–214.

(6) Kahane, E.: *L'Action de l'Acide Perchlorique, Actualités Scient. et Indust. No. 167 and 168, Paris*, 1934.

(7) Kidson, E. B. and Askew, H. O.: *New Zealand J. Sci. Techn.*, 1940, *21*, 178B–189B.

(8) Lingane, J. J. and Kolthoff, I. M.: *Chem. Rev.*, 1939, *24*, 1–94.

(9) Marmoy, F. B.: *J. Soc. Chem. Ind.*, 1939, *58*, 275–276.

(10) Marston, H. R. and Dewey, D. W.: *Aust. J. Expt. Biol, and Med. Sci.*, 1940, *18*, 343–352.

(11) Maunsell, P. W.: *New Zealand J. Sci. Techn.*, 1940, 22, 100B–111B.

(12) McNaught, K. J.: *New Zealand J. Sci. Techn.*, 1938, *20*, 14A–30A.

(13) Parker, W. E. and Griffin, F. P.: *Can. J. Res.*, 1939, *17B*, 66–70.

(14) Sylvester, N. D. and Lampitt, L. H.: *Analyst*, 1935, *60*, 376–382.

(15) Walkley, A.: *Aust. Chem. Inst. J. and Proc.*, 1938, *5*, 291–311.

(16) Walkley, A.: *Aust. Chem. Inst. J. and Proc.*, 1942, *9*, 29–34.

(17) Walkley, A.: *Aust. J. Expt. Biol. and Med. Sci.*, 1942, 20, 139–147.

(18) Wichmann, H. J.: *Ind. Eng. Chem. (Anal. Ed.)*, 1939, *11*, 66–72.

(19) Willard, H. H. and Greathouse, L. H.: *J. Amer. Chem. Soc.*, 1917, *39*, 2366–2377.

INDEX: SOIL ANALYSIS

INDEX: PLANT ANALYSIS